United Nations
Department of
Technical
Co-operation for
Development

Ground Water in Western and Central Europe

Natural Resources
Water Series No. 27

United Nations New York, 1991

NOTE

Symbols of United Nations documents are composed of capital letters combined with figures. Mention of such a symbol indicates a reference to a United Nations document.

ST/TCD/12

UNITED NATIONS PUBLICATION

Sales No. E.90.II.A.25

ISBN 92-1-104355-7

CONTENTS

CONTENTS (continued)

CONTENTS (continued)

CONTENTS (continued)

CONTENTS (continued)

CONTENTS (continued)

CONTENTS (continued)

CONTENTS (continued)

CONTENTS (continued)

CONTENTS (continued)

CONTENTS (continued)

CONTENTS (continued)

CONTENTS (continued)

CONTENTS (continued)

Figures

(Maps except as noted)

CONTENTS (continued)

CONTENTS (continued)

CONTENTS (continued)

Note: With effect from 3 October 1990 the Federal Republic of Germany and the German Democratic Republic united to form one sovereign State under the designation "Germany".

ABSTRACT WITH KEYWORDS

This volume is the ninth in a series giving encyclopedic coverage of the hydrogeology, use and resources potential of groundwater world-wide. For Austria, Belgium, Denmark, France, the German Democratic Republic, Germany, Federal Republic of, Ireland, Italy, Luxembourg, Malta, the Netherlands, Portugal, Spain, Switzerland and the United Kingdom of Great Britain and Northern Ireland is given a summary of the physiography and climate, the geologic setting, hydrogeology and stratigraphy, history of groundwater use, principal aquifers, characteristics of the groundwater (including mineral and thermal springs), development practices, training and research facilities, and extent of international cooperation in the field. Includes maps, tables of data and lists of supplementary reference sources.

PREFACE

The Economic and Social Council, in its resolution 675 (XXV) of 2 May 1958, requested the Secretary-General to take appropriate measures for the establishment, within the Secretariat, of a centre to promote co-ordinated efforts for the development of water resources. It also singled out groundwater problems as one of the priority subjects in the development of a programme of studies. Large-scale Groundwater Development, published in 1960, 1/ was the first study prepared in this field by the Water Resources Development Centre (now the Water Resources Branch of the Natural Resources and Energy Division, Department of Technical Cooperation for Development of the United Nations Secretariat).

The Advisory Committee on the Application of Science and Technology to Development, in its World Plan of Action, 2/ gave priority to groundwater exploration and development. In fact, in the course of the First and Second United Nations Development Decades, more than 100 projects assisted by the United Nations Development Programme (UNDP) and other United Nations technical cooperation programmes were entirely or partly devoted to groundwater prospecting, assessment or pilot development. (A list of groundwater projects in the subregion sponsored by UNDP is contained in annex I to the present publication.)

While such operational activities were developing, the need for a comprehensive review of the results of the projects and for the dissemination of relevant information became more evident. As a result, the Economic and Social Council, in its resolution 1761 B (LIV) of 18 May 1973, requested the Secretary-General to take the necessary measures, within budgetary limitations, to improve and strengthen the existing United Nations services for the analysis, evaluation and dissemination of world-wide data on natural resources, including water resources. The recommendations of the United Nations Water Conference (1977) 3/ placed added emphasis on these aspects.

With respect to groundwater, a first comprehensive review of the African continent was published in 1972 and 1973 under the title Groundwater in Africa 4/ as a synthesis of material available in the records and files of the United Nations. The material of the second volume in this series, Groundwater in the Western Hemisphere 5/ (1976) was drawn from country papers prepared by hydrogeologists and groundwater engineers in the countries concerned; this was also the case for material presented in a third volume, Groundwater in the Eastern Mediterranean and Western Asia 6/ (1982), which includes Turkey in Europe, and a

---

1/ United Nations publication, Sales No. 60.II.B.3.

2/ United Nations publication, Sales No. E.71.II.A.18.

3/ Report of United Nations Water Conference, Mar del Plata, 14-25 March 1977 (United Nations publication, Sales No. E.77.II.A.12), chap. I.

4/ United Nations publication, Sales No. E.71.II.A.16.

5/ United Nations publication, Sales No. E.76.II.A.5.

6/ United Nations publication, Sales No. E.82.II.A.8.

fourth volume, _Groundwater in the Pacific Region_ 7/ (1983) which covers also the major Asian countries off the coast of continental Asia, including Japan, the Philippines and Indonesia, and a fifth volume _Groundwater in Continental Asia_ 8/ (1986). The first volume _Groundwater in Africa_ being out of print and largely outdated, it was decided to publish an updated and expanded review in two volumes: _Groundwater in North and West Africa_ 9/ (1987) and _Groundwater in East, Central and Southern Africa_ 10/ (1988).

It is hoped that the two volumes on groundwater in Europe, which are the last in this series on groundwater in the world's regions, will be useful to developing countries. In addition to a description of groundwater occurrences and availability in European countries, they provide information on the conditions of groundwater development and related problems, information on technologies and methods used in the investigation, assessment and development of groundwater, and on international cooperation among the countries concerned.

In particular this volume on Western and Central Europe deals with extensive groundwater development in densely populated and industrial areas where water resources are threatened by depletion, pollution and also sea-water intrusion in coastal areas. As a result national water resources management policy involving both surface water and groundwater at the scale of a river basin had to be developed which involved the design and operation of mathematical models. The same problems are experienced in fast-growing metropolitan areas of developing countries which may draw from the experience gained in Western Europe.

Water resources management policies had also to be developed in the semi-arid areas of the south of the subzone especially in southern Spain, southern Italy and Sicily where the best is to be made from scarce groundwater resources, from the irrigation of valuable cash crops. The experience gained there may benefit many arid areas of the world.

Many populated islands with water problems occur in the North Sea, the Atlantic and the Mediterranean off the shores of Western Europe. They are of quite varied geologic types: volcanic, sandy, limestone and others. Similar problems and conditions are encountered in many Caribbean and Pacific Islands and islands off the coast of Africa.

Central Europe is a mountainous area. In the densely populated Alpine valleys where groundwater resources are extensively developed from alluvium, proper management methods had to be devised as to avoid depletion and prevent pollution. Similar conditions are found in developing countries, especially in the Asian subcontinent.

7/ United Nations publication, Sales No. E.83.II.A.12.

8/ United Nations publication, Sales No. E.86.II.A.2.

9/ United Nations publication, Sales No. E.87.II.A.8.

10/ United Nations publication, Sales No. E.88.II.A.5.

For the 15 countries reviewed in this volume, abundant and valuable information has been provided by governmental organizations and by scientists, consultants and specialists whose assistance is gratefully acknowledged by the United Nations. In particular, thanks are due:

Mr. L. J. Anderson, Dr. P. A. Aureli, Mr. H. Aust, Mr. H. Bender, Mr. J. Bintz, Dr. E. Boeckh, Dr. D. Burdon (deceased), Dr. E. Custodio, Mr. R. A. Downing, Mr. P. L. Ferreira, Mr. P. Frederiksen, Pr. H. Hetzer, Mr. N. Kelstrup, Pr. Dr. R. Llamas, Mr. C. R. Meinardi, Mr. J. Margat, Ms. N. Miguens, Mr. J. D. Rodrigues, Mr. A. C. Skinner, Dr. W. Struckmaier, Mr. H. Vierhuff, Dr. P. H. Zahner, Dr. J. G. Zotl and also Administration des Ponts et Chaussées (Luxembourg), the British Geological Survey, the Bureau de Recherches Géologiques et Minières (Orléans, France), the Central Geological Institute (Berlin, German Democratic Republic), the Federal Institute for Geosciences and Natural Resources (Hannover, Federal Republic of Germany), the Geological Survey of Denmark, the Institute for Geothermics and Hydrogeology (Graz, Austria), the National Institute of Public Health and Environment Hygiene of the Netherlands (Bilthoven) and the Water Works Department (Valletta, Malta).

EXPLANATORY NOTES

The following symbols have been used in the tables throughout the report:

Two dots (..) indicate that data are not available or are not separately reported.

An em dash (--) indicates that the amount is nil or negligible.

A hyphen (-) indicates that the item is not applicable.

A full stop (.) is used to indicate decimals.

A slash (/) indicates a crop year or financial year, e.g., 1976/77.

Use of a hyphen (-) between dates representing years, e.g., 1975-1978, signifies the full period involved, including the beginning and end years.

The word billion means the number 1,000,000,000 or 10^9.

Reference to "dollars" ($) indicates United States dollars.

Details and percentages in tables do not necessarily add to totals because of rounding. Some of the data series are not homogeneous; they have been taken from various reviews and publications.

The designations employed and the presentation of the material in this publication do not imply the expression of any opinion whatsoever on the part of the Secretariat of the United Nations concerning the legal status of any country, territory, city or area or of its authorities, or concerning the delimitation of its frontiers or boundaries.

Regional or local terminology practice may differ from the forms used in the report. The following equivalencies should be noted:

mg/l = ppm (parts per million)

1 km^2 = 100 hectares

1 hm^3 = 1 million m^3

1 km^3 = 1 billion m^3

Abbreviations used

AC	Asbestos-cement
AICA	Asociación de Ex-participantes en los Cursos del Instituto de Hidrología
ACSAD	Arab Centre for Studies of Arid Zones and Dry Lands
BP	before the present
BRGM	Bureau de Recherches Géologiques et Minières
CEC	Commission of the European Communities
CIEH	Comité Inter-africain d'Etudes Hydrauliques
CIHS	Curso Internacional de Hidrología Subterranea
CILSS	Comité Inter-africain de la lutte contre la Sécheresse au Sahel
CNA	Comissao Nacional do Ambiente
CNRS	Centre National pour la Recherche Scientifique
DGOH	Dirección General de Obras Hidráulicas
DTCD	Department of Technical Co-operation for Development
ECA	Economic Commission for Africa
EEC	European Economic Community
ESCAP	Economic and Social Commission for Asia and the Pacific
ESB	Electricity Supply Board
FAO	Food and Agriculture Organization of the United Nations
F/m	Francs per cubic metre
GTI	Institute for Geotechniques
IAEA	International Atomic Energy Agency
IAH	International Association of Hydrogeologists
IAHS	International Association for Hydrological Sciences
ICW	Institute for Land and Water Management Research
IGH	Institute for Geothermics and Hydrogeology
IGME	Instituto Geológico y Minero de España
IHE	International Institute for Hydraulic and Environmental Engineering
IHP	International Hydrological Programme
INE	Instituto Nacional de Estadística
IRYDA	Instituto de Reforma y Desarrollo Agrario
ITC	International Institute for Aerial Survey and Earth Sciences
IUGS	International Union of Geological Sciences
KIWA	Testing and Research Institute of the Waterworks Association
KNMI	Royal Meteorological Institute
LN	log normal

MAPA	Ministerio de Agricultura, Pesca y Alimentación
meq	milliequivalent
MIE	Ministerio de Industria y Energia
MOPU	Ministerio de Obras Públicas y Urbanismo
MPAT	Ministerio do Planeamento e da Administraçëo do Territorio
mV	millivolt
m.y.	million years
ODA	Overseas Development Administration
PGCA	Programa de Gestión y Conservación de Acuíferos
PNANU	Plan Nacional de Abastecimientos a Núcleos Urbanos
PVC	Polyvinylchloride
REPO	Comisaría de Aguas del Pirineo Oriental
RGD	Netherlands Geological Survey
RIVM	National Institute of Public Health and Environmental Hygiene
RWS	Rijkswaterstaat
SGOP	Servicio Geológico de Obras Públicas
SN	short normal
SP	Self potential; spontaneous potential
SUWT	Symposium on Underground Water Tracing
TDS	total dissolved solids
UNDP	United Nations Development Programme
UNESCO	United Nations Educational, Scientific and Cultural Organization
UNICEF	United Nations Children's Fund
WHO	World Health Organization
WL	Delft Hydraulics Laboratory
WVC	Ministry of Welfare, Health and Cultural Affairs

OVERVIEW

Introduction

Europe is generally described as a large peninsula, located at the western end of Eurasia, the largest landmass in the world. To the east, Europe is separated from Asia proper by the Ural Mountains (1,885 m at Mount Narodnaya) and the Ural River, and to the south by the Caucasus Mountains. From Nordkapp to the north in Norway to Cape Taínaron in the south (Greece), the distance is about 3,850 km. From Cape Roca, in Portugal to the west, to the Urals in USSR to the east, the distance is about 5,300 km.

By Sea, Europe is separated from Asia by the Caspian Sea, the Bosporus, the Sea of Marmara and the Aegean Sea; from Africa by the Mediterranean Sea and the Strait of Gibraltar. To the west and north, Europe ends at the Atlantic Ocean, the Norwegian Sea and the Barents Sea.

With only 7 per cent of the landmass of the world, Europe contains about one fifth of the world's population and accounts for half of the world's trade. The influence of the sea is considerable over the continent, both from the point of view of climate and the closeness to the sea of most of the territory which contains islands and peninsula. There are few places in Europe more than 800 km from the sea.

The physiography of Europe can be summarized as follows (figure 1). A vast plain in the north crosses from west to east comprising much of France, the Low Countries, Germany, Poland and the USSR. The plain is flanked to the south by plateaus covering central France, the Ardennes, the Black Forest, the Harz and a mass of highlands extending through Czechoslovakia, Poland and the Ukrainian SSR. A huge mountainous mass crosses Europe, mainly in the west and south, which includes the Pyrenees, the Alps, the Carpathian Mountains, the Balkan Mountains and the Caucasus Mountains. The highest point is Elbrus in the Caucasus (5,630 m). In Western Europe the highest peak, in the Alps, is Mont Blanc (4,807 m).

A great variety of climates from arctic (northern USSR) to sub-tropical (Canary Islands, Madeira) are encountered in Europe which is drained by a number of rivers, some counting among the major rivers in the world (Volga, Danube, Rhine). Geologic conditions are also extremely diverse. All the geologic time periods are represented in Europe. The first geologic descriptions were made in Europe; consequently, stratigraphic terms have been chosen on the basis of the European locations where they were first described. Major orogenies, or mountain-building tectonic phases were also named after the massifs of Europe (Caledonian, Hercynian, Alpine).

As a result of the great diversity of conditions affecting groundwater, morphology, climate and geology, all kinds of hydrogeologic situations are found in Europe. Also, much European expertise in managing groundwater has been put at the disposal of developing countries in the form of training, transfer of technology, assignment of experts and supplying of equipment.

Such cooperative activities are described in the country papers that follow. A list of groundwater projects benefiting from participation of the United Nations Development Programme (UNDP) is mentioned in annex I.

Figure 1

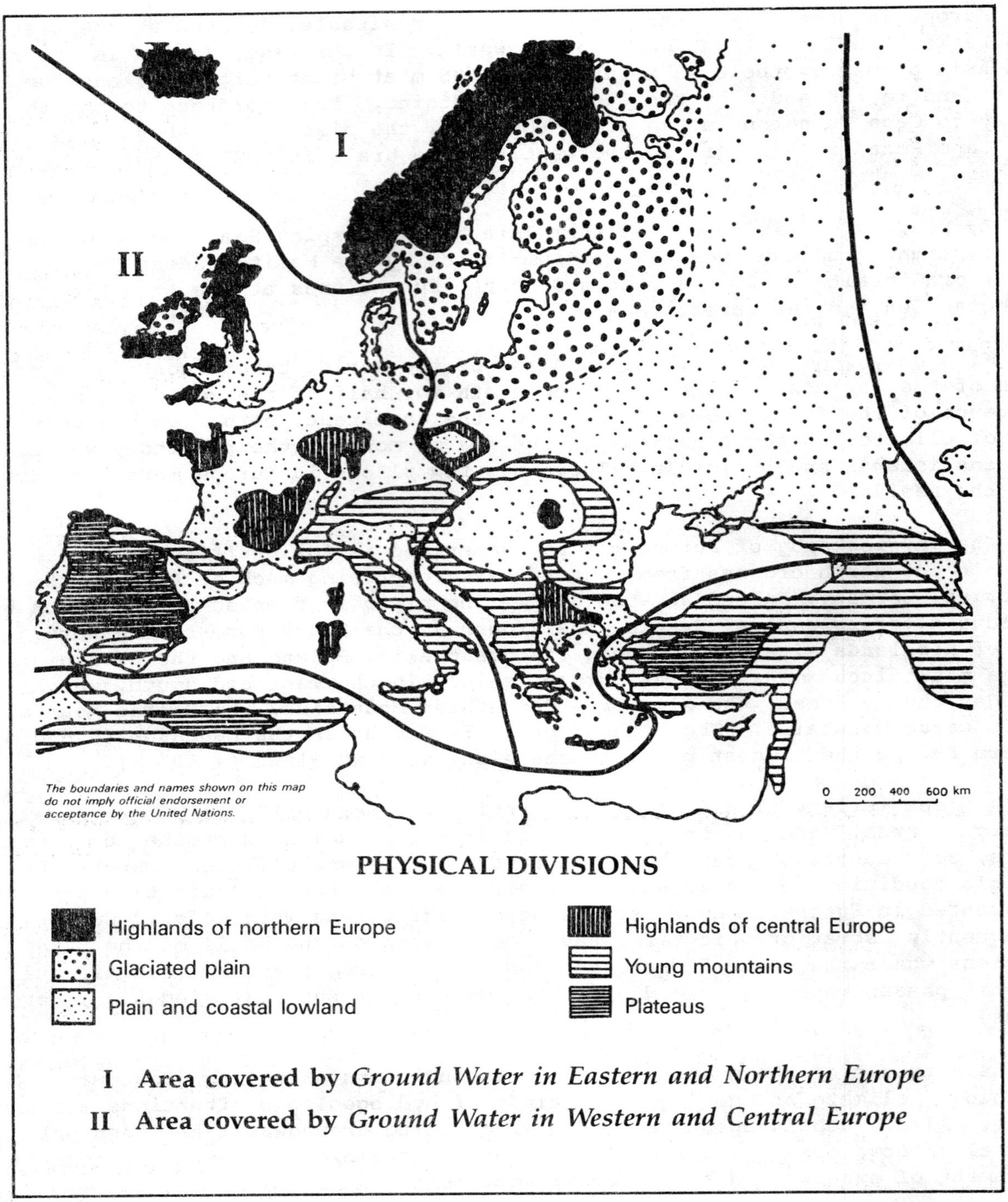

3546.1x

Source: Norman J.G. Pounds, *Europe and the Mediterranean*, New York: McGraw-Hill, 1953.

This overview of groundwater resources, knowledge and development in Europe may appear rudimentary as compared with the abundant literature available and the extensive cartographic effort carried out in recent years, such as the joint International Association of Hydrologists (IAH) and UNESCO Hydrogeological Map of Europe at the scale of 1:1,500,000 and the publication entitled Ground Water Resources of the European Community in 10 volumes and nine sets of maps at the scale of 1:500,000.

The present work is intended, within the framework of a world-wide review of the subject, and primarily directed at water resources professional engineers, planners and decision makers of developing countries, to serve as a reference document, bringing to completion the series initiated in 1972 with the volume Ground Water in Africa.

For reasons of format convenience, coverage of Europe is in two volumes (see figure 1): I. Eastern and Northern Europe; II. Central and Western Europe. This division is not political but is rather related to geographic considerations with the dividing line including the Adriatic Sea, the Alps, the Bohemian Forest, the Oder-Neisse Rivers, and the Skagerrak and Kattegat Straits at the entrance to the Baltic Sea.

A. Physiography

The area covered by this volume comprises the following four physiographic subregions:

1. To the north-east: The western end of the northern European "Baltic" plain, west of the Oder River, which extends across northern Germany, Denmark, the Netherlands, Belgium and northern France with practically no relief feature. A substantial part of the Netherlands is below sealevel but kept emerged through an extended system of dykes.

2. To the north-west: The British Isles and Ireland composed mainly of rolling hills and plains and also plateaus of moderate elevation mainly in the north and the west of Great Britain (Scotland and Wales) and in the north and the west of Ireland (a few hundred metres elevation). The highest point is Mount Snowdon in Wales (1,085 m).

3. The other major plains and lowlands of western Europe extend mainly in northern and western France from the shores of the English Channel and the Atlantic Ocean to 200-300 km inland. The plain of the Po River and the Bavarian plateau flank the Alps to the south and to the north. The central part of the subregion, located between the northern plains and the mountainous region to the south, includes mainly massifs at moderate elevation: 500 to 1,000 m in the Massif Central with some peaks above 1,500 m (highest point: Plomb du Cantal, 1,858 m). The "Slate uplands" of central Germany, north of the Bavarian plateau, with elevations mostly 200 to 500 m and some higher zones: the Harz (1,142 m), the Thüringer Wald - Frankenwald (1,051 m at Fichtel Gebirge) with their extension to the western Rhine Valley: the Ardennes. Other important physiographic features of this central part include the valleys of the Rhone and Rhine rivers, and the mountainous ridges of Vosges (1,424 m), Black Forest (1,493 m) and Jura (1,693 m).

4. To the south are two large mountainous peninsulas, the Iberian and Italian, flanked by narrow coastal plains and dissected by many river valleys, large in Spain and Portugal, small in Italy with most areas at elevations between 500 and 1,000 m. Between the central and southern part, the extensive mountainous systems of the Pyrenees (3,298 m at Mount Vignemale) and the Alps (4,807 m at Mont Blanc).

The major islands of the region are to the north, in addition to the British Isles, the Danish Islands (about 50 per cent of the country's land area), the Zeeland Islands of the Netherlands and the Faroe Islands (Denmark). To the south, in the Mediterranean Sea, the mostly rugged or mountainous, Balearic Islands, Corsica, Sardinia and Sicily. Malta is an island country, small in size but densely populated, with gentle topography. To the west, in the Atlantic Ocean, are the volcanic archipelagos of the Azores, the Canary Islands and Madeira.

B. Climate

Most of the region enjoys moderate climatic conditions as a result of the vicinity of the ocean to the west and to the Mediterranean Sea in the south.

Two main types of climatic zones are encountered in the area:

a. The Mediterranean zone with cold winters and hot dry summers divided into two subzones:

i. To the south: temperate Mediterranean with mild winter, some rain in summer and average rainfall exceeding 600 mm. The subzone includes most of Portugal, the coastal Mediterranean areas of Spain, the coastal Mediterranean of France and their hinterland up to 100 km, peninsular Italy, Corsica and Sardinia.

ii. To the north, an area still with Mediterranean characteristics but with continental influence and sharper climatic contrasts with summer rainfall, severe winters and hot summers. The subzone includes the northern parts of Italy between the Alps and the Italian peninsula and the central plateau of Spain.

b. The temperate cool zone where average temperature is above 18° C for eight months of the year. Summers may be hot. Also divided in two subzones:

i. The oceanic subzone which includes the British Isles, the shores of the North Sea and the Atlantic Ocean from Denmark to Cap Finesterra and their hinterland west of a line Copenhagen-Toulouse. In the subzone, rainfall is well distributed. The climate is both mild and moist. Average temperature never goes below 0° C. Cloudiness is frequent with heavy fog in winter. Summers are seldom hot.

ii. The dryer central Europe subzone east of the line Copenhagen-Toulouse and north of the line Toulouse-Trieste with greater seasonal temperatures ranges. Rainfall is maximum in the summer, winters are cold.

Temperature distribution is related to altitude, latitude and distances to the sea. Average annual temperatures are as follows:

a. Less than 7.5° C in the Alps, the Pyrenees-Cantabric chain, central Spain, the Apennines,

b. From 7.5° to 10° C in most of the British Isles, central and north-eastern France and Germany,

c. From 10° to 15° C in continental Spain, the Italian plains (Po), France west of line Rouen-Marseille,

d. From 15° to 17.5° C on the Atlantic shores south of Coimbra and the Mediterranean shores. Also, peninsular Italy south of the Pisa-Ancona line.

The coldest month is February with monthly temperatures averaging 0° C and less in:

a. Northern and eastern Denmark,

b. The Germanys east of Lubeck-Dresden line,

c. Germany, France, Italy, Switzerland, Austria: a vast area within a triangle Dresden-Montelimar-Graz (Austria). The area corresponds to the Alpine zone and the highlands flanking it to the north,

d. Massif Central in France,

e. Northern Highlands of Spain,

f. The Pyrenees mountains,

g. The higher parts of the Apennines and Corsica.

The warmest month is July or August with mean temperatures above 20° C south of the Pyrenees-Cantabric chain and south of the Alps. Also as warm is the Aquitaine plain in France.

The mean annual range of temperatures is found along the Atlantic and North Sea shores, in the British Isles and in mountainous areas. The highest range is found in central Spain and the Po plain of Italy.

C. <u>Precipitation</u>

The highest annual values (1,000-1,500 mm) are found in areas facing the Atlantic, i.e., western Ireland, western Scotland, Wales, northern Portugal, north-western Spain (see figure 2). Also with high precipitation are the Alps (including snow), western Pyrenees and the northern Apennines. The driest areas (400-500 mm or less rainfall) are found in most of Spain (exception made of the southern tip and the plateaus north of Seville and Madrid), the coastal areas of France west of Marseille and south of Bordeaux. In Italy, the central part of the plain of the Po, the south-eastern part of the peninsula, southern Sicily, the Rhone Valley, and south-eastern England are also dry.

From the above, it can be concluded that most of the subregion gets 500 to 1,000 mm of precipitation. The Atlantic islands are subjected to an oceanic humid climate in the north (Azores) and to a subtropical mild or semi-arid climate in the south (Madeira, Canary Islands).

Figure 2. Europe: Precipitation

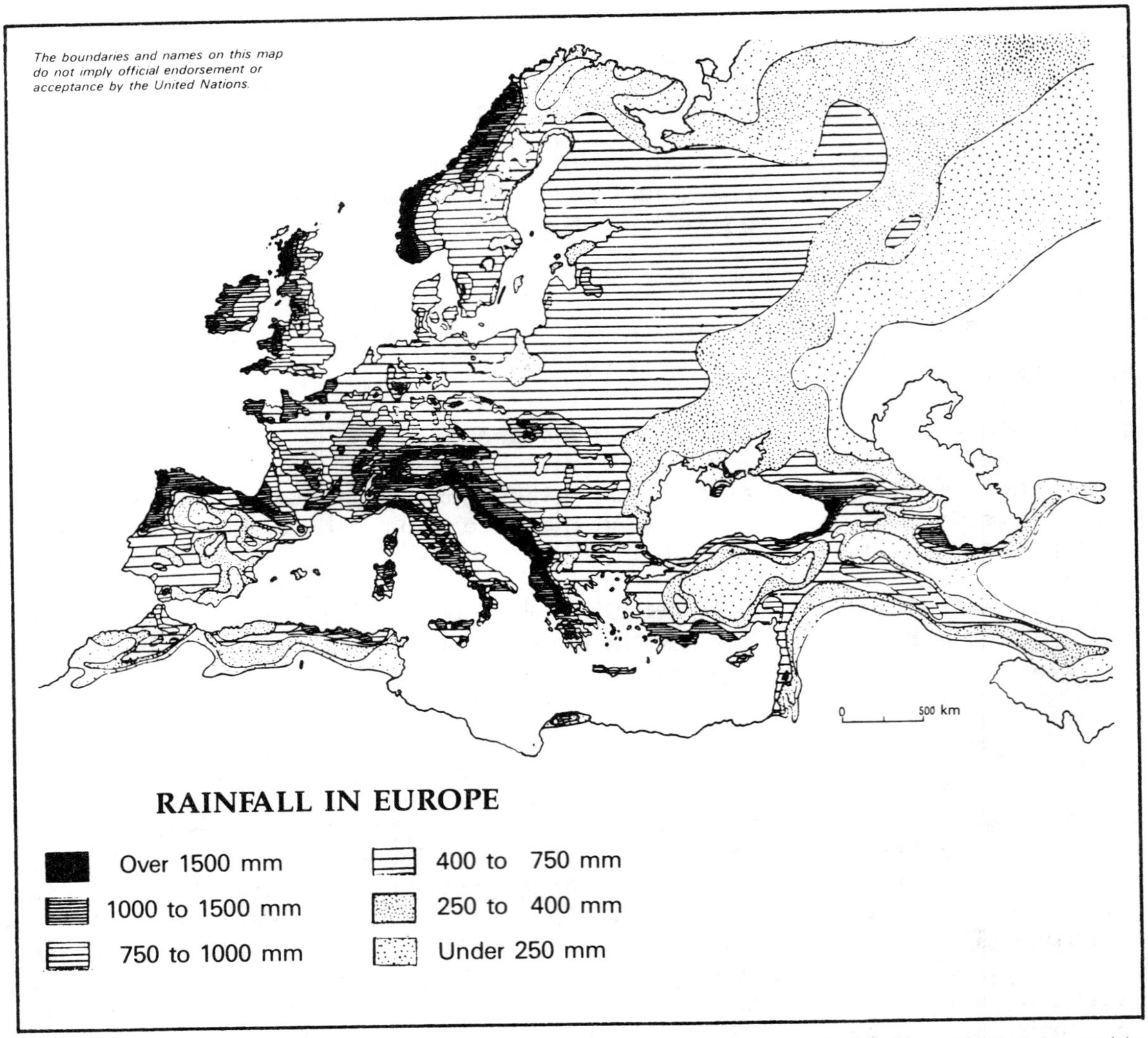

3546.2x

Source: Norman J.G. Pounds, *Europe and the Mediterranean*, McGraw Hill, 1953, p. 876.

D. Geology

The landscape of Europe is marked by three orogenic periods:

a. The Caledonian folding of Devonian age,

b. The Hercynian folding of Permian age,

c. The Alpine folding of late Cenozoic age when an enormous mass of Mesozoic rocks which had been deposited in the Tethys sea south of Europe were buckled against the highlands of central Europe, resulting in the Alpine system.

In the area considered here, the main geologic formations of the four physiographic subregions mentioned above are as follows (see figure 3).

1. Northern Highlands

In the Northern Highlands (A) are included:

a. Great Britain (A1) with north to south:

i. The highlands of Scotland: mostly Precambrian metamorphic rocks,

ii. The central valley trough: upper Paleozoic rocks with basaltic lava,

iii. The Southern Uplands of Scotland: Paleozoic rocks,

iv. The Pennine chain with an escarpment up to 750 metres: it is an anticline of Lower Carboniferous rocks with locally exposed karstified limestone, capped elsewhere by the Millstone grits (sandstone),

v. The lake region (Cumbria), an Ordovician-Silurian dome of both fine- and coarse-grained sandstone and slate,

vi. Wales with Precambrian and Paleozoic formations: including Carboniferous limestone,

vii. Cornwall, a Paleozoic syncline with geologic features similar to those of Wales.

b. Ireland (A2) with the continuation to the west of the above formations:

i. Precambrian metamorphics (north-west),

ii. Massive lavas flows (Antrim),

iii. Paleozoic formations intruded by granite in the north-east (continuation of Southern Uplands),

iv. In south Wexford to the south-east, continuation of Wales formations: Wicklow granitic mountains, Devonian sandstone, Carboniferous-age limestone,

v. The central part of Ireland is a plain of Carboniferous limestone.

Figure 3

3593.1x

c. Brittany (A3) Paleozoic formations, folded, intruded by granite with Appalachian structure comprising elongated ridges to the north and south with, in between, the Rohan plateau covered by Cenozoic deposits which extend to the north-east, the east and the south-east as to include Normandy and Le Maine.

2. Great Central Plains

The Great Central Plains (B) comprise from north to south:

a. The western end of the Baltic plains (B1) which extends across the Northern Germanys, Denmark, the Netherlands and Belgium, along three belts parallel to the coast comprising successively from the shores and inland:

i. Dunes, marshes, polders,

ii. Glacial sands up to 70 m thick (from Belgium to the Weser River),

iii. The foreland of the plateau of Belgium and northern Germany (B2).

b. The English lowlands (B2) that is the coastal plains of south-east England, the main feature being the London syncline. A complete stratigraphic sequence is present here from Paleozoic through Cenozoic (London chalk and greensands).

c. The Paris basin (B3): the geology is similar to that of the London basin from Cenozoic (Oligocene) in the central part to Triassic. The geomorphology is marked by cuestas, especially to the east towards the Vosges mountains.

d. The Aquitaine basin (B4):

i. To the north towards the Central Massif; Cenozoic to Jurassic formations,

ii. To the east in Montagne Noire; Cenozoic belts,

iii. To the south towards the Pyrenees, the Lannemezan Plateau; congregation of alluvial fans comprising Tertiary deposits covered with glaciofluvial outwash,

iv. To the west: Les Landes; alluvium and Cenozoic beds; sand dunes in the coastal region.

3. Central Massifs

The Central Massifs (C) include:

a. The bulk of the Iberic peninsula (C1) with:

i. The horst of the Spanish Meseta: a plateau flattened by erosion comprising Paleozoic sedimentary beds with, to the north-west, Precambrian schist, gneiss and granite, the whole being covered by Cenozoic formations,

ii. The Andalusia lowlands between the Meseta and the Sierra Nevada: a Cenozoic-Quaternary syncline,

iii. The Ebro valley: depression between the Meseta and the Pyrenees mountains with Oligocene and continental Cenozoic formations,

iv. The Portuguese lowlands including the lower Tagus River, mostly Neogene and Quaternary formations.

b. Also included is the Central Massif of France (C2): Uplands with mean elevation of 1,000 metres, steeply rising in its southern part and dipping gently to the north and south-west. They are part of an Hercynian mountain system with schist, gneiss, intrusive granites and volcanic rocks which occurred following Alpine uplifts.

c. The Rhone-Saone corridor (C3) is filled with thick Cenozoic and alluvium beds.

d. The Vosges and Black Forest mountains (C4) with, in between, the Rhine graben (Cenozoic floor covered by Rhine sediments) are Hercynian massifs mostly constituted of granite and metamorphics covered by Triassic beds.

e. The Slate Mountains of Germany (C5): Uplands consisting of folded Devonian slate, rhyolite sandstone, quartzite and crossed by the Hohe Venn anticline of older rocks; also volcanics.

f. The Highlands of Central Germany (C6): Flat shaped, Cenozoic area including uplifted Thüringer Wald and Harz mountains (Paleozoic schist and rock) intruded by granite.

g. The Bavarian Plateau (C7) located between the Alps, the Black Forest and to the north, mostly Triassic-Jurassic limestone, and to the south, mostly Cenozoic flysch (between the Danube and the Alps).

4. Alpine system

The Alpine system (D) includes:

a. The Alps (D1), an arcuate chain of mid-Cenozoic age developing from Nice on the Mediterranean through France, Switzerland, northern Italy and Austria, to Vienna: crystalline rocks, massifs and nappes.

b. The Apennines (D2), backbone of peninsular Italy and extension of the Alps to the south, with clay and sand to the north, limestone, granite, gneiss to the south (Calabria) and volcanoes including Vesuvius, Etna and Stromboli.

c. The Sierra Nevada of Spain (D3), or Betic Chain from Gibraltar to Catalonia and Balearic Islands. An arch of micaceous schists and Mesozoic formations.

d. The Pyrenees (D4) with Paleozoic formations metamorphozed by granitic injections. To the north are Mesozoic formations and Cenozoic nappes towards the Aquitaine basin.

e. Corsica and Sardinia (D5) are remnants of a Tyrrhenian massif, mostly granite.

f. The Jura (D6) is a Mesozoic arch, mostly limestone, 300 km long, a tableland to the north, folded to the south.

g. The Swiss plateau (D7) is a syncline between the Jura and the Alps with upper Oligocene and Miocene formations and alluvium.

h. The Po Basin (D8), continued by the Adriatic coastal plain of the Italian peninsula, is a vast and deep trough filled with deposits from the Alps and the Apennines.

E. Groundwater

The main aquifers are related to:

a. Among unconsolidated formations, the alluvium fill of most rivers in mountain valleys and in plains. Also included are the glacial and periglacial gravel formations in the great central plains of the north, also sandy areas, including coastal dunes and low-lying islands in the North Sea area (the Netherlands, northern Germany, Denmark),

b. The complex major sedimentary basins of London, Paris, Aquitaine and others. Also in troughs of the Ebro valley, plain of the Po and Guadalquivir rivers,

c. Karstic areas particularly in the Jura, Austria, Bavaria, Sicily, peninsular Italy, southern France and mountainous and coastal areas of Spain including Majorca,

d. Volcanics, for example in the Etna volcanic complex and in the Canary Islands.

In several areas of Europe, groundwater management has reached high levels of sophistication through system analysis. Such is the case in the Rhine valley, in several basins of south-east England, like the Thames basins, and in Alpine valleys where surface water and groundwater are conjunctively developed, also in the metropolitan areas of Hamburg and Barcelona, the lower Aquitaine basin of France and the lower Rhone area, and the Netherlands.

Considerable experience has also been acquired in the monitoring and the control of sea-water intrusion in coastal areas (the Netherlands), karstic hydrology, and cleansing of polluted aquifers.

Groundwater is extensively used for community water supply and industrial use in a number of metropolitan areas (Barcelona, Hamburg) and for village water supplies, throughout all Europe.

Among the EEC member countries, Ireland is where groundwater is the least developed due to the easy availability of low-cost surface water.

In such densely populated islands as the Canary and Balearic islands, groundwater supplies are a most valued commodity which is threatened by overdraft and sea-water intrusion.

With the exception of certain Mediterranean regions, groundwater is not commonly used for irrigation. Such usage is found mainly in Spain (Andalusia, the Mediterranean shores south of Valencia), also in France (lower Rhone, Languedoc area), in southern Italy and Sicily.

A problem of concern in Europe is the pollution of groundwater - a major source for drinking water - with industrial pollutants and chemical fertilizers. The activities of the Committee on Water Problems of the Economic Commission for Europe of the United Nations served by the Water Unit of the Environment and Human Settlements Division are largely directed at this type of problem.

For the last 15 years or so, several sheets of a hydrogeologic map of Europe at the scale of 1:1,500,000 have been printed, accompanied by detailed explanatory booklets. The map is published jointly by the International Association of Hydrogeologists (IAH) and by UNESCO, with the support of the Commission for the Geological Map of the World. The printing is done by the Bundesanstalt für Geowissenschaften und Rohstoffe, Hannover.

The sheets covering the area are the following:

A-5, La Coruña	B-3, Edinburgh	C-3, Oslo
	B-4, London	C-4, Berlin
	B-5, Paris Sud	C-5, Berne
	B-6, Madrid	C-6, Rome

Sheets A-6, Lisbon, and D-6, Athinai, are expected (1989) to be published in the near future.

I. AUSTRIA

Area: 83,855 km^2 Population: 7.6 million

A. Geology

North of the Danube, which crosses the north-eastern part of the Republic of Austria from west to east, the country's portion of the Bohemian massif takes up the most space. This area is constituted of crystalline rocks (granite and gneiss) (figure 4) and along the Danube shows remnants of broad erosion terraces formed during the late Tertiary period. The moderate precipitation (700-900 mm per year), which mostly falls in the growing season and summer, is mainly consumed by evapotranspiration and not much is left to infiltration, storage and surface runoff. The uppermost areas of the Bohemian massif are an old peneplain at 800-1,000 m elevation, above which rise some inselbergs (1,100-1,300 m). Conifer forest predominates here. Towards the north-west, annual precipitation reaches 1,000 mm.

The Bohemian massif has a rugged climate. Its eastern portion is among the zones in Austria with the least precipitation. Surface streams (the largest is the Kamp, with a drainage area of 1,490 km^2) are mostly tributaries of the Danube.

1. Alps

The Alps cover the central area of Austria or about 60 per cent of the country.

The geologic structure of the mountains was thoroughly complicated by nappe tectonics and overthrusts beginning in the Mesozoic and uplifts during the Tertiary. Tectonics and exogenous evolution formed the present topographic and morphologic features that permit a simplified subdivision into the Northern Limestone Alps, the Greywacke Zone, the Central Alps and the Southern Alps.

(a) Northern Limestone Alps

The rugged Northern Limestone Alps extend from south-west of Salzburg to the eastern part of the Alps as massifs with individual underground drainage.

Triassic and Jurassic limestone dominates, in places as layers more than 1,000 m thick. The plateaus show a mean altitude of 2,000 m and the highest summits reach 3,000 m. Since the tree line on the northern flank is at about 1,700 m, large areas of the plateaus are free of vegetation and highly karstified. There is practically no surface runoff in these areas. In summer, even heavy rainfall infiltrates completely and feeds underground karstic flow.

(b) Central Alps

In the central zone of the Alps, mica schist, orthogneiss and paragneiss dominate. Phyllites extend into the greywacke zone on the northern flank which divides the Northern Limestone Alps from the Central Alps.

In the central and western part of the Alps are glaciers of remarkable extent in the Otztaler and Stubaier Alps, covering an area of 485 km^2.

Figure 4

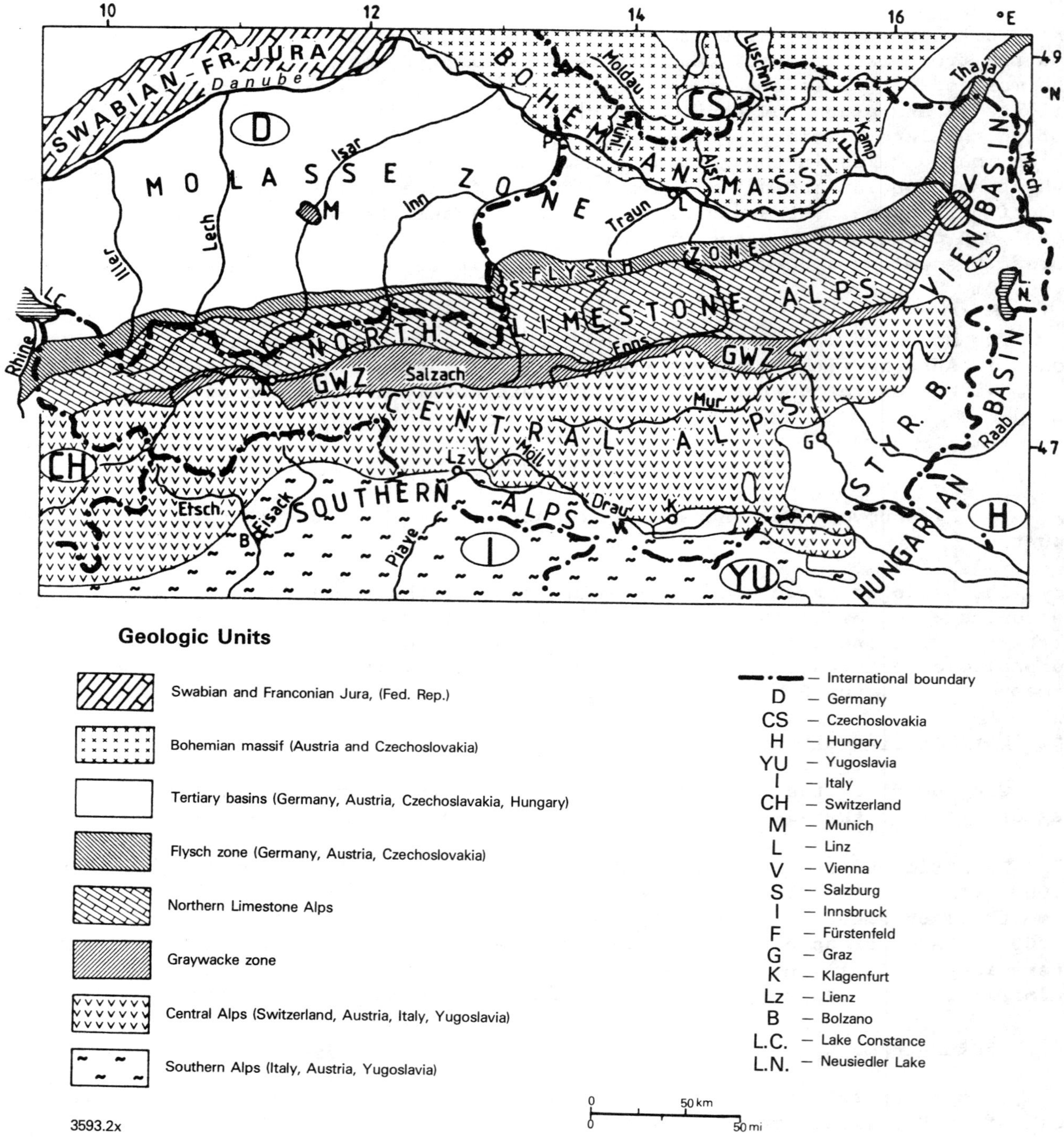
10
12
14
16
°E
49
°N
47
SWABIAN-FR.JURA
Danube
D
MOLASSE ZONE
M
Isar
Inn
Lech
Iller
BOHEMIAN MASSIF
CS
Moldau
Luschnitz
Mühl
Aist
Kamp
Thaya
March
Traun
L
FLYSCH ZONE
V
VIENN BASIN
L.N.
L.C.
Rhine
S
NORTH LIMESTONE ALPS
Enns
GWZ
Salzach
CENTRAL ALPS
Mur
STYR. B.
BASIN
Raab
G
CH
Etsch
Eisack
B
SOUTHERN ALPS
Lz
Möll
Drau
K
Piave
I
YU
HUNGARIAN
H
Geologic Units
Swabian and Franconian Jura, (Fed. Rep.)
Bohemian massif (Austria and Czechoslovakia)
Tertiary basins (Germany, Austria, Czechoslavakia, Hungary)
Flysch zone (Germany, Austria, Czechoslovakia)
Northern Limestone Alps
Graywacke zone
Central Alps (Switzerland, Austria, Italy, Yugoslavia)
Southern Alps (Italy, Austria, Yugoslavia)
— International boundary
D — Germany
CS — Czechoslovakia
H — Hungary
YU — Yugoslavia
I — Italy
CH — Switzerland
M — Munich
L — Linz
V — Vienna
S — Salzburg
I — Innsbruck
F — Fürstenfeld
G — Graz
K — Klagenfurt
Lz — Lienz
B — Bolzano
L.C. — Lake Constance
L.N. — Neusiedler Lake
0 50 km
0 50 mi
3593.2x

In the western part of the central zone, precipitation reaches 2,000-3,000 mm per year; towards the east it comes under the wind shadow of the Northern Limestone Alps.

The Southern Alps are mainly located outside of Austria. Only the Gailtaler Alps between Lienz and Villach are entirely within Austria.

2. Basins and valleys

(a) Molasse basin

In Austria, the Molasse basin is the area between the flysch in the south and east and the Bohemian massif in the north. The landscape is that of gentle hill country with Quaternary terraces along the Inn, Salzach, Traun and Enns, the main tributaries of the Danube. In the south, moraines from Pleistocene glaciers influence the topography and hydrogeology. Average elevations range from some 200 m in the east to 400 m in the west.

(b) Viennese and Styrian basins

After the molasse zone north of the Alps, the most important Tertiary basins are the Viennese and the Styrian basins on the eastern edge of the Alps. In both of these basins, Tertiary sediments also reach a thickness of about 1,000 metres. The flatland between the Danube and March (Marchfeld, altitude 150-170 m) grades northwards into a Tertiary hill country with moderate relief (170-300 m). This part of Austria has the least precipitation and of which most is consumed by evapotranspiration at high air temperatures (monthly mean for June and July 19°-20° C).

Central-Alpine rocks (mainly gneiss) form the base of the Styrian basin. The topography of the basin is formed by ridges of Tertiary sediments, overshadowed by cones and residual mountains from Miocene (trachyte and trachy-andesite) and Pliocene (basalt) volcanoes.

Last but not least, tectonics have also preserved Tertiary deposits in some inner-alpine valleys.

B. Groundwater

1. History of research

Until recently, hydrogeology in Austria was not dealt specifically apart from general geology. In the standard work, Der geologische Aufbau Osterreichs (1980) hydrogeology is covered in only 14 pages as a subchapter of applied geology, in a book of 700 pages. Karst hydrogeology is an exception, due to the circumstance that the classical karst areas of Europe were within the borders of the former Austro-Hungarian empire.

The periodical Geologie und Bauwesen since the early 1930s covered hydrogeology and its publisher, J. Stini, was the author of a standard work Die Quellen (1933). Numerous hydrogeologic works by various authors in almost all the geologic journals were followed by individual reports on the hydrogeology of landforms and provinces (K. Vohryzka, 1973; T. E. Gattinger, 1980), and others.

The geological survey has one department for hydrogeology.

The founding in 1962 of the association for hydrogeologic research in Graz by A. Winkler-Hermaden gave considerable impetus to the advancement of systematic hydrogeologic research. The expansion of the Research Center Joanneum in Styria included an Institute for Geothermics and Hydrogeology. With 16 staff members, well-equipped laboratories and field equipment, this is the primary institution working in hydrogeology in Austria.

2. Non-carbonate aquifers

The hard rocks of the Bohemian massif, and the Paleozoic greywacke and Mesozoic strata of the Central Alps include all non-carbonate hard rocks with the exception of those of the flysch zone and Cretaceous sediments. It also includes magmatic rocks and their metamorphic products (granitoids, granite gneisses and porphyroids of the greywacke zone; diabase, metadiabase and green schist of the Paleozoic series and serpentine).

The permeability of non-karstifiable hard rocks depends upon their cohesion, their degree of weathering and the nature of the weathering products. Shale, phyllite, green schist, mica schist and slaty, mica-rich gneiss show few wide joints.

The more compact schist groups, and especially granite and related rocks, generally unchanged and massively developed porphyroids and diabase as well as serpentine can carry fracture water in faulted areas, allowing for the occurrence of springs. This is seen in the north-west part of the Bohemian massif and in areas of pure quartzite, as for example in the Mesozoic of the Central Alps and within the greywacke zone.

More significant water outlets may be related to tectonically predetermined zones with more jointed rock groups (granite to quartzite), or to intercalated or wedged-in accidentals (e.g., karstified marble ranges in the Niedere Tauern, Koralpe and Gleinalpe, see figure 5).

Water supply for individual dwellings is generally not a problem but larger amounts of water for communities and industry are in most cases not available.

Larger glacier areas form a special topic. They are located above the permafrost line, and daytime temperatures and precipitation have different effects than in low-lying areas; these factors result in especially low amounts of runoff which for the 1970s was 2-5 $l/s/km^2$.

3. Carbonate aquifers

Karstifiable rocks, mainly Paleozoic limestone and dolomite, form a large part of the Graz and Mur regions. Mesozoic carbonate rocks are the dominant element in the Northern Limestone Alps and in large areas of the Southern Alps.

The hydrogeologic map (figure 5) shows no stratigraphic division for these areas. Limestone and dolomite are both karstifiable, but in different ways. The pre-Alps, with large areas of dolomite at altitudes averaging 1,700 m have

Figure 5. Austria: Hydrogeology

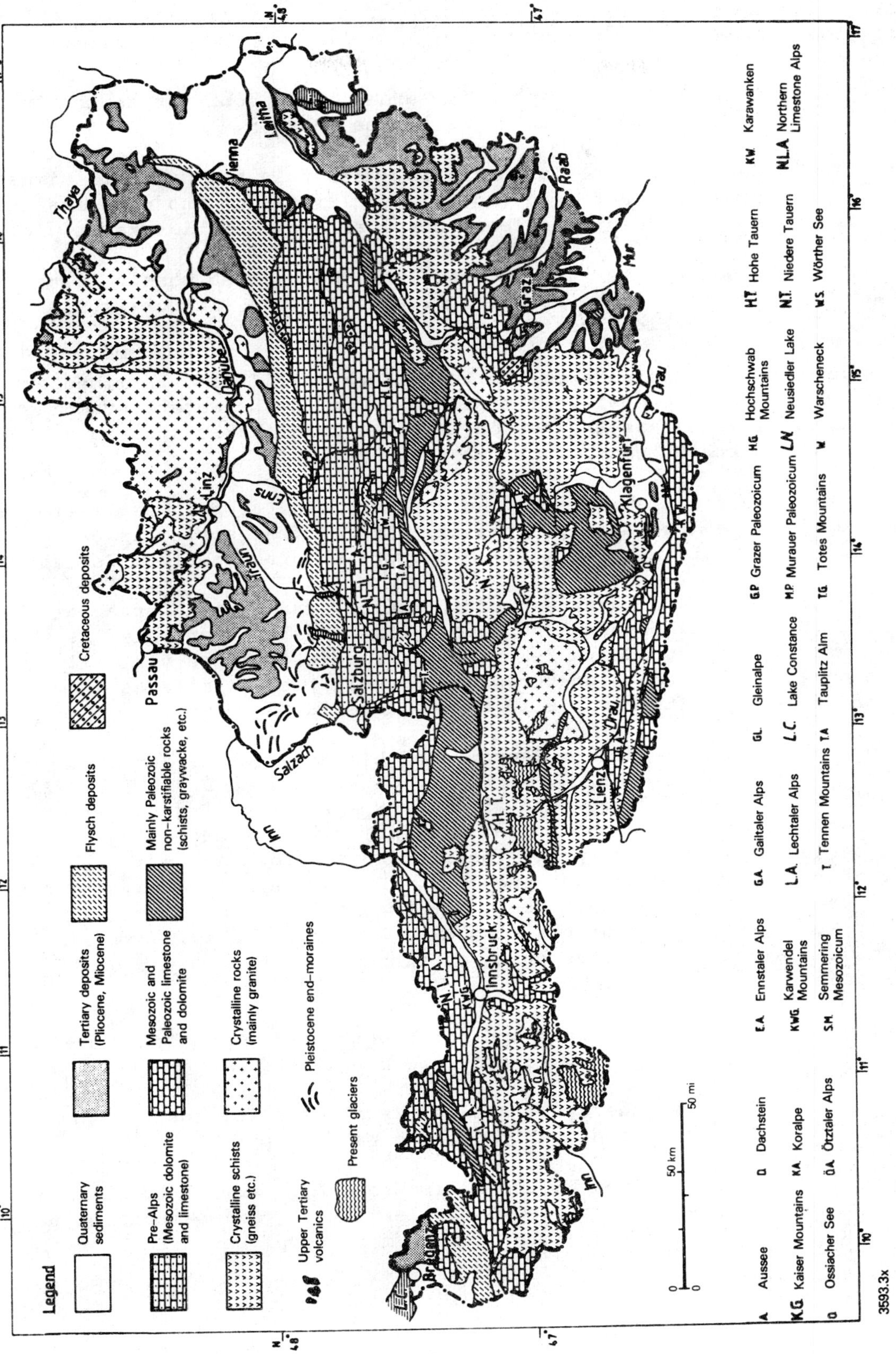

therefore been separated from the High Limestone Alps. North and east of the river Salzach, the High Limestone Alps are more or less individualized massifs. Of these, the Dachstein massif is among the most thoroughly studied in Austria. The banked, readily karstifiable Dachstein limestone (Triassic) reaches a thickness of 1,000 m.

The lack of water in the high-lying areas is in contrast to the large karst springs in the valleys and on the edges of the massifs.

The Totes Gebirge is the largest karst massif in the Northern Limestone Alps (590 km^2). It comprises Jurassic and Triassic limestone and dolomite and the plateau is among the wildest and least hospitable of the high karst plains.

As far as useable water is concerned, the Hochschwab massif is the most important karst massif in the Northern Limestone Alps. It consists mainly of Middle and Upper Triassic limestone and dolomite.

The underground drainage of the Hochschwab massif is determined by the geologic structure; the east-west trending domes of the water-retaining base (Werfen schist layers) are the reason that the large karst springs are mainly located on the northern edge of the massif.

The most important springs on the northern edge of the massif are tapped for the Second Viennese High Spring Pipeline. The Kläffer spring is one of the largest karst springs in Austria. Its maximum output is more than 10 m^3/s (see figure 6).

Owing to the severe surface karstification, the plateau areas of the limestone massifs of the High Limestone Alps have a high infiltration rate, as there is no surface runoff. The considerable precipitation infiltrates virtually without evaporation loss. In the higher regions, precipitation falls as snow from October to May and there is little water in the valleys in the winter.

Measurements of the stable isotopes of hydrogen (deuterium) and oxygen-18 in the Totes Gebirge (T. Dincer, B. R. Payne, C. K. Yen and J. Zötl, 1972) showed that the low amount of water discharged from karst springs during the winter consists only of summer rainfall stored in the massif. With the beginning of snowmelt, for a short period in March and April, the water discharged at lower altitudes is a mixture of old storage water and snowmelt. During May and June, the discharge is only meltwater from the plateau area. At the beginning of July, melt- and rainwater are mixed, and from the middle of August on, only rainwater is discharged in decreasing amounts until the onset of winter when precipitation is only in the form of snow.

Owing to the prevailing west and north-west winds, precipitation is high in the Northern Limestone Alps, up to and exceeding 2,500 mm per year.

Farther to the east there are smaller karst massifs; their large springs were the basis for the First Viennese High Spring Water supply.

West of the Salzach, the Northern Limestone Alps turn into mountain chains (Kaiser Gebirge, Karwendel Gebirge, Lechtaler Alps). Owing to steep nappes, faults and folding, the tectonic structure is much more complicated than in the eastern massifs.

Figure 6. Austria: Kläffer and Brunngraben springs discharge curves

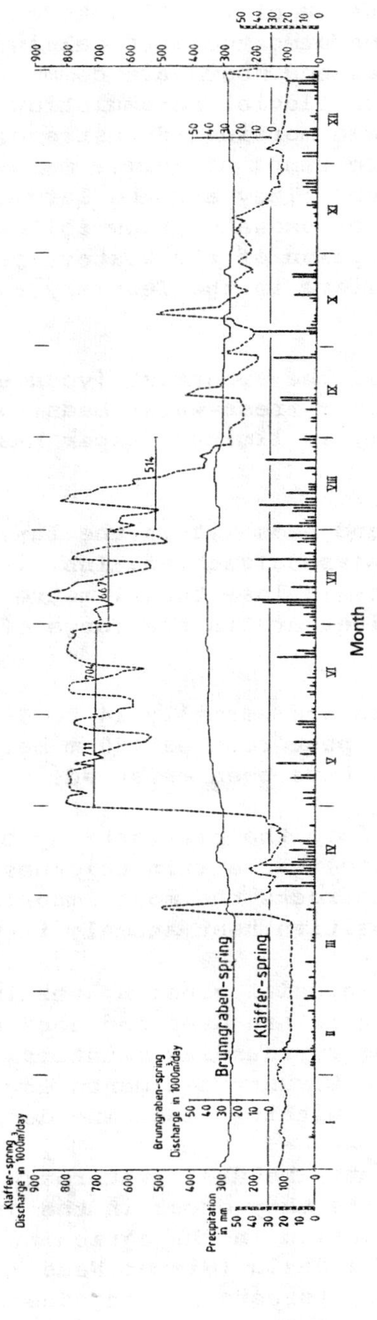

(The recharge area of the Kläffer Spring is limestone, that of the Brunngraben Spring mainly dolomite.)

4. Molasse and Tertiary basins

Between the hydrogeologically unfavourable flysch zone and the crystalline rocks of the Bohemian massif in the north, there lies the hydrogeologically complicated and diverse sedimentary body of the vast molasse zone consisting of Quaternary deposits, including moraines of the Pleistocene alpine glaciers that extended far into the foreland. The unsorted material in the moraines does not hold extensive aquifers; small lakes and moors are common. Unlike the hydrogeology of the glacial outwash plains, glaciofluvial accumulation fans and terraces along the meltwater channels follow the end moraines downstream. Gravel and sand whose clayey material was washed out by torrents of summer meltwater contains large quantities of unconfined groundwater. They are the largest and most productive aquifers in the molasse zone. The groundwater flow follows the present river systems (Vohryzka, 1973). A description of the western part of the molasse zone (table 1) gives information on aquifers in the Tertiary and pre-Tertiary formations.

Of the molasse sediment phases, the uppermost (youngest) cycle contains mainly coarse granular sediments of the upper fresh-water beds, which contain mainly unconfined shallow groundwater; only in limited deeper basins do artesian conditions occur.

The next-lower aquifers are sand beds within the Innviertel formation. The underground flow is towards the rivers Salzach and Inn, into which the artesian waters discharge. The flow velocity is less than 1 metre per year and the residence times close to the Inn River are in the range of 30,000 to 40,000 years (Andrews and others, 1985).

The depths of confined aquifers are normally 14 to 40 m below land surface. Individual wells are deeper; the deepest reaches 480 m below land surface. Discharges between 1 l/s and 35 l/s have been measured.

Jurassic and Cretaceous rocks form the pre-Tertiary basin floor in which the Malm (Jurassic limestone and dolomite) can attain thicknesses up to 500 metres. Karstification makes these carbonate rocks the most important deep aquifers. The uplift zones along faults show a positive heat anomaly in this area.

Owing both to tectonics and to glacial erosion, the inner-alpine valleys of the Rhine, Inn, Salzach and Enns rivers are deep and considerable groundwater reserves are known, but springs from surrounding mountains supply most of the villages. In the Salzburg basin, Quaternary sediments down to a depth of 300 m contain thick sand and gravel layers within lacustrine deposits.

In the Viennese basin north of the Danube, Quaternary deposits are of minor thickness with little groundwater. As they occur in the driest part of Austria, groundwater must be pumped for irrigation in the agricultural "Marchfeld" area. In the southernmost part of the Viennese basin (Wiener Neustadt), the land surface is also dry and supports only large pine forests serving the turpentine industry. Underground, however, there is a graben 40 km long and 2-8 km wide filled to a depth of 50-150 m with Pleistocene gravel of high permeability. Rainwater and surface runoff from marginal Alpine mountains infiltrate into this graben. The north-western margin of the Viennese basin follows large faults along the spur of the eastern Alps. Here, the karstified limestone along the so-called "thermal line" may reach considerable depths and, rising thermal water led to the development of spas whose history goes back to Roman times.

Table 1. Austria: Stratigraphy of the Molasse basin

			EPOCH	STAGE	LITHOLOGY	FORMATION	AQUIFERS	OIL GAS	THICKNESS IN METRES
	QU.		HOLOCENE PLEISTOCENE		SAND, GRAVEL, BLOCK	MORAINES and TERRACES	IMPORTANT	--	0 - 300
MOLASSE SEDIMENTS	TERTIARY	NEOGENE	PLIOCENE	UPPER PLIOCENE	SANDY GRAVEL	. .	MINOR	--	0 - 300
			MIOCENE	PANNONIAN-KARPATIAN	GRAVEL, MARL, SANDSTONE	COAL-BEARING FRESH-WATER BEDS	MINOR	--	
				OTTNANGIAN	MARL, GRAVEL, SAND	INNVIERTEL FORM.	MAJOR	--	0 - 700
				EGGENBURGIAN	MARL, SHALE, SAND	HALL FORM	MINOR	GAS	0 - 800
		PALEOGENE	OLIGOCENE	EGERIAN	GRAVEL, MARL, SAND	UPPER PUCHKIRCHEN FORM	MINOR	GAS	0-1050
						LOWER	MINOR	GAS	0-1000
				RUPELIAN	CLAY, MARL, DOLOMITE, LIMESTONE	SHALE STAGE BANDED MARL LIGHT MARLY LIMEST.	--	--	0 - 550
				LATTORFIAN	CLAY, MARL	FISH-BEARING SHALE	--	--	0 - 40
			EOCENE	UPPER EOCENE	LIMESTONE, MARL, SANDSTONE	SANDSTONE STAGE LIMNIC SERIES	--	OIL, GAS	0 - 120
BASIN FLOOR SEDIMENTS	MESOZOIC	CRETACEOUS	UPPER CRETACEOUS	CAMPANIAN-TURONIAN	SHALE, CLAY, SANDSTONE	. .	MINOR	OIL	0-1000
				CENOMANIAN	SANDSTONE	. .	IMPORTANT	OIL	
		JURASSIC	MALM		LIMESTONE, DOLOMITE, SANDSTONE	CARBONATE-GROUP BASAL SANDSTONE	IMPORTANT	OIL	0 - 500
			DOGGER		OOLITIC LIMESTONE	. .	--	--	. .
		PERMO-TRIAS			SANDSTONE, CLAYSTONE	. .	--	--	0 - 230
	PAL	CARB	UPPER CARB WESTFAL		SANDSTONE, CONGLOMERATE	. .	--	OIL	0 - 40
CRYSTALLINE BASEMENT					MASSIVE HARDROCK	. .	--	--	. .

The Styrian basin south-east of the Alps is divided by step faults into smaller basins, some deeper than 3,000 m, just 40 km from the Alps. The substratum of the Tertiary basin fill includes Paleozoic carbonate rocks and phyllite over the crystalline basement rocks of the Central Alps (mainly gneiss).

The influence of the Pleistocene glacier runoff on the permeability of the Quaternary valley fill is expressed drastically in the Styrian basin. Unlike the northern margin of the Alps (molasse zone), in the south-eastern Alps, the Pleistocene glacier tongues reached only into the inner Alpine valleys (end moraine Mur Valley, figure 5). The Quaternary gravel terraces of the Mur River on the margin of the Alps and in the Styrian basin (from Graz to the Yugoslavian border) thus were formed by meltwater originating from the Würm glaciers. Their fine clay material having been washed out, they have good permeability and yield abundant groundwater. This situation is different for the Raab River and its tributaries. The recharge area of this river was not covered by a glacier in the Pleistocene. The climate was periglacial and the daily fluctuation between freezing during the night and thaw during the day resulted in solifluction, i.e., soil flow that transported fine clayey soil particles down the slopes to the valley floor. The high percentage of clay in the valley fill resulted in poor effective porosity and low permeability. As a consequence, the yield of wells is low and the quality of unconfined shallow groundwater poor because of pollution by flood and high natural content of iron and manganese. Artesian wells supply water to villages and small towns.

5. Water quality

The physical and chemical quality of water depends on natural (geologic and environmental) and man-made factors.

The detailed chemical data in table 2 provide basic information. No. 1 is typical for waters low in mineralization in crystalline rocks of the Bohemian massif and Central Alps. These waters generally have total mineralization of 100 to 120 mg/l and pH of only 5.6 to 6.2; they must be treated (marble grain solution filter) to protect household equipment (washing machines, dishwashers, hot-water heaters). Low potassium content can lead to deficiency symptoms as well. These waters taste stale and flat.

The mineralization of waters from carbonate rocks (limestone and dolomite) depends on their degree of karstification, and not on the formation to which they belong. Water flows quickly through highly karstified massifs with caves and wide joints and has little opportunity to interact with rock surfaces. Thus, total mineralization is often a scant 300 mg/l, but the pH is always around or above 7. It is a common misconception that karst water must have a high total (in this case carbonate) hardness. 1/

1/ It is no longer customary to give the hardness of water in German, French or English degrees of hardness (German degree of hardness or °dH = meq Ca + meq Mg X 2.8). Older essays and instruction manuals, however, still use this concept, which is also shown in table 2.

Table 2. Austria: Chemistry of water from different geologic units

No. and locality	Key rock	pH	el. Conductivity uS/cm, 20° C	Ca^{2+} mg/l	Mg^{2+} mg/l	Na^+ mg/l	K^+ mg/l	HCO_3^- mg/l	Cl mg/l	NO_3^- mg/l	SO_4^{2-} mg/l	CO_2 mg/l	°dH*	Remarks
1. Bohemian massif	Granite	5.8	..	12	4.0	9.2	..	16	6.3	32.0	17.5	22	2.6	2/ Radon 42.6 nCi/kg
2. Prealps	Jurassic marl and limestone	8.4	197	40	3.4	1.5	0.3	128	0.4	2.8	8.4	..	5.4	1/
3. Northern Limestone Alps	Paleozoic limestone	7.7	250	49	5.8	1.0	0.6	163	0.8	4.0	12.6	..	8.2	1/
	Triassic limestone	8.1	177	39	2.3	0.2	0.1	122	0.4	3.3	5.9	..	6.5	1/
	Triassic dolomite	8.2	199	26	12.6	0.1	0.1	121	0.2	1.4	16.7	..	6.6	1/
4. Wells in Salzach valley and Molasse Zone	Quaternary gravel and sand	..	..	104	2.4	..	..	323	9.1	6.0	..	..	10.5	1/
5. a) Southern Alps	Gravel and sand	..	..	59	9.8	10.1	0.6	184	20.1	..	30.1	..	10.5	1/
b) Gailtal Alps	Gypsum	..	..	517	103.0	3.7	2.1	189	2.7	..	1,500.0	..	68.4	1/
c) Gail valley	Moraines	..	..	45	19.0	0.6	0.5	177	0.9	..	45.3	..	10.7	1/
6. Mur valley a) Polluted well	Quaternary sand and gravel	..	..	119	30.2	64.7	74.2	387	104.7	140.0	74.0	..	8.5	1/
b) Grazer water work		..	..	81	17.1	5.4	1.9	270	10.5	8.0	33.5	..	15.0	1/
7. Southern Vienna basin	Gravel	..	..	99	39.7	10.4	1.8	307	17.2	6.0	147.5	..	23.0	2/
8. Molasse Zone	Sand of fresh-water molasse	7.2	354	99	24.7	5.3	1.4	405	9.6	..	20.3	55.0	19.3	1/
Artesian waters		7.8	569	97	24.3	6.1	1.3	397	4.0	0.1	22.1	..	19.1	1/
9. Styrian basin a) Art. water Blumau 1	Sand, Pannon.	7.7	385	60	18.8	11.0	1.9	299	3.0	0.0	8.0	..	12.9	1/ Well, horizon in 60-65 m depth
b) Blumau 2	Sand, Sarmat.	7.8	385	49	9.3	45.6	2.6	305	5.0	..	7.0	..	9.0	175-177 m depth
10. Styrian basin	Tertiary sediments	..	..	263	197.4	106.2	6.8	2,050	22.7	..	6.4	2,000	81.0	2/ Mineral water H_2SiO_3 66 mg/l
11. Molasse basin formation water	Upper Puchkirchen formation	..	..	254	159.0	7,740.0	31.6	256	13,000	..	..	..	..	2/ Bad Hall spa; information; artesian flow

* Degree of total hardness.

1/ Analyses by Research Center Graz.

2/ Local information.

Less karstified solid carbonate rocks provide waters with average values for dissolved substances (calcium-hydrogen-carbonate waters, table 2, Nos. 2 and 3), and with pH values between 7 and 8 they are pleasant tasting. Groundwater in gravel terraces has higher carbonate mineralization. The degree of mineralization depends on the percentage of carbonate gravels in the terrace (table 2, Nos. 4, 6b, 7). In some cases, these waters are too highly mineralized for use in household equipment and pipes, and must be treated to prevent precipitation. There are considerable differences between the shallow confined aquifers of the Tertiary fresh-water molasse (No. 8) and similar aquifers in the Styrian basin (No. 9). The former, like the Quaternary deposits, are mainly of Alpine dolomitic limestone, while the Styrian basin mainly received deposits from the surrounding crystalline mountains with less carbonate rock content.

Both natural and man-made factors can exert major chemical and physical influences. Layers of gypsum in the underlying beds of Triassic limestone in the Gailtaler Alps give the water (No. 5b) a high sulphate content of up to 1,500 mg/l that makes it unfit for human consumption. On the other hand, the high Cl and NO_3 content of No. 6a is clear evidence of human pollution. When surface waters are dammed up, this sort of human pollution may also occur. This factor must be considered when hydroelectric power plants are built on rivers.

C. Groundwater exploration

1. Shallow unconfined groundwater

Preparatory studies are necessary if a large amount of water is to be tapped in an unstudied area where individual domestic wells suggest plentiful supplies of unconfined shallow groundwater.

Test-boreholes are drilled to include sampling for grain size classification, study of permeability and porosity, collection of water samples, information on water table fluctuation and flow direction, electrical and geophysical logging, and application of the point dilution method.

Refraction seismic studies are made to determine the morphology of the underlying bedrock of the aquifer, or the thickness of the water-bearing beds.

Resistivity logs of the boreholes show especially well the alternation of sand layers and clay. These measurements should always be taken when large wells with output of more than 100 l/s are planned.

In Austria, pumping tests are made for periods of up to six months for high-yielding wells. In general, the tests are continued until the water levels have stabilized. Water samples are collected repeatedly, and the area to be protected against pollution is, in the end, determined. The water pumped out must be transported far enough away that it cannot reinfiltrate the area from which it was taken.

Mathematical models are used in Austria mainly to calculate water balances for shallow aquifers and to predict the effects of man-made activity (e.g., damming of rivers) on the groundwater body. Two-dimensional computer models are usually used, in which stationary and non-stationary cases are calculated either according to the method of finite elements or the so-called difference method.

The following example shows the results of calculations for a groundwater field in the Mürz valley in Styria (the Mürz River is a tributary of the Mur River). The groundwater in the sandy gravel, permeable aquifer communicates with the Mürz River. Detailed studies were necessary because the waterworks of the town of Kindberg (population about 10,000) are located too near the river and water taken includes a high percentage of bank filtrate. The water, however, is not treated in any way. The model predicted how damming of the river for a small hydroelectric power plant would affect the groundwater situation and consequently the waterworks. The calculations assumed a maximum input to the waterworks of 40 l/s and showed that as the waterworks increased the situation would worsen.

2. Deep confined groundwater

Deep bores to tap confined aquifers for water supply are possible only to a certain limited depth owing to increasing mineralization and temperature of the water. Depths in excess of 250 m are only occasionally tapped in both the molasse zone and the Styrian basin.

Seismic data and borehole logs are especially important in geothermal water exploration. The most common measurements include: SP (self potential) log, i.e., measurements of specific electric resistance using probes with different spacing and radius of investigation, sonic logs for porosity measurements (in special situations, also suitable for recognizing joints and cavities in hard rock), and measurements of natural gamma radiation (gamma ray log). Together with the petrographic record of cuttings and cores, these methods provide information on the aquifers and estimates of their permeability and the quantity of stored water.

A boring in the Styrian basin at depths of 1,540-1,675 m showed clay-free water-bearing sand beds separated by thick clay and marl layers and covered by a clay and marl series with thick sandstone layers.

Log correlations with neighbouring borings show that these rather porous sands (25 per cent porosity according to the sonic log) are quite extensive and thus form an aquifer important as a source of geothermic energy. The temperature in the formation is about 70° C.

3. Research technology

The research technology for water supply from shallow wells or springs issued from rock formations includes hydrogeologic survey of the area under investigation with regard to the geology, discharge from springs, measurement of pH value, water temperature, and collection of samples for physico-chemical measurement. Representative springs are selected for periodic or episodic (after rainy or dry seasons, thunderstorms, floods, snowfall or snowmelt) physico-chemical measurements and discharge control by limnigrams. Periodic or episodic measurements are made of environmental isotopes in rainwater, spring and wells. Groundwater tracing experiments use environmental isotopes, both radioactive and non-radioactive, and other substances.

Tracing experiments are not limited to karst regions alone. They were also successful in solving problems related to a magnesite mine in the crystalline rocks of the Central Alps, in identifying groundwater flow beneath basalt flows in Kenya

(Austria-sponsored project) and in investigating flow direction and velocity of shallow groundwater in gravel and sand of alpine valley fill.

Groundwater tracing techniques have special importance in karst hydrology because of the rapid movement of water in open joints, the lack of a filter effect in karst rocks, and the leakage of surface water into the karst water body. They are also used to identify recharge and discharge areas. The following properties are required for substances used for groundwater tracing: they must be non-toxic, water-soluble, traceable in small quantities, chemically resistant, incapable of absorption by soil and rocks, not susceptible to ion exchange, suitable for simple and rapid analyses and inexpensive.

Salt (NaCl) and potassium chloride (KCl) were the first substances to be used successfully in karst water tracing. Lithium chloride (LiCl) has been used successfully in some cases.

A disadvantage of these salts is that large quantities are necessary for tracing over longer distances and for labelling large quantities of water. An injection of 500 kg or more may be necessary for distances of only 3 to 5 km.

An advantage is that exact quantitative calculation of the output of the injected material is possible with titration of chloride and flame photometry for sodium.

Fluorescein dyes, i.e., sodium fluorescein (uranin), rhodamine and sulphorhodamine in various commercial forms, are commonly used in tracing experiments. Trace amounts of these dyes in solution were first detected with an ultraviolet lamp. Currently, spectrofluorometers are used for exact quantitative measurements.

The fluorescence of uranin solutions up to a dilution of 10 mg/m^3 is visible without any equipment. A spectrofluorometer can measure the fluorescence of concentrations as small as 0.02 mg/m^3. As the fluorometer detects different wave lengths, various dyes can be injected simultaneously into several sinkholes. Such dyes include uranin, eosin (sodium tetrabromide fluorescein), rhodamine B (tetraethylrhodamine chlorhydrate), pyranin (1-hydroxyprene 3, 6, 8 trisulphonic acid-trisodium salt), naphtion salt (1-naphthylamine- 4 sulphonic acid salt) and tinopal.

The "activated charcoal method" with uranin has become popular for karst water tracing.

Lycopodium spores (lycopodium clavatum) are pollen used for karst water tracing. The diameter of a single spore is 30-35 micrometers. The spores are added to the water as suspended (not floating) matter. Because they are solid drifting bodies, though microscopic in size, they are limited in application to the exploration of the qualitative distribution of groundwater in fissured, mainly karstic rocks. These spores may become the most important tracer in such areas, since they are larger than the typhus bacterium (length approximately 8 microns) and a positive result indicates that the area must be protected. Use of lycopodium spores in karst regions increased when a method was developed to dye the spores different colours.

When springs are to be studied, plankton nets are installed to filter the spores. The water emerging from many springs around a karst massif may be checked continuously and the experiment is economical as it is possible to conduct simultaneous examinations under identical hydrological conditions without additional observation time and equipment, as dyed spores can be injected simultaneously at several seepage points.

Only the most commonly used artificial tracers have been discussed above. For special studies in a limited area, for example leakage from a reservoir, artificial radioisotopes have been used successfully, with limitation of the amount of material injected and of human exposure to radiation. The exact measurability of the tracer is an advantage. However, it is not possible to use artificial radioactive tracers for experiments over long distances and involving large areas.

With the so-called neutron activation analysis method for groundwater tracing, a non-radioactive compound is injected. The samples collected from springs are then irradiated with thermal neutrons in a nuclear reactor.

Bacteria (Serratia marcescens) have been used as artificial tracers, particularly when a bacteriological threat to a water supply is under study. Other methods, such as the use of odoriferous substances are little more than curiosities.

A hydrologic model of the karst system can be made with data from isotope, discharge and tracing experiments. The model system consists of "cells" describing water quantity and tracer concentration depending on an input-output function. A given quantity of water with a given concentration of tracer enters a cell representing a fictive recharge area, and leaves the cell with the next input.

4. Data banks

Data banks for surface and groundwater data in Austria are kept at the Central Hydrographic Office in Vienna. This institution publishes the Hydrographisches Jahrbuch von Osterreich (most recent volume 89/1981, Vienna, 1984). The yearbook includes all data on precipitation, air temperature, surface runoff and groundwater. Monthly means and extreme values for groundwater level and temperature are published, along with a registry of groundwater measuring stations.

The Hydrogeological Department of the Geological Survey in Vienna can provide general information on inventories of wells and boreholes. Detailed information on inventories and data banks as well as monitoring networks is available at the governmental offices of the individual provinces in Eisenstadt (Burgenland), Klagenfurt (Carinthia), Linz (Upper Austria), Salzburg (Salzburg), Graz (Styria), Innsbruck (Tirol) and Bregenz (Vorarlberg).

D. Groundwater development

In Austria, virtually all drinking water originates from groundwater sources (locally supplemented by surface water). Owing to the different types of plants used, and the different factors regarding the size of protected areas and the yearly variations in yield, it is useful to distinguish spring water, shallow unconfined groundwater exploited by wells, and artesian wells exploiting confined groundwater. Table 3 shows some examples of the most important sources of water supply for towns in Austria.

Table 3. Austria: Municipal water supply in various regions

City	Inhabitants supplied by central supply (1983-1985)	Percentage of central supply through:			Length of pipe (km)	Litres per person per day	Price (Schillings per cubic metre) a/
		Spring water (mainly karst-water)	Shallow ground-water	Artesian water			
Vienna	1 500 000	83.4	15.4 b/	--	3 194 c/	128	9.80
Graz	250 000	--	100 b/	Few private wells (brewery, etc.)	661 c/	242	9.24
Innsbruck	115 000	99 (karst tunnel)	1	--	22 d/	383	4.08 to 7.30
Klagenfurt	83 900	7	93	--	424 c/	267	7.50
Linz	248 000	--	100	--	500 c/	250	6.05
Salzburg	136 000	50	45	--	370	381	8.25
Fürstenfeld	6 800	--	--	100	62 c/	192	12.0

a/ Rate of exchange for 1987-1988 was 12 Austrian schillings per United States dollar.

b/ With surface water recharge.

c/ Pipeline plus distribution network.

d/ Pipeline only.

1. Springs

The first central water supply for the city of Vienna was water pumped from a drainage system near the city (10,000 m^3 per day, or about 16 l per day per person). The quality was bad and cholera and typhus epidemics occurred. The geologist Eduard Suess suggested the use of spring water from the Northern Limestone Alps and this proposal was accepted in 1866 by the Vienna city council. Work on the channel, aqueducts and tunnels was begun in 1869. The distance from the largest spring, the Kaiserbrunn, to the reservoir in Vienna is 90 km; this was the First Viennese High Spring Water pipeline. Construction was completed in 1893. The initial capacity of the channel was 138,000 m^3 per day; later it was increased to 200,000 m^3 per day. A tunnel was built in 1967-1969 to feed more karst water into the transport pipeline.

In March 1900, it was decided to build the Second Viennese High Spring Water pipeline with a capacity of 200,000 m^3 per day (Kochanek and Donner, 1967). This second supply transport is 200 km long and went into operation in December 1910.

Supply of the city of Innsbruck is an especially good example of utilization of karst water. As early as 1890, a modern drinking-water supply was combined with a power station through the construction of a tunnel in the the Nordkette (figure 7). The Nordkette shows a distinct nappe structure. The apron on the south slope of thick Höttinger breccia and the clay content both of the rauhwacke and the Bunter sandstone lead with the northern dips of the layers to a backwash of water in the karstified Muschelkalk (limestone). A tunnel was made through the backwash layers into the karst-water body in the Muschelkalk. The karst water yield is 800 l/s in winter and 1,500 l/s in summer. The difference in elevation of 445 m between the tunnel and the valley floor is used to produce electrical power averaging 400 kW yearly.

2. Shallow unconfined groundwater

Figure 8 shows the distribution of large groundwater fields in Austria. As with spring water, the most important water consumers are large cities and industry. An example is the water supply of the city of Graz, capital of the province of Styria and second largest city in Austria. Its water supply originates exclusively from exploitation of shallow groundwater in the Mur valley. Today, in Graz there are two horizontal filter wells each in the waterworks Graz-South, Graz-Andritz and Graz-North (Friesach in the Mur valley north of Graz). It is estimated that an adequate water supply is now secured for many years to come. Average consumption is 58,000 m^3 per day (peak 76,000 m^3 per day or 880 l/s).

With regard to the material used for pipelines, asbestos-cement (AC) pressure pipes have become increasingly popular; at present, both such piping and also synthetic polyethylene and polyvinylchloride (PVC) are in use.

The newest additions to the Third Viennese Water supply system can draw on large groundwater occurrences in the southern part of the Viennese basin. This area is called the Mitterndorf fault trough. After large-scale pumping tests in 1963-1966, it was decided in 1970 to include two existing horizontal filter wells in the Mitterndorf fault trough with capacities up to 800 l/s (69,120 m^3 per day). These horizontal filter wells (Ranney wells) are closed casing wells made of reinforced concrete with an inner diameter of 4 m and walls 40 cm thick. They

Figure 7. Austria: Water supply of Innsbruck

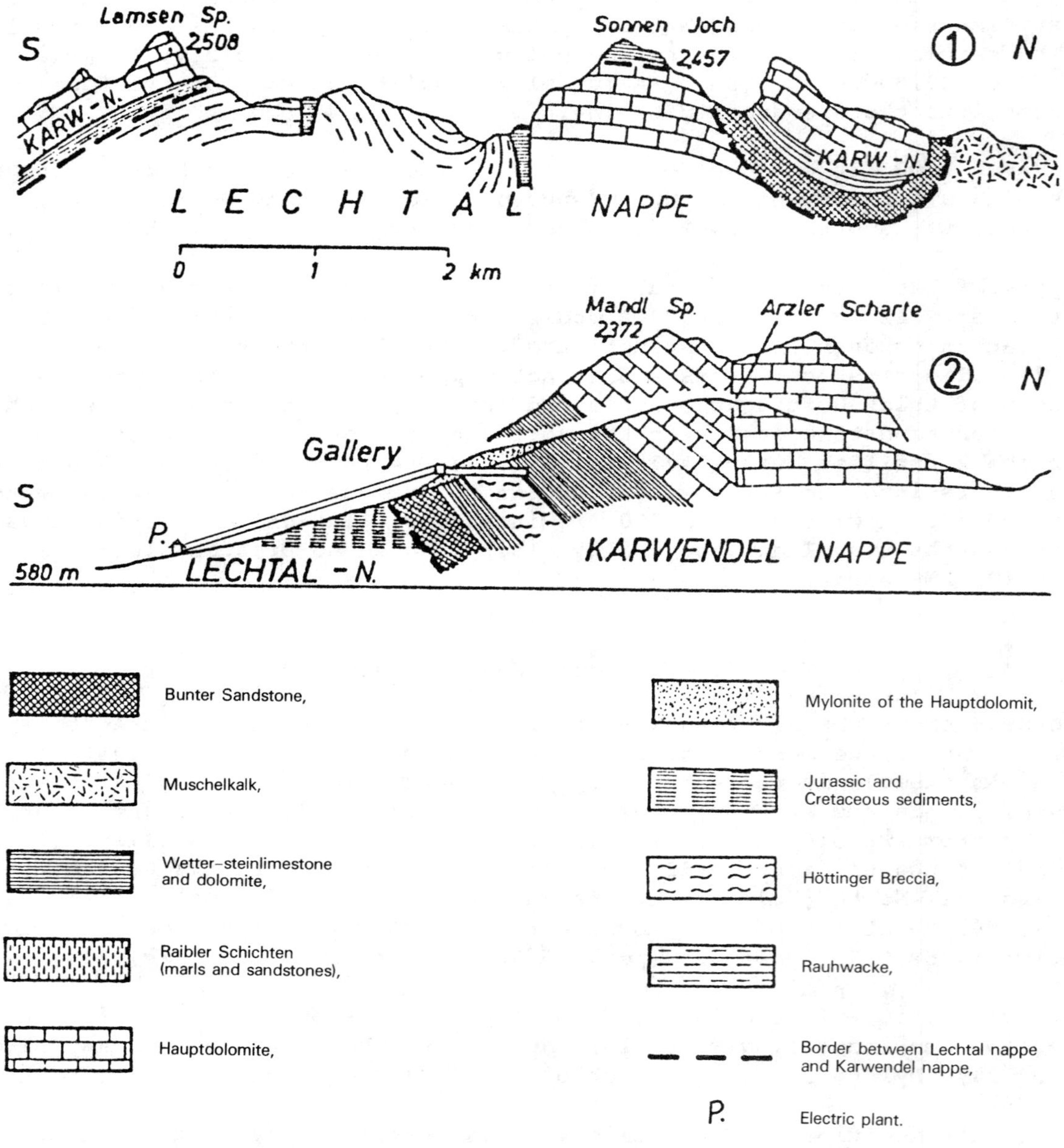

extend to a depth of 25 m under the land surface and are sealed at the bottom by a reinforced-concrete sole about 2 m thick. One-and-a-half metres above the well sole, 12 horizontal filter pipes with lengths of 25-30 m extend from the well shaft in a star pattern into the groundwater body. After completion of the wells and conduits, pump tests of up to 400 l/s were performed over a period of several months. These and other tests showed a groundwater reserve of 1.2 km^3 for the Mitterndorf fault trough. The pump station was built into the northern end of the Mittelfeld tank.

In sum, the Quaternary river deposits in the Viennese and Graz basins in south-east Austria and the Pleistocene gravel channels in the Molasse basin north of the Alps are the largest and most reliable sources for safe, potable unconfined shallow groundwater in Austria.

3. Artesian wells

Problems of exploitation arise in regions where hydrogeologic conditions demand the use of confined aquifers through artesian wells. In the molasse zone or Enns valley, in most cases water supply is available from springs or areas with unconfined shallow groundwater. In the Styrian basin, however, confined aquifers have to be developed. In this area, artesian wells supply three small towns and a large number of villages (see table 2, No. 9); the largest of the towns has six artesian wells. The aquifers are located at depths between 60 and 258 m. The deepest boreholes discharge 12 and 15 l/s, respectively, and the total discharge is 35 l/s. The problems are a slow decrease in output over the course of years and high water temperature from deep wells (water temperature of the deepest well is 20° C and is used with a heat pump to heat the town hall). 2/

4. Water associations

Water associations have been formed in places where there is a substantial water potential close to small localities having insufficient water supply. An improved water supply is often sought when demand increases as a result of the development of tourism or industrial and agricultural operations in an area. Requirements include reasonably favourable orographic conditions that do not involve costly large-scale pumping stations, and also favourable circumstances for the laying of pipelines. In these water associations, users and wells are usually widely scattered and construction of a complex network of pipeline system with different diameters is necessary.

One example is the Water Supply Association of Northern Burgenland around Neusiedler Lake. The transport pipeline, not including the local networks in the village, has a length of some 410 km and serves an area of some 1,900 km^2 with 53 communities, 130,000 residents and 50,000 vacationers. Groundwater extraction is from 11 wells with an output of 40-50 l/s. It supplies some 30 reservoirs (50-12,500 m^3) with a total storage of 94,000 m^3 and 30 pumping stations.

2/ For additional information on mineral and thermal waters see Kupper and Wiesbock, 1966, and Zotl, 1980-1985.

Figure 8. Austria: Large groundwater fields and mineral and thermal springs

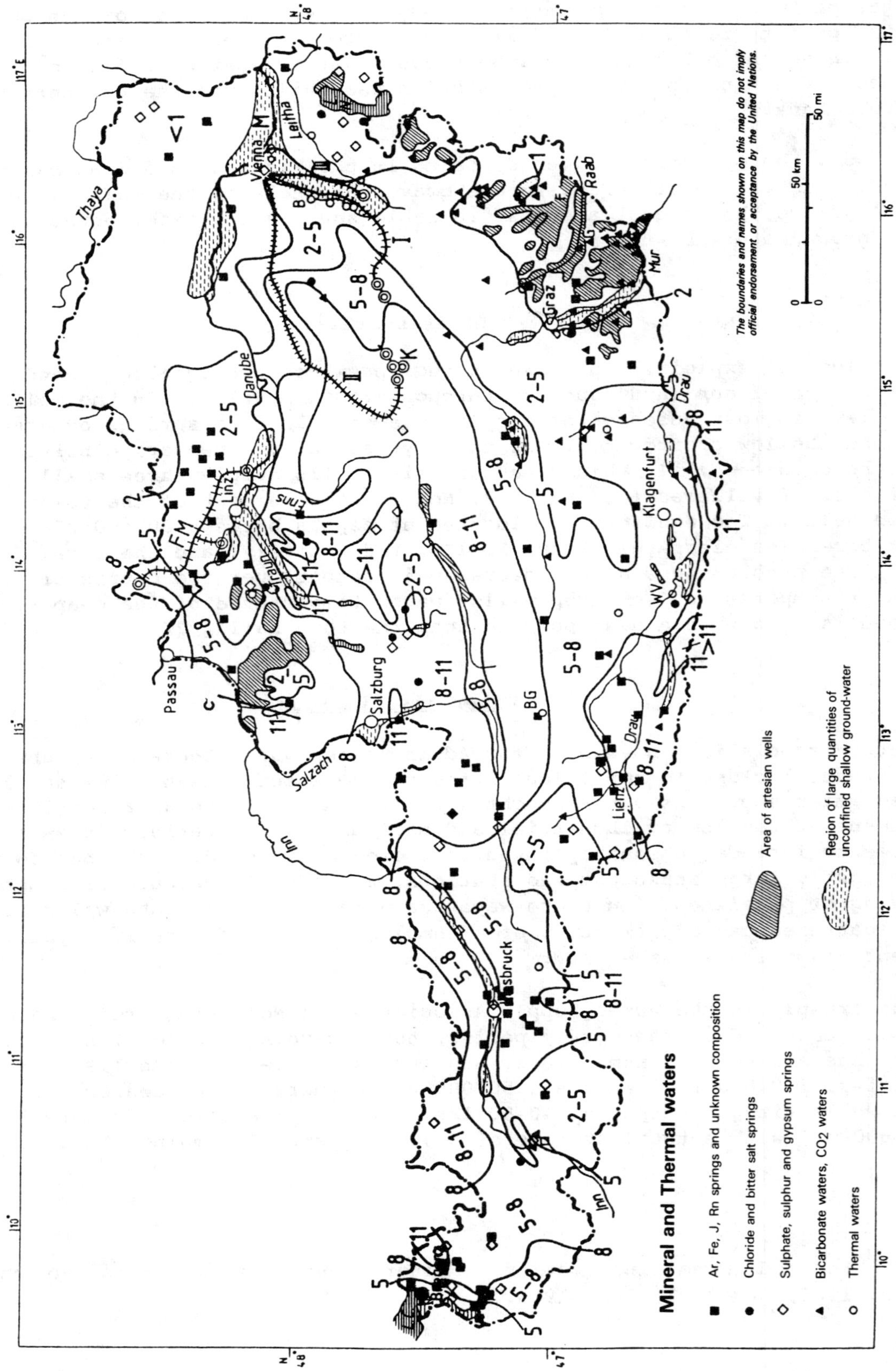

Large cities and areas with permanent lack of water are supplied from distant sources. One example is the Mûhlviertel Regional Water Supply System, which covers most of Upper Austria north of the Danube on the Bohemian massif. Water is supplied here from three marginal areas, springs in the extreme north-west and two well areas in gravel terraces along the Danube west and east of Linz to the middle of the mountain region. The main difference as compared with a water association is that, in addition to the external supply to the area, little consideration can be given to orographic and geological conditions (relief) for the transport pipelines, and a relatively large number of high-capacity pump stations is necessary. Depending on the pressure situation, AC pressure pipes or nodular-cast pressure pipes must be used. The pumping stations are located at altitudes between 243 and 869 m.

Although this regional supply system does not cover a large area, the difficult terrain with sharp orographic contrasts renders it expensive. It consists of a 108,400-m pipeline (91,918 m of which is transport pipeline), including 71,787 m of AC pressure pipe, 26,990 m of nodular-cast pressure pipe, 9,375 m of PVC pipe and 250 m of steel pipe. The eight water towers (500-1,600 m^3) have a relatively small volume of 7,500 m^3. This small storage volume must be compensated for with high transport performance of the pumps of 1,200 m^3/h and yearly power requirements of some 1,200,000 kW. This is needed to overcome a total transport height of 630 m and a maximum individual transport height of 207 m. The pumps are usually run after 10 p.m.

Differing chemical composition of water has led to special problems in exploitation. The water taken from springs in the north-western corner of the Austrian part of the Bohemian massif shows a pH value between 5.6 and 6.1. The total hardness is only 0.7 to 1.8° dH, or 0.25 meq Ca plus Mg to 0.7 meq Ca plus Mg. The corrosive CO_2 lies between 9.7 and 21.9 mg/l. Water of this corrosiveness requires treatment with antacidification equipment, which generally has four chambers with 15 m^3 of filter gravel each (marble). In contrast, the groundwater of the two wells located in the Danube has 10° and 20° dH, respectively (i.e., 3.5 meq Ca plus Mg and 6.4 meq Ca plus Mg, respectively). Without treatment, the mixing of these two water types would lead to technical problems.

5. Irrigation

In Austria, the exploitation of groundwater for irrigation is generally limited to small areas and nurseries. The Marchfeld east of Vienna is an exception. There, 640 km^2 are cultivated intensively (that is, 4 per cent of the total agricultural area of Austria); it produces 10 per cent of the cereal grains, 17 per cent of the sugar beets and includes 2,100 ha of vegetable crops. With the lowest annual rainfall in Austria and low mean runoff, this area is irrigated with pumped groundwater. Over the years, an increase of the area under intensive cultivation has led to overdraft of groundwater resources and a slow but steady lowering of the water table by 2.5 m in the period from 1951 to 1982. The uneven bedrock morphology at the base posed the danger that the groundwater body could break up into individual basins with stagnant water whose quality would deteriorate from infiltrating waste water. These problems could be solved only with groundwater recharge by continuing infiltration of surface water. An artificial channel south of Klosterneuburg brings Danube water to intermittent natural streams, which were widened to improve infiltration.

For this purpose, some 100 km of channels were built or improved not only for groundwater recharge but also to improve surface irrigation and remove waste water. The main channel has an average width of 34 m, water depth of 0.9-1.8 m, depth of the sole under land surface of 4-6 m and maximum flow-through of 15 m^3/s (Kaupa, 1985). When completed the project will have cost some 2 billion Austrian schillings (about $US 170 million).

E. Agencies involved in exploration and development

Universities in Austria do not offer a hydrology curriculum. The technical schools in Vienna and Graz have institutes of hydrology. Training in hydrogeology and hydrology may be obtained at the Department of Geology and the Department of Water Economy at the Agricultural University in Vienna.

The Geological Survey of Austria and the Institute for Geotechniques (GTI) at the Bundesversuchs- und Forschungsanstalt Arsenal in Vienna have programmes for groundwater exploration and exploitation.

The Styrian provincial government founded the Joanneum Research Centre (FGJ) in Graz to make up for deficiencies in research and education. After the Reactor Centre in Vienna, it is, with 26 departments, the second-largest paragovernmental research institution in Austria. Unlike the universities, the Centre's departments work according to the economic principle of supply and demand. The Centre's Institute for Geothermics and Hydrogeology (IGH) has a well-trained staff educated in modern techniques for field and laboratory work and for the development of models for practical projects. The Institute has technical equipment at its disposal including a Punch Pinzgauer all-terrain vehicle, a motor pool for transport of equipment and personnel, and a mobile drilling rig. The Institute's modern laboratories are equipped with an atomic absorption device, a spectrofluorometer and a microcomputer terminal connected with the Centre's main computer. New devices developed at the Institute include a computer-driven pressure measuring device for maintenance-free automatic evaluation of long-term observations of confined aquifers, and a device for computer-driven discharge measurements in open channels. This HYDRODAT computer program can also calculate the quantity of discharge for turbulent flow on the spot (Benischke and Harum, 1984). Current projects at the Institute are concerned with the calculation of groundwater resources and their use and protection, reservoir leakage, hydroelectric power plants and their influence on groundwater, hydrogeologic problems caused by tunneling and mining, and exploration and exploitation of geothermal waters.

IGH is the only Austrian institution in this field that maintains ongoing exchange of know-how with developing countries. Since 1969, a postgraduate training course on groundwater tracing techniques has been held every other year at the Institute; it is the result of a contractual agreement between UNESCO and the Technical University in Graz, and is financed by the Austrian Fund for Developing Countries. The involvement of the Technical University is limited to administration. Training and field work are directed by IGH staff members, foreign experts from GTI Arsenal, the Geological Survey and the International Atomic Energy Agency in Vienna, and an expert from Poland (Institute of Physics and Nuclear Techniques, University of Mining and Metallurgy, Krakow).

Postgraduate participants have come from Latin America (Argentina, Brazil, Bolivia, Cuba, Haiti, Honduras, Mexico and Venezuela), Africa (Algeria, Burundi, Egypt, Ethiopia, Ghana, Kenya, Lesotho, Libyan Arab Jamahiriya, Nigeria, Senegal, Sierra Leone, Somalia, Sudan, United Republic of Tanzania, Tunisia, Uganda, Zambia and Zimbabwe) and Asia (China, Cyprus, India, Indonesia, Islamic Republic of Iran, Iraq, Jordan, Kuwait, Lebanon, Malaysia, Pakistan, Saudi Arabia, Syrian Arab Republic, Thailand, Turkey). These courses emphasize hydrogeology, hydrochemistry, geophysical methods in hydrology, environmental and artificial isotopes, water tracing techniques, models and computer evaluation; field work with tracing experiments and water sampling; laboratory techniques and data interpretation. For maximum effectiveness, the number of participants is limited to 16 per course.

Special hydrogeological courses have been held in Kenya and Mexico. Cooperative hydrogeologic research projects have been carried out with Saudi Arabia, with assistance from the Austrian Academy of Sciences.

Continuing cooperation with European specialists began with the first Specialist's Conference on Tracing of Subterranean Waters in Graz in 1966, sponsored by the Association for Hydrogeological Research in Graz. A small core of enthusiasts at this meeting formed a working committee with the objective of keeping up working contacts and holding symposia every four to six years. The Austrian members of this group known now as SUWT (Symposium on Underground Water Tracing), include the Institute for Geothermics and Hydrogeology, the Geotechnical Institute of Arsenal in Vienna and the Federal Institute for Water Economy in Karst Areas, Vienna, which assists in the fluorescein measurement of water samples. Members from other European countries include Swiss institutions (Geographic Institute, University of Bern, Department of Hydrology; Centre d'Hydrogeologie, University of Neuchâtel; Geographic Institute, University of Zürich; Physics Institute, University of Bern), institutions in the Federal Republic of Germany (Geologisches Landesamt, Baden-Württemberg; GSF Institute for Radiohydrometrics, Munich; Institute for Geology, University of Karlsruhe); Greece (Institute for Geology and Mineral Exploration, Athens) and Yugoslavia (Institut za raziskovanje krasa SAZU, Postojna; Zvezar Vodnih Skupnosti Slovenije Hidroloska sluzba, Ljubljana). Meetings have been held in 1966 in Graz, Austria, in 1970 in Freiburg im Breisgau, Federal Republic of Germany, in 1976 in Bled, Yugoslavia, and in 1981 in Bern, Switzerland. The Fifth International Symposium on Underground Water Tracing was held 22-27 September 1986 in Athens, Greece. The symposia differ from the first meeting in that they are divided into preparation and reports. Preparation involves long-term scientific work in a previously neglected area (detailed hydrogeologic mapping, measurement of precipitation and surface runoff, chemical and isotopic measurements of surface or groundwater and a final tracing experiment to identify underground drainage). These studies are performed in cooperation with the above-mentioned institutions at their own expense and with contributions in the form of tracers and personal involvement. New tracers have also been tested in the course of this work. As these studies benefit the country involved, the group representing the host country endeavours, whenever possible, to provide food and lodging during field work.

References

Andrews, J. N., J. E. Goldbrunner and others. A radiochemical, hydrochemical and dissolved gas study of groundwaters in the Molasse basin of Upper Stria. Earth and Planetary Science Letters, vol. 73, 1985.

Back, W. and J. Zötl. Application of Geochemical Principles, isotope methodology, and artificial tracers to karst hydrology. Hydrogeology of Karstic Terrains. IUGS series B, No. 3. IAH, Paris, 1975.

Bauer, F. and W. Perlega. Ergebnisse des Fluoreszenztracernachweises mittels Aktivkohle. Steirische Beiträge zur Hydrogeologie, vol. 32, Graz, 1980.

Benischke, R. and T. Harum. Computergesteuerte Abflussmessungen in offenen Gerinnen nach der Tracerverdünnungsmethode (Integrationsverfahren). Steirische Beiträge zur Hydrogeologie, vol. 36, Graz, 1984.

Bub, F. B. and H. Hötzl. Ergeonisse der Dünnschicht-chromatographischen Auswertung von Direktproben. Steirische Beiträge zur Hydrogeologie, vol. 32, Graz, 1980.

Dincer, T., B. R. Payne, C. K. Yen and J. Zötl. Das Tote Gebirge als Entwässerungstypus der Karstmassive der nordöstlichen Kalkhochalpen (Ergebnisse von Isotopenmessungen). Steirische Beiträge zur Hydrogeologie, vol. 24, Graz, 1972.

Gattinger, T. E. Hydrogeologie. In: R. Oberhauser. Der geologische Aufbau Osterreichs. Geologische Bundesanstalt, Wien, 1980.

Goldbrunner, J. E. Zur Hydrogeologie des Oberösterreichischen Molassebeckens. Steirische Beiträge zur Hydrogeologie, vol. 36, Graz, 1984.

Katzer, F. Karst und Karsthydrographie. Zur Kunde der Balkanhalbinsel, vol. 8, Sarajevo, 1909.

Kaupa, H. Der Marchfeldkanal. Planungsgesellschaft Marchfeldkanal. Wien, 1985.

Kochanek, F. and J. Donner. Die Wiener Wasserversorgung. Der Aufbau, Heft Februar/März, Wien, 1967.

Küpper, H. and I. Wiesböck. Erläuterungen und Index zur Ubersichtskarte der Mineral- und Heilquellen in Osterreich. Geologische Bundesanstalt, Wien, 1966.

Maurin, V. and J. Zötl. Karsthydrologische Untersuchungen im Toten Gebirge mit besonderer Berücksichtigung der versorgungswasserwirtschaftlichen Belange im Tauplitzgebiet. Osterreichische Wasserwirtschaft, vol. 16, Wien, 1964.

Nemecek, E. P. Grundwassererschliessung mittels Horizontalfilterbrunnen. Wiener Mitteilungen Wasser-Abwasser-Gewässer, vol. 41, Wien, 1981.

Przewlocki, K. Interpretation of Environmental isotope data by means of digital simulation. International Report, IAEA, Section of Isotope Hydrology, Wien, 1973.

Vohryzka, K. Hydrogeologie von Oberösterreich. Amt der o.ö.landesregierung, Linz, 1973.

Zötl, J. G. Karsthydrogeologie. Springer-Verlag, Wien-New York, 1974.

__________. Information Geothermie. FGJ-Institut für Geothermie und Hydrogeologie, Graz. Heft 1, 2, Graz, 1980; Heft 3, Graz, 1981; Heft 4, Graz, 1982; Heft 5, Graz, 1985.

Zupan, M. Ergebnisse der Fuoreszenzmessung von Direktproben. Steirische Beiträge zur Hydrogeologie, vol. 32, Graz, 1980.

II. BELGIUM

Area: 30,513 km^2 Population: 9.9 million

A. Physiography and geology

The Kingdom of Belgium has three main physiographic regions. The coastal region, or lower Belgium, along the North Sea shores, for about 70 kilometres in length is low lying and consists mainly of sand dunes and polders protected by dykes. It extends 15 to 50 km inland with elevations ranging from slightly below sealevel to 20 metres above and presents itself as flat pastureland crossed by drainage canals, as in the south-western Netherlands.

The rolling central plains of middle Belgium range in elevation from 20-30 m in the north to 200 m in the south. They are crossed by a number of waterways and contain fertile valleys.

The rocky Ardennes plateau, or upper Belgium, is densely wooded, averages 300 to 500 metres in elevation to a maximum of 694 m north of Malmedy. The climate in lower Belgium is the temperate humid type. Temperatures increase from the shores inland. In the Ardennes, the climate contrasts sharply between hot summers and cold winters. Mean annual temperature in Brussels is 10° C (2° C in January, 18° C in July), average temperature for the country is 8° C. The yearly rainfall increases from 700 mm in the coastal region to 1,400 mm on the Ardennes plateau.

The territory of Belgium is unevenly distributed among five hydrographic basins. The Meuse basin covers 14,630 km^2 and has the Sambre and Ourthe as main tributaries. The Scheldt (Escaut) basins, covering 12,240 km^2 has the Lys, Dender, Senne and Rupel rivers as main tributaries. The Scheldt is the main waterway of Belgium. Coastal basins cover 2,230 km^2; the Mosel-Rhine basin, 871 km^2, and the Vise sur Seine basin, 73 km^2.

Lower and middle Belgium are densely populated and have experienced extensive agricultural and industrial development. Coal is mined both in the north (Campine) and south (Charleroi, Mons, Liège, Namur). The population density of Belgium is one of the highest in Europe, 324 persons per square kilometre.

Two main geologic structures are found in Belgium (see figure 9). A sedimentary basin in the north consists mainly in a flat-lying Cenozoic and Mesozoic series of unconsolidated or consolidated layers with a great number of relatively thin aquifers.

Roughly south of a line from Tournai through Namur to Liège is a folded Paleozoic area with, to the north, a mainly limestone belt of Carboniferous and Middle Devonian terranes, and to the south, the shale and limestone massif of the Ardennes, which is Devonian to Cambrian in age. Quaternary formations include the Flemish valley, and alluvium of the Escaut and Meuse basins.

Figure 9. Belgium: General map

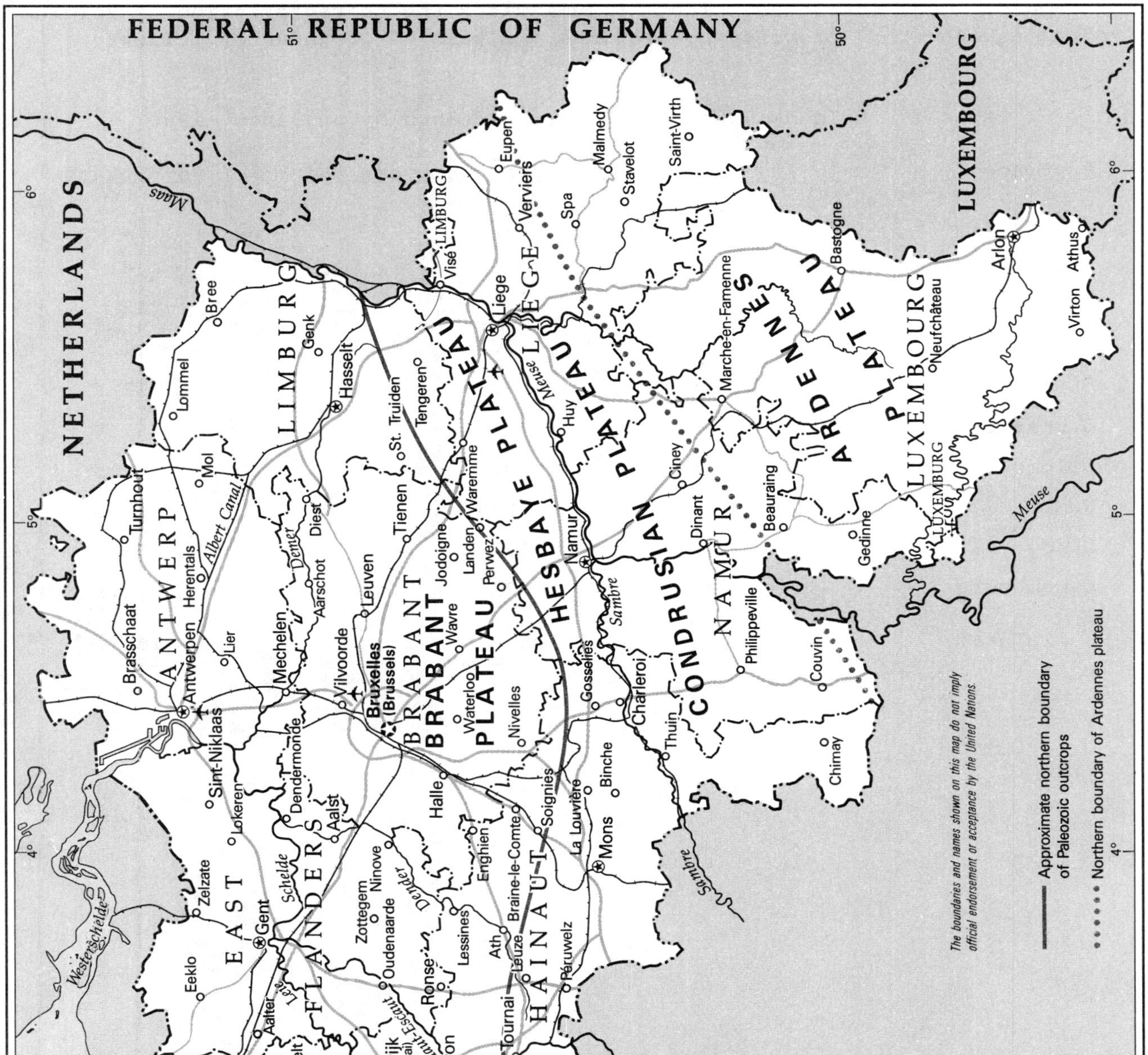

B. Groundwater development

Table 4 gives the extraction of groundwater in 1980 in terms of amount and percentage of total extracted. Table 5 summarizes the characteristics of Belgium's principal aquifers. A province-by-province analysis of groundwater follows.

Table 4. Belgium: Groundwater extractions by province, 1980

Province	Million m^3 per year	Percentage
West Flanders	17	2.5
East Flanders	25	4
Antwerp	70	11
Limburg	70	11
Brabant	76	12
Brussels	5	1
Hainaut	200	31.5
Namur	67	10.5
Liège	82.5	13
Luxembourg	22.5	3.5
Total	635	100

Table 5. Belgium: Characteristics of the principal aquifers

Aquifer layers are as follows:

Stratigraphy		Aquifer and/or area	Type		
Quaternary	Holocene	Dunes, Coastal plain	P	A	X
	Pleistocene	Flemish Valley, Valleys of Escaut, Meuse	P	A	X
		Campion Sands	-	B	-
Pliopleistocene		-	P	M	-
Cenozoic	Pliocene	Lillo sand	P	M	-
1)	Miocene	Antwerp and Diest sand	P	AM	XX
2)	Oligocene	Rupelian sand	P	AM	-
	Eocene	Ledo-Bruxellian, Bruxellian Ypresian sands	P	A	X(Brux)
3)	Paleocene	Landenian sands	P	A	-
Cretaceous		Flanders	F	A	-
		Brabant	F	A	-
		Hesbaye (Liège)	F	A	X
		Herve	F	A	XX
		Mons	F	A	X
		Maastrichtian - Geer	F+P	A	XX
		Maastrichtian Artesian	F+P	A	XX
Jurassic	Middle Jurassic	Bajocian Limestone	F	A	X
	Lower Jurassic	Virtonian "Macignos"	F+P	A	X
		Sinemurian limestone, sandstone, sand	F+P	A	X
Triassic		Sand, conglomerate	P	A	-
Permian		Stavelot conglomerate	F+P	A	-
Carboniferous	Dinantian	Tournai	Karst	A	XX
		Namur, Dinant	Karst	A	X
Devonian	Middle Devonian	Devonian limestone of Namur and Dinant basins	F	A	X
	Devonian Lower Middle Upper	Massif of shales and sandstone of the Ardennes	F	A	-
Silurian Ordovician Cambrian		Flemish and Ardennes Basement	F	A	X

Main impervious layers
1) Boom clay
2) Asse clay
3) Ypresian clay

P: Porous
F: Fracture
A: Monolayer
M: Multilayer

XX major
X important

1. West Flanders

In the coastal area, which includes De Panne, Ostende and Brugge (see figure 10), groundwater is exploited for water supply purposes by means of batteries of wells 10 to 20 metres deep. Extraction is at a rate of 4 million cubic metres per year from sand dunes and 4 million m^3 per year from Eocene and Paleocene sands. An additional 1.5 million m^3 per year could be exploited in the Ostende area. Transmissivity is in the order of 10^{-2} m^2/s for dune sands, 10^{-3} m^2/s for Cenozoic sands (Eocene and Paleocene sands are semi-artesian under a clayey layer). Salinity is moderate and the aquifers are vulnerable to pollution.

In the sector which includes Diksmuide, Torhout, Ieper, Poperinge and Roeselare is an artesian aquifer of Paleocene sands. Its transmissivity is 2×10^{-5} m^2/s. The fractured basement rock is tapped by wells 200 m deep but water salinity is 200 mg/l and up. Total extraction is less than 2 million m^3 per year. All groundwater resources have been developed.

In the sector around Courtrai, a confined basement aquifer is tapped by wells down to 300 m. Salinity is 120 mg/l. All potential sources have been tapped.

2. East Flanders

In the area around Ghent, Eeklo, Aalter and Dendermonde, Pleistocene sediments of the Flemish valley to the north provide water. The aquifer is shallow, less than 10 m thick, and tapped by batteries of wells. Total extraction is about 10 million m^3 per year and only about half of its estimated potential. To the south, the aquifers consist of separate blocks of Eocene sand from which about 10 million m^3 per year are extracted.

The same aquifers provide water to the areas around Oudenaarde, Aalst, Ninove and Zottegem. About 1.5 million m^3 per year are taken from the Pleistocene sediments and about 1 million m^3 per year from the Eocene sands. There is also a fractured basement from which about 3 million m^3 per year is extracted. Salinity, however, is high (500 mg/l) and the water is mainly used for industrial purposes. All potential resources have been developed.

The aquifers in the Sint Niklaas area are mostly Eocene and Oligocene sands, producing 1.2 million m^3 per year of water whose salinity is 500 mg/l. All potential resources have been developed.

3. Antwerp

In the Antwerp area, Miocene sands overlie the Cretaceous Boom clay. Transmissivity varies from 10^{-2} to 10^{-3} m^2/s depending on grain size and other factors. The groundwater in some cases is ferruginous and corrosive, in others polluted by industrial effluents. Locally, the water is brackish, being up to 1,000 mg/l in salinity. Extraction is 70 million m^3 per year over an area of 2,800 km^2. It is estimated that an additional 30 million m^3 per year could be developed.

Figure 10. Belgium: Principal aquifers

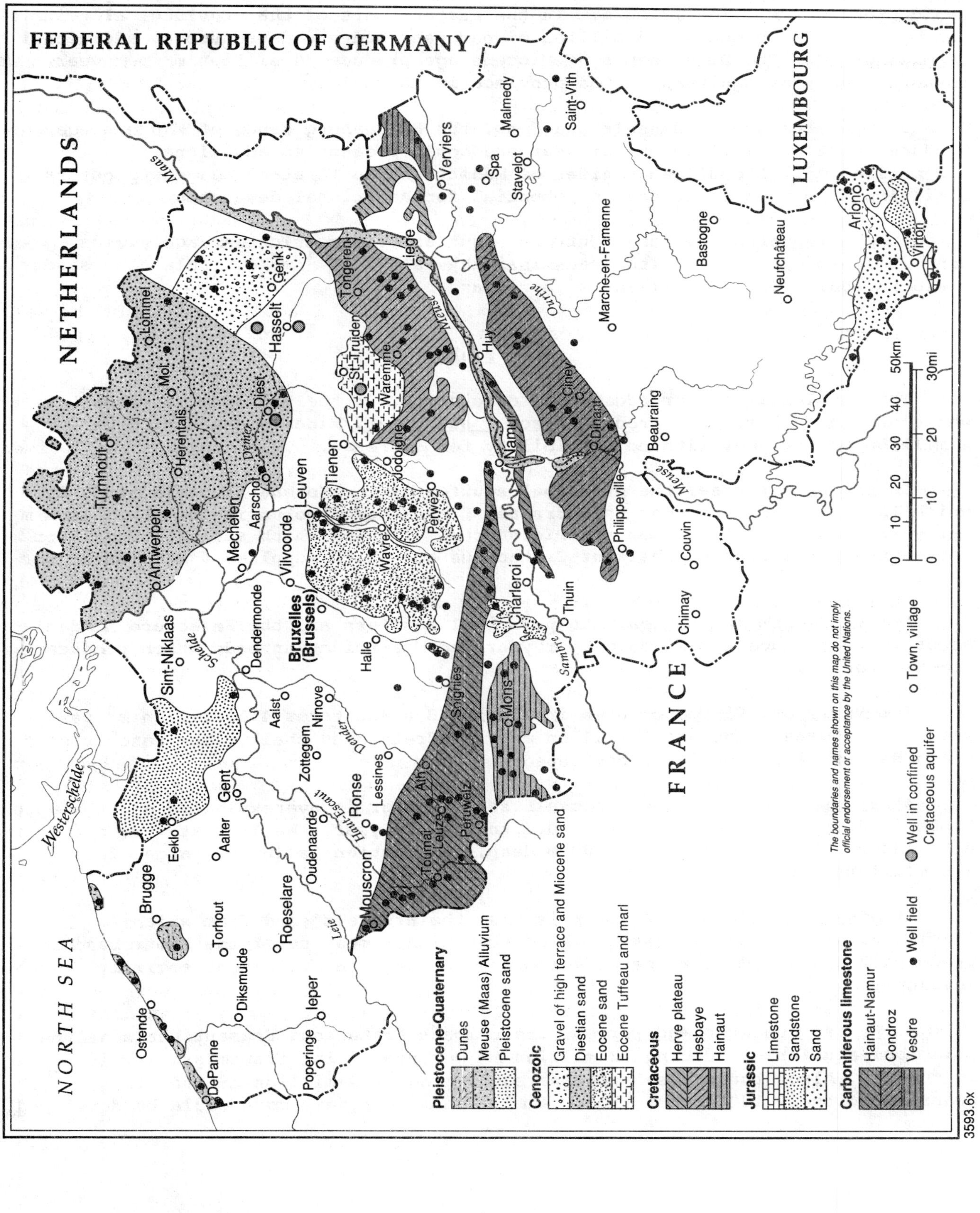

4. Limburg

Two principal aquifers provide water in Limburg. In the western part of the province, the Campine sands of Pleistocene age produce 9 million m^3 per year and could produce 9 million m^3 more. In the eastern part of the province, alluvium of the Meuse valley produces 15 million m^3 per year and could produce an additional 14 million m^3. The Diest sands of Miocene age produce 20 million m^3 per year. The transmissivity of aquifers in the province is about 10^{-2} m^2/s.

In the area around Hasselt and Genk, the Bolderberg sands of Miocene age produce about 3.5 million m^3 per year and could produce an additional 3.5 million m^3. A confined aquifer of Maestrichtian (Cretaceous) age produces 5 million m^3 per year but has no potential for additional development.

In the Tongeren area the "Tuffeau" aquifer of Maestrichtian age yields 8.5 million m^3 per year. Its transmissivity is 4×10^{-3} m^2/s. Paleocene sands yield an additional 2.5 million m^3 per year.

5. Brabant

The possibilities for additional development in the province of Brabant as a whole are quite limited. Around Aarschot and Diest, Miocene sands produce 3 million m^3 per year with no possibility for increase.

In the Leuven area, the principal aquifer is the Eocene Brussels sands from which 13 million m^3 per year is extracted from a total potential of 18 million m^3 per year. Additional sources are Pleistocene alluvium which supplies 2 million m^3 per year and the Brabant chalk of Cretaceous age which supplies 4 million m^3 per year.

Around Tienem and Landen, the principal aquifers are the Paleocene Landenian sands which produce 4 million m^3 per year and Maestrichtian beds which produce 2.5 million m^3.

The Halle and Vilvoorde area in Flemish Brabant takes 4.5 million m^3 per year from Eocene sands and 3 million m^3 from Cretaceous chalk. The resources of the area are almost entirely developed.

At Brussels the confined Brussels sand aquifer is overexploited and the water table is dropping at rates of from 0.7 to 2 m per year. Wells that were once artesian now have water levels 100 m deep. Extraction is in the range of 2.5 million m^3 per year.

At Nivelles, the main aquifer is also the Brussels sand from which 13 million m^3 per year is taken. Also, Devonian limestone of the Namur basin produces 2 million m^3 per year. Resources in the area are almost entirely developed.

At Wavre, Jodoigne and Perwez, the compact Cretaceous Brabant chalk is the principal aquifer, producing 12 million m^3 per year. Its transmissivity is 10^{-2} m^2/s. Also, the Brussels Eocene sands produce 9 million m^3 and its transmissivity is 10^{-4} m^2/s. About 4 million m^3 per year more could be developed.

6. Hainaut

In the area encompassing Mouscron, Comines, Tournai, Leuze and Peruwelz, Carboniferous-age limestone of the Namur basin produces 9 million m^3 per year. From 1945 to 1973, the water table dropped 65 metres due to overdraft. Extraction volumes were then strictly regulated and the levels stopped dropping. The land, however, subsided near the Escaut River and also in the middle of its bed. The resultant recharge from river water raised the water table in the limestone so that by 1978 it was back up to the 1971 level.

The groundwater is generally hard because of sulphates in variable amounts. Flourine occurs in confined parts of the aquifer. In the Tournai area, quarrying draws off 10 million m^3 per year. Total extraction amounts to 50 million m^3 per year with no prospects of developing more.

Around Ath, Lessines and Enghien, Carboniferous-age limestone is compartmented into independent blocks caused by faulting. Permeability varies from block to block and water levels may differ by 40-50 metres. Extraction amounts to 6.5 million m^3 per year. Additional resources may be developed up to an amount yet unknown, including about 2 million m^3 per year from the Silurian and Cambrian basement rocks.

At Mons and its surrounding Borinage coal-mining district, the Mons chalky aquifer was studied by mathematical modelling. Total extraction amounts to 80 million m^3 per year and 8 million m^3 more could be developed. Also, Carboniferous-age limestone of the Namur basin produces 8 million m^3 per year.

At La Louvière and Soignes, mainly Carboniferous-age limestone of the Namur basin produces 15 million m^3 per year. Additional resources, in unknown amounts, could be developed. Here also, the Mons chalk produces 7 million m^3 per year, and the Brussels sands, 2 million m^3.

At Charleroi, Carboniferous-age limestone of the Namur basin produces 20 million m^3 per year. Its transmissivity is 10^{-2} m^2/s.

In the Thuin and Chimay area, mostly Devonian limestone of the Dinant basin produces 2.5 million m^3 per year. About 3.5 million m^3 per year additional could be developed from Ardennes shale and sandstone aquifers.

7. Namur

At the city of Namur, Carboniferous-age limestone of the Dinant and Namur basins produces 30 million m^3 per year. An additional 5-10 million m^3 could be developed. Pleistocene alluvium of the Meuse valley produces 8.5 million m^3 per year, and Devonian limestone, 1.2 million m^3. Transmissivities are: Carboniferous limestone, 10^{-3} m^2/s (after acidification); Meuse alluvium, 10^{-2} m^2/s.

At Philippeville and Couvin, mostly Carboniferous-age limestone of the Dinant basin produces 6.5 million m^3 per year; Devonian limestone, 1 million m^3 and alluvium, 0.8 million m^3. The Ardennes shale and sandstone aquifers are not exploited but have the potential to produce 3 million m^3 per year.

In the Dinant and Cirey area, mainly Carboniferous-age limestone of the Dinant and Namur basins produces 12 million m^3 per year and 3-5 million m^3 additional is available.

At Beauzaing and Gedinne, the Devonian limestone has been exploited to the limit, 200,000 m^3 per year. The Ardennes shale and sandstone aquifers, however, produce 1.2 million m^3 and an additional 3.6 million m^3 per year is available.

8. Liège

In the Huy, Waremme and Liège area, at least 10 million m^3 per year is available from the main Carboniferous-age limestone aquifer of the Dinant basin. At Verviers, several Cretaceous aquifers produce 700,000 m^3 per year, and Carboniferous-age limestone, 600,000 m^3. Most of the aquifers are underdeveloped and an additional 4 million m^3, at least, is available.

Around Stavelot, Malmedy and Saint Vith, the Ardennes shale and sandstone aquifers produce about 5 million m^3 per year and an additional 2 million m^3 more may be available. Alluvium provides 1.5 million m^3 per year. Transmissivity in the area ranges from 10^{-5} to 10^{-3} m^2/s.

9. Luxembourg

At Marche-en-Famenne, the main aquifer is the Carboniferous-age Dinant limestone. It produces 5 million m^3 per year and is exploited to only about two thirds of its potential. The Ardennes Formation could be developed for an additional 6 million m^3 per year.

At Bastogne, the main aquifer is the Ardennes Formation which produces 4 million m^3 per year. An additional 5 million m^3 more could be developed.

At Neufchateau, the Ardennes Formation produces 5 million m^3 per year and is exploited to only a third of its potential. Its transmissivity is 10^{-4} m^2/s.

Around Azlon and Virton, the "Macignos" give way to many springs yielding up to 500 m^3 per day. The Ardennes Formation yields 4.5 million m^3 per year which is about 25 per cent of its potential. Its transmissivity is 10^{-3} m^2/s.

C. Overall status and future prospects

The current status of groundwater development in Belgium is summarized in two following tabulations. The overall conclusion is that groundwater is developed in Belgium up to 75 per cent of potential, renewable, resources. The principal aquifers rank as follows:

Aquifer	Percentage
Carboniferous Namurian limestone	18.6
Miocene sands	15.9
Carboniferous Dinantien limestone	11.4
Cretaceous chalk and limestone of Mons basin	9.7
Brussels Eocene sands	7.9
Cretaceous Hesbaye (Liège) chalk-limestone	5.0
Meuse valley alluvium	4.0
Maestrichtian "Tuffeau" limestone	3.0
All other	24.5

On a regional basis, Flanders requires 120 million m^3 per year for domestic use, 100 million m^3 for industry and agriculture for a total of 220 million m^3 per year; Wallonia requires 300 million m^3 for domestic use, and 110 million m^3 for industry and agriculture for a total of 410 million m^3 per year.

The quality of groundwater in Belgium is threatened by the high concentration of industries and population, by the fact that in northern Belgium groundwater levels are at relatively shallow depth, and by the very high permeability and transmissivity of the karstic limestone of Carboniferous and Middle Devonian age.

The potential for future development of groundwater in Belgium can be inferred from the following tabulation showing the estimated amounts of additional groundwater that could be extracted annually as needed.

Province	Amount (in million m^3 per year)
West Flanders	2
East Flanders	10
Antwerp	30
Limburg	28
Brabant	15
Hainaut	50
Namur	35
Liège	15
Luxembourg	35
Total	220

Most of the data contained in this paper is extracted from the EEC publication of 1982 Bilan des ressources en eau souterraine which, in turn, contains an extensive listing of publications on groundwater in Belgium.

III. DENMARK

Area: 43,075 km^2 Population: 5,100,000

A. Physiography and climate

The Kingdom of Denmark consists of a peninsula, Jutland, a few large islands and approximately 400 small islands separated by straits and channels, and surrounded by the North Sea, the Skagerrak, the Cattegat and the Baltic Sea.

Topographically the country consists of two major relief zones. Western Jutland has slightly undulating old moraine landscapes from the penultimate glaciation in a network of flat meltwater valleys and outwash plains from the last glaciation. Eastern and northern Jutland and the islands have an undulating moraine landscape from the last glaciation incised by several inlets and, in places, cut by deep valleys, particularly in eastern Jutland. Most of the country lies less than 100 metres above sealevel.

Denmark is situated in the northern temperate climatic zone but, as it is surrounded on three sides by large water bodies, the climate has a maritime character with mild winters and cool summers. The average annual precipitation calculated from measured values from 1930 to 1960 is 662 mm. According to more recent studies, however, such measurements of precipitation may be 10 per cent too low because certain wind effects had been overlooked. Considered region by region, annual precipitation in western Jutland averages 700-800 mm; in northern Jutland, 600-700 mm and in eastern Jutland and the islands, 500-650 mm.

Precipitation is highest in July and August (70-80 mm per month) and least in February and March (30-40 mm). Low intensity rainfall associated with frontal passages can occur any time during the year, whereas high intensity precipitation, associated with convection, occurs mostly in the summer. Snow constitutes less than a tenth of the yearly precipitation. On average, relative humidity is 82 per cent, but ranges from 72 per cent in May to 91 per cent in December.

Mean annual air temperature is 7.9° C, ranging from -0.4° C in February to 16.6° C in July. Mean monthly maxima range from 2° C in January to 21.1° C in July and the mean monthly minima range from -3° C to 12.2° C, respectively. The highest and lowest temperatures recorded are 36.4° C and -31.2° C, respectively. In general, differences between mean summer and winter temperatures increase inland and from west to east.

Potential annual evapotranspiration is in the range of 500 mm with highest values in June. Actual annual evapotranspiration, estimated from the hydrologic balance using measured precipitation, is 380 mm ranging from 360 mm in western Jutland to 400 mm in the rest of the country. The actual annual evapotranspiration, calculated on the basis of corrected precipitation values, would be 430 mm for western Jutland and 460 mm for the rest of the country.

The prevailing winds blow from the west at an annual average speed of 4.9 m/s with little seasonal variation. Strong winds of hurricane force mainly occur in winter in connection with the passage of low-pressure systems.

The root zone soils consist of sand and loamy sand in western Jutland, and of loamy sand, sandy loam and clay in the rest of the country. Silty soils occur only in tidal and non-tidal low-energy coastal environments.

Groundwater recharge is highest in western Jutland compared to the rest of the country. This is because precipitation is higher, actual evapotranspiration is lower, and the soils are coarser with a lower water-retention capacity.

The infiltration capacity of the soils is normally high enough to prevent overland flow. Exceptions may occur in winter during thawing of frozen soils, and on poorly managed fields without contour ploughing during high-intensity precipitation events outside the growing season. Even then, much of the surface runoff will be collected in depressions and slowly infiltrate down as groundwater instead of flowing towards river courses.

The differences in deep percolation to groundwater, 280-380 mm per year in western Jutland and 140-250 mm per year on the islands, result in significantly different drainage density and river discharge pattern for these two parts of the country. The drainage network density is lower and the difference between the maximum winter and minimum summer discharge is smaller in western Jutland.

B. Geology

The main geologic structures of continental Denmark include units partly of the Danish-Norwegian basin, partly of the Danish-German basin and the Ringkøbing-Fyn high, which separates the two basins. The two basins are connected by minor grabens which cross the bedrock high. Two thirds of Denmark lies north of the high one third south of it. The high itself trends west-north-west.

From the point of view of lithology and stratigraphy, the two basins are quite similar because they were connected and both received material from the Fennoscandian shield. The thickness of the deposits in the Danish-Norwegian basin ranges from 3 to 10 km while in the Danish-German basin the thickness ranges from 2 to 4 km.

Precambrian granite and gneiss constitute the substratum of the two sedimentary basins and the dividing high (see figure 11). They outcrop on the island of Bornholm, east of the basins.

Sediments from Cambrian to Silurian and from Triassic to Senonian are found on Bornholm either close to the surface or as outcrops. In the rest of the country, pre-Senonian rocks are found only in deep wells.

The outlines of the three structural elements were formed during the Permian period. In marginal parts of the basins, i.e., on either side of the high limestone in the southern and central Denmark, and evaporites in the central part of northern Jutland were developed.

During the Zechstein, evaporites of rock salt were deposited north and south of the Ringkøbing-Fyn high in thicknesses of up to 1 km. South of the Ringkøbing-Fyn high, the evaporites are present as salt beds at depths of 2 km. North of the high the Zechstein rock-salt formation appears as a number of diapirs or interdomal structures. The highest diapirs have the caprock only a few hundred metres below the surface.

Figure 11. Denmark: Pre-Pleistocene geology

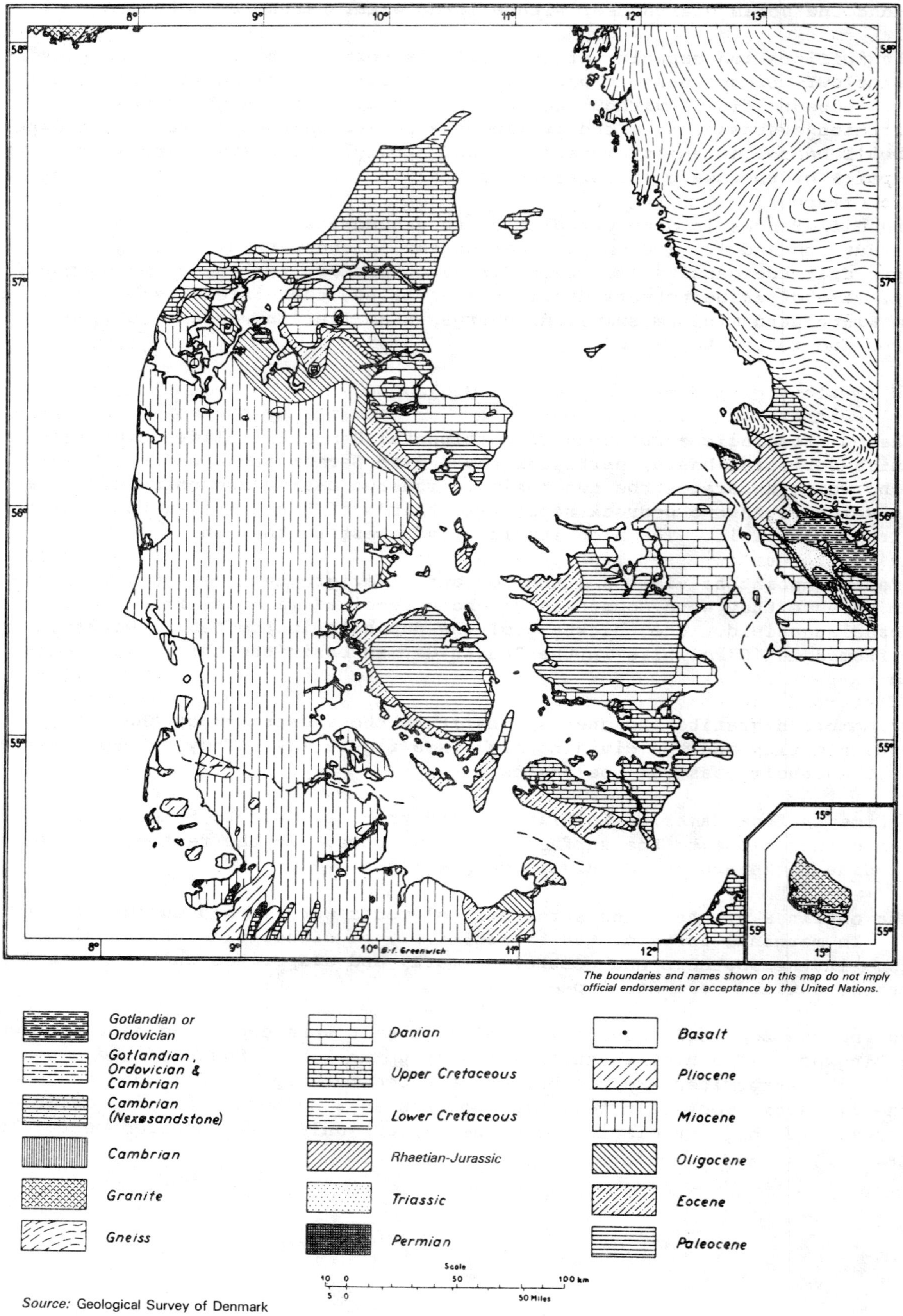

Source: Geological Survey of Denmark

3593.7x

The lower Triassic period was characterized by deposition of material from the north-east. These deposits have, therefore, a decreasing grain size, from gravel over sand to clay and with evaporites on either side of the high towards the south-west. The thickness of these deposits ranges from less than 200 m over the high to 4 km in the graben traversing the ridge. During the late Triassic, sandy deposits covered the entire area.

During the Jurassic and Early Cretaceous periods, clay and sand were deposited in the Danish-Norwegian basin - sand at the margins and organic-rich clay in the central parts of the basin. These deposits are now found as sandstone, claystone and shale, respectively. Such deposits are not found in the southern Danish-German basin.

Upper Cretaceous and lower Tertiary formations include rather homogeneous limestone structural units several kilometres thick. The porosity of the limestone decreases with depth from approximately 50 per cent to 15 per cent owing to compaction due to the weight of the overlying sediments and by cementation.

The Tertiary period was characterized by deposition of limestone, clay, silt, sand, volcanic ash and lignite. During the late Tertiary, the coastline crossed the area and material was washed out from the Fennoscandian shield from east to west into the basins. This, together with a more pronounced glacial erosion in the eastern part of the area, have generally resulted in increased thickness of Tertiary deposits towards the south-west.

During the Quaternary period, the territory was covered by icecaps at least four times. The glacial dynamics together with the production of meltwater left a complex system of clayey till and sand. The fact that the glaciers came from different directions, and moved over areas with various pre-Quaternary deposits with different lithologies, render correlation difficult in most cases. Only in the western part of Jutland, where late Quaternary outwash plains and valleys were left undisturbed, is correlation possible over larger areas. The sediments laid down during the transgression due to melting of the icecaps is also easy to correlate, but these deposits are mainly restricted to the northern and coastal parts of the country. The Quaternary landscape is shown on figure 12.

Figure 12. Denmark: Features of the Quaternary landscape

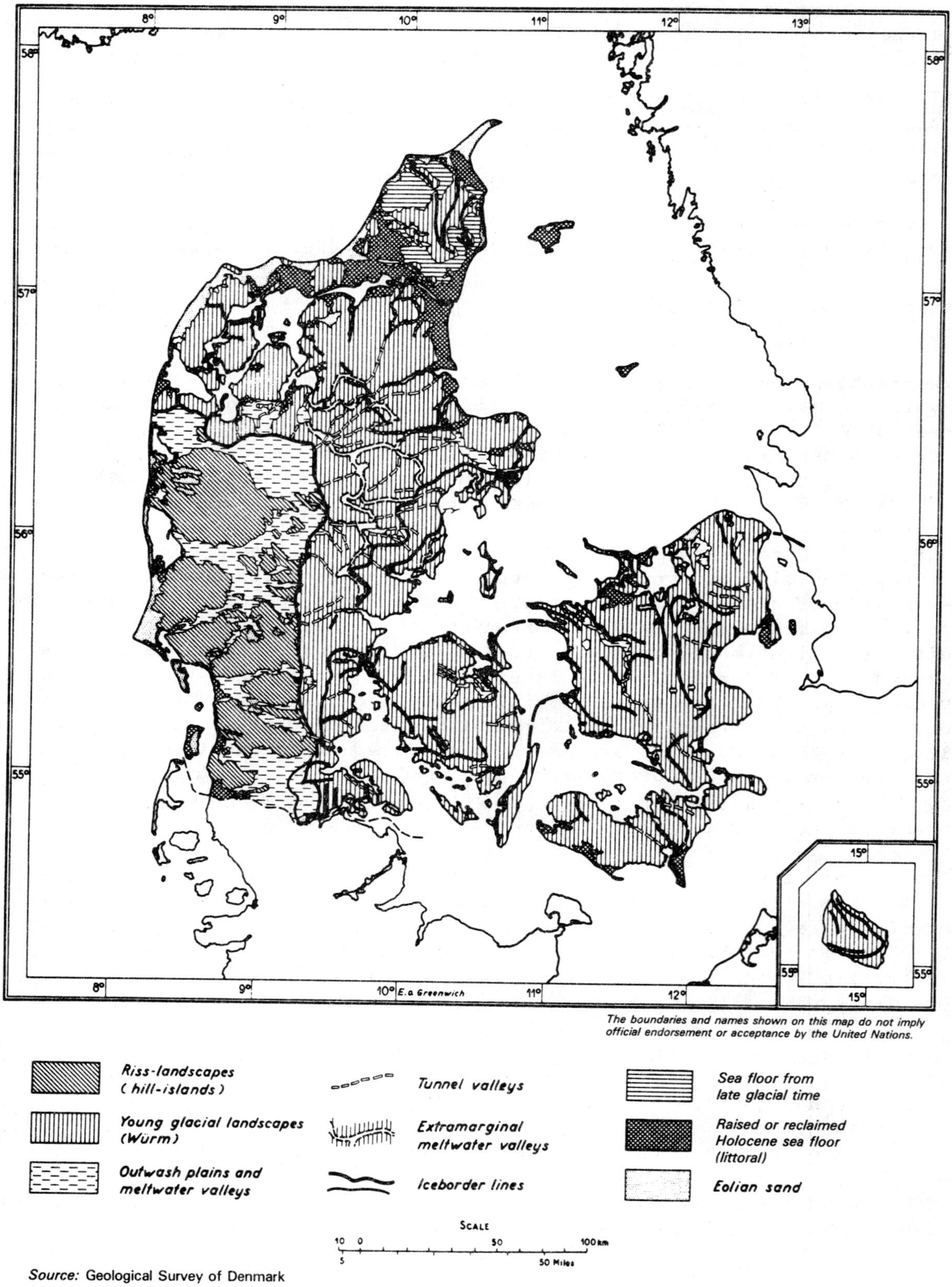

Source: Geological Survey of Denmark

3593.8x

C. Principal aquifers

1. General

Water supply in Denmark originates almost entirely from groundwater sources: 95 per cent of drinking water is from wells, the remaining 5 per cent from lakes.

The principal Danish aquifers are described in table 6, which includes age, type of aquifer material, permeability type, indication of confined or unconfined conditions, depth to the top of aquifer and the importance of the aquifer as water supply.

The major aquifers, together with the age of the aquifer formation and an estimate of the contribution to the total water supply, are listed below:

Type and age	Percentage
Glaciofluviatile sand and gravel, Quaternary	50
Danian limestone, Cenozoic	15
Sand and gravel, Cenozoic	10
Greensand limestone, Cenozoic	10
White chalk, Cretaceous	10
Other	5

2. Cenozoic

(a) Glaciofluviatile sand and gravel

Approximately 50 per cent of the water supply comes from Quaternary sand and gravel. In spite of the fact that these aquifers can be found all over the country, there are considerable regional differences regarding their importance for the water supply. In western Jutland outwash plains and valleys cover large areas. The aquifers are unconfined, up to 25 m thick and the water table is 0-25 m below the surface. Transmissivity ranges from 10^{-3} to 10^{-2} m^2/s.

In the moraine landscape of eastern and northern Jutland and on the islands, aquifers occur either as extensive or as minor discontinuous sand layers in till. The aquifers are typically confined and their thickness ranges from few metres to about 20 metres for the discontinuous and extensive aquifers, respectively. Of particular importance are the coarse-textured meltwater deposits, which fill pre-Quaternary valleys. Transmissivity ranges from 10^{-3} to 10^{-2} m^2/s.

(b) Miocene sand and gravel

The Miocene aquifers consist of alternating, marine and limnic layers of mica clay, mica sand, quartz sand and gravel and lignite. The aquifers in the quartz sand layers are 10-20 m thick and have a transmissivity of 10^{-3} to 10^{-2} m^2/s.

The top of the aquifers can be at depths ranging from a few metres to more than 200 m. These aquifers are predominantly confined.

Table 6. Denmark: Characteristics of the principal aquifers

Era	System	Subsystem	Aquifer	Permeability	Aquifer type	Top and bottom of aquifer (in metres)	Significance of aquifer
Cenozoic	Quaternary	Holocene	Marine and eolian sand	Intergranular	Phreatic and artesian	0-15	Local
		Pleistocene	Meltwater deposits of sand and gravel	All intergranular			All major
			(1) Outwash plains		Phreatic	0-50	
			(2) Extensive sand-layers in till		Artesian	0-220	
			(3) Discontinuous sand-layers in till		Artesian	0-200	
	Tertiary	Miocene	Quartz sand and gravel, mica sand	Intergranular	Phreatic and artesian	0-120	Major
		Paleocene	Greensand limestone	Intergranular and fissures	Artesian	5-150	Both major
			Danian limestone	Fissures	Phreatic and artesian	0-125	
Mesozoic	Cretaceous	Upper Cretaceous	White chalk	Fissures and karst	Phreatic and artesian	0-150	Major
			Bauneodde greensand	Intergranular	All artesian	35-90	All local
			Arnager limestone	Fissures		0-25	
			Arnager greensand	Intergranular		5-80	
		Lower Cretaceous	Sandstone	Fissures	Artesian	5-90	Local
			Robbedale sand	Intergranular	Phreatic and artesian	5-90	Moderate
	Jurassic	..	Sand and sandstone	Intergranular and fissures	Artesian	0-75	Local
Paleozoic	Silurian to Middle Cambrian	..	Shale and limestone	Fissures	Artesian	0-150	Minor
	Lower Cambrian	..	Sandstone and quartzite	Fissures	Artesian	0-125	Moderate
Precambrian	..	..	Granite and gneiss	Fissures	Artesian	0-150	Local

(c) Greensand limestone

The upper Paleocene greensand and limestone aquifers occur only on the islands and consist of a glauconitic limestone characterized by different degrees of cementation and clay content, interbedded with clay layers. The transmissivity ranges from 10^{-3} to 10^{-2} m^2/s, depending on the clay contents and the degree of fracturing. The porosity and permeability are caused by interstitial pores as well as fissures. The thickness ranges from a few metres to 150 m and the top aquifer lies from few metres to 75 m below the surface. Everywhere, the aquifer is confined by clayey till. In the deeper parts of the aquifers the water quality is often influenced by mineral water.

(d) Danian limestone

The lower Paleocene Danian limestone aquifer occurs in different parts of the country as bryozoan limestone, hardened limestone, lime sand and lime sludge. Flint layers and differences in the degree of hardening produce hydraulic heterogeneities. The porosity is mainly of intergranular origin, whereas the permeability is caused by fissures, particularly where the limestone has been crushed by glacial activity. Typically, the aquifers are confined, with transmissivities ranging from 10^{-3} to 10^{-2} m^2/s.

3. Mesozoic and older

(a) White chalk

Upper Cretaceous white chalk aquifers occur below Quaternary deposits in northern Jutland and on the south-eastern islands. In Jutland the thickness of these water table aquifers, developed in karst topography, ranges from 10 to 100 m, and the transmissivity values are high, up to 10^{-1} m^2/s.

On the south-eastern islands the aquifers are found from a few metres to 30 m below the surface. These aquifers are confined and are characterized by interstitial porosity and fissure permeability. Fissures produced by glacial erosion and pressure are especially productive. Transmissivity is typically around 10^{-3} m^2/s.

(b) Other aquifers

The aquifers on Bornholm comprise older, mainly fissured rocks:

a. Confined limestone aquifers, Upper Cretaceous;

b. Confined sand and gravel aquifers, Lower Cretaceous-Jurassic;

c. Shale and limestone aquifers, Middle Cambrian-Silurian;

d. Sandstone and quartzite aquifers, Lower Cambrian; and

e. Precambrian gneiss and granite.

D. Groundwater quality

1. Overall assessment

Groundwater quality is generally good. In most cases only basic treatment, if any, is needed. Hydrochemical data for various aquifers are presented in table 7, with the following terms used in the description:

a. Hardness: Ca plus (Mg X 40/24) unit: mg/l;

b. Groundwater types: Dominant anion/cation composition on an equivalent basis considering only Ca, Mg, Na, SO_4, Cl and HCO_3;

c. Percolation time: The time a tracer, moving as natural water, needs to reach from the land surface to the top of the aquifer.

Hardness is clearly more pronounced in the limestone than in the sand aquifers for obvious reasons. Groundwater in a limestone aquifer is mainly of the Ca-HCO_3 type with more local occurrences of Na-HCO_3, NaCl and mixed types. The totally dissolved solid contents increase, from sandy aquifers not related to limestone deposits (western Jutland), to sandy aquifers related to limestone (the rest of the country where limestone became part of till and meltwater sand), to the limestone aquifers.

The nitrate content (NO_3) of the shallow groundwater has risen from a level of a few milligrams per litre (mg/l) around 1950 to more than 50 mg/l today with large regional and local variations, however. In western Jutland the groundwater is under water table conditions and more polluted by nitrate than on the islands with artesian water. The nitrate content increases up to 2-4 mg/l per year. Today, 69 per cent of the population drink water with NO_2 content less than 5 mg/l, 14 per cent drink water with more than 25 mg/l and 5 per cent drink water with more than 50 mg/l.

The major source of nitrate pollution is the leaching of high nitrate fertilizers from the root zone - averaging 70-80 kg of nitrates per hectare per year. Since the percolation time from the root zone varies much, some aquifers will be more polluted than others in the near future. However, confined aquifers covered by clayey till with a high reduction capacity related to nitrate are well protected even on a long-term basis. Conversely, water table aquifers with low reduction capacity to nitrate even with a long percolation time will be affected and, at present, several measures are taken to prevent this.

2. Saline intrusion

In some regions of Denmark water quality is affected by sea-water intrusion, fossil salt water, upconing mineral water and by pollutants from above.

Salt-water problems due to sea-water intrusion or fossil salt water occur in areas with marine sediments of late Pleistocene and Holocene age and in areas with Recent tidal marshes.

Table 7. Denmark: Hydrochemistry of the principal aquifers

Aquifer	Hardness (Ca in mg/l)	Groundwater type	Totally dissolved solids (mg/l)	Percolation time to aquifer (years)	Special substances Cl, areal; <u>Cl</u>, point importance, sources
Late and Postglacial (northern Jutland)	40-120	$Ca-HCO_3$ Na-Cl Mixed	<500	1-20	Cl
Quaternary melt-water sand and gravel (throughout country)	<40 Central West Jutland 40-120 East and West Jutland scattered parts of major islands 120-200 Scattered parts of major islands	Mixed, West Jutland $Ca-HCO_3$, rest of the country	<500, Jutland 1,000-1,500, islands	1-20, West Jutland 1-20, rest of the country >20	NO_3 + (Cl) Jutland Cl + (NO_3) the islands
Miocene sands (West Jutland)	<40 West Jutland 40-120 Central Jutland	Mixed, West Jutland $Ca-HCO_3$, Central Jutland	<500	1-20 >20	<u>NO_3</u>
Upper Paleocene	<40 (Fuen)	$Ca-HCO_3$, Na-HCO, mixed (Fuen)	1,000-1,500	>20	<u>Cl</u>
Greensand, limestone (two major isles)	40-120 (Zealand)	Na-HCO (Zealand)			Cl <u>F</u>
Lower Paleocene Danian limestone (North Jutland and Zealand)	40-120 (North Jutland) 120-200 (Zealand)	$Ca-HCO_3$ (locally NaCl and mixed)	1,000-1,500	1-20 >20	F Cl
Upper Cretaceous limestone	40-120 (North Jutland) 120-200 (South-east Denmark)	$Ca-HCO_3$	<500 (North Jutland) 1,000-1,500 (South-east Denmark)	1-20	Cl

Sea-water intrusion may locally take place along the coast as a result of heavy pumping. Mineralized water is a problem at great depths everywhere in the country. At shallow depths in connection with subsurface structures, such as fault zones and diapirs, mineral water problem is known in some of the limestone aquifers.

3. Pollution

The pollution of groundwater caused by point sources like leachate from waste disposal sites and industrial areas, oil from storage tanks and pipelines, leakage from sewers, together with the pollution resulting from agricultural activities (nitrate, pesticides, herbicides), pose a serious threat to the groundwater.

Until the early 1970s there were practically no restrictions on deposition of waste and chemicals in the soil or ground. Control and monitoring measures on existing dump sites were not required. As a result, more than 1,000 sites may constitute potential sources of groundwater pollution. Many of the dump sites were placed on locations with almost direct contact to groundwater, e.g., abandoned gravel pits with the water table immediately below the surface. The Ministry of Environment and local authorities are entrusted with defining the size of the problem and, in some cases, preparing remedial action.

There are about 500,000 private oil storage tanks, many of them in unsatisfactory condition. These tanks, together with oil pipelines, may cause widespread pollution of groundwater.

Recent investigations of some sewers indicate that a high percentage of them leak and pollute the groundwater. The replacement of several thousand kilometres of pipe will be necessary to eliminate this source of pollution.

Much attention is placed on pollution originating from agricultural activities. A significant nitrate increase in shallow, water-table aquifers has occurred in Jutland. The Parliament has introduced several measures intended to reduce by 50 per cent the input of nitrate to the environment. Another major problem being considered is the leaching of pesticides and herbicides. The difficulties in assessing the magnitude of these problems are the high cost of analysis and insufficient knowledge of the mechanisms governing the transport of the pollutants in the soil and groundwater.

A national programme to monitor the water table has existed since 1950. A similar programme to monitor groundwater quality is being organized through a network of about 1,200 observation wells.

E. Groundwater development

1. Exploration

Site selection of new wells is first and foremost based on the information contained in 200,000 well records kept in the archives of the Geological Survey of Denmark including, in particular, hydrogeologic maps, reports on geoelectric surveys and records of pumping tests.

The most commonly used drilling method is percussion drilling, but during the 1980s different rotary drilling methods, especially airlift and reverse circulation, became more common. Normally, the temporary steel casing used during the drilling process is replaced by a permanent polyvinylchloride (PVC) casing and a PVC screen. Steel casings are used only with wells completed without screens in limestone aquifers where the temporary initial casing is left as a permanent, protective casing.

In unconsolidated formations, wells are completed with screen and gravel pack whereas wells in consolidated rocks, such as limestone, chalk, sandstone or granite, generally are completed without screens. After completion, the wells are pumped clean by increasing the pumping rate. Pressure acidizing of new and old wells in limestone aquifers is in many cases carried out to increase well capacity.

Submersible pumps are the most common type used, but ejector pumps are used when the demand is small. Centrifugal pumps are installed in shallow wells.

2. Legal and administrative

The Parliament has passed several acts, such as the Water Supply Act and the Environmental Protection Act, which establish the legal framework of the water supply structure. In principle, the Minister of the Environment is responsible for the issue of licences and regulations, for the settlement of disputes and for supervision and control.

In practice, many of these duties are delegated to the National Agency of Environmental Protection. In addition the Agency acts as adviser and submits recommendations to the Minister.

According to the Water Supply Act, regional authorities are responsible for the assessment and management of the surface- and groundwater resources in cooperation with municipalities.

The regional council considers how a region is to be supplied with water, issues licences and decides where the water is to be taken and which priority should be accorded the different consumer groups.

The Geological Survey of Denmark registers most of the wells constructed in the country. It has (1989) a geologic data base of more than 200,000 well records, each giving information about position, depth, geology, screened section, water table, drilling, pumping tests, and other data. The geochemical data base contains information on chemical composition of groundwater from more than 15,000 wells. Apart from that, the Survey produces various hydrogeologic and hydrochemical maps, including basic data maps depicting most well records as cyclogrammes. The Geological Survey coordinates the mapping and assists the regional authorities in mapping the groundwater resources.

Private consulting companies specializing in different fields of groundwater hydrology and chemistry play important roles in assisting the authorities with the hydrologic mapping, aquifer analysis and protecting the aquifers against pollution. Further, private companies are responsible for the construction of most public waterworks.

3. Current status

After the use of surface water and shallow dug wells for thousands of years, the first deep well was drilled in the mid-1800s. Today, groundwater accounts for 95 per cent of the country's water supply.

Water supply schemes may be divided into three groups (with number of units and percentage of population served as of 1970):

1. Single wells (250,000) supplying a large number of farms and isolated houses (17 per cent);

2. Small, private waterworks (3,000) constructed from the beginning of the 1900s on a cooperative basis in rural districts and in recreational areas (25 per cent);

3. Municipal waterworks (124) which serve the larger cities and their surroundings (58 per cent).

The distribution and trend of demand vary between consumers (see table 8).

Table 8. Denmark: Trends in demand by consumers, 1970-1982

Consumer group	1970	1977 (in million cubic metres)	1982
Households	340	445	564
Institutions	..	40	41
Agriculture, animals	105	90	81
Agriculture, irrigation	25	330	374
Industry	200	265	295
Fish farming	50	40	37
Total	720	1 210	1 392

This corresponds to a per capita household consumption of more than 100 m^3 in 1982, the largest figure in the water budget. The use of groundwater for irrigation has risen drastically, especially on the sandy soils in western Jutland in dry years. All wells for irrigation are private. Industry receives approximately 60 per cent of its water from municipal waterworks and 40 per cent from private ones.

4. Future trends

The need for water during the rest of the century is difficult to estimate. Among other factors climate conditions loom large since a major part of the water is used for irrigation.

The balance between the renewable potential of each main aquifer and the quantity of water abstracted varies from region to region and from aquifer to aquifer. The variations depend on the hydrogeological characteristics on the geochemical state of each aquifer, on population density and the requirements on quality and quantity of the water.

In western Jutland, where population density is low and recharge to the sandy aquifers is high, there is a surplus of groundwater. Shortages of water occur only in dry years when irrigation demand is high. Nitrate pollution may, however, become a serious threat to the quality of the ground water.

In eastern and northern Jutland and on the islands, the regional pattern is more complex because of their higher population densities.

In general, the sand and limestone aquifers either have a surplus or are fully exploited (equal distribution between the two). A significant number of reservoirs are of local importance only. In the metropolitan area of Greater Copenhagen, some of the resources have been overdeveloped. Salt water and mineral water is a local problem only. The aquifers are better protected against nitrate pollution here than in western Jutland but they are more exposed to pollution from chemical dump sites and landfill due to industry and higher density population in eastern Denmark.

F. International cooperation

As groundwater is the only source of water supply in the country, Denmark has a highly experienced team of geologists, hydrogeologists and engineers that are able to offer services to other countries within the following fields:

a. Hydrogeologic surveys of all kinds,

b. Groundwater chemistry,

c. Groundwater modelling,

d. Drilling technology,

e. Design of national data bases (national well record archives, water quality archives),

f. Computerized groundwater mapping,

g. Supervision and control of pollution,

h. Assessment of pollution hazards,

i. Pollution countermeasures,

j. Design of computerized waterworks,

k. Safety of water supply,

l. Groundwater treatment (deferrification, clearing, filtering, desalination),

m. Legal and administrative aspects,

n. Planning of water supply,

o. Postgraduate training,

p. Remote sensing.

References

Akademiet for de tekniske Videnskaber, "Vand", ATV-beretning, No. 44, Copenhagen, 1963.

Andersen, L. J. Cyclogramme technique for geological mapping of borehole data. Geological Survey of Denmark, Series III, No. 43, 25 p., 1 map., Copenhagen, 1973.

Bartman, J. C. In L. B. Rasmussen: Geological aspects of the Danish North Sea sector. Geol. Surv. of Denmark. Series III, No. 44, Copenhagen, 1978.

Forureningsrådet. Vandressource, Publ. No. 14, Copenhagen, 1971.

Hansen, J. M. Geologi for enhver. Danmarks Geologiske Undersøgelse, 88 p., 1984.

Kommissionen for de europaeiske faellesskaber. Grundvandsressourcer i Danmark, 74 p., 1982.

__________. Grundvandsressourcer i de europaeiske faellesskaber. Grundvandets sårbarhed og kvalitet, Danmark, (inkl. bilag), 1983.

Miljøstyrelsen. Vandindvinding og vandforsyning. NFM 13, Copenhagen, 1977.

Sorgenfrei, Th. and O. Berthelsen. Geologi og vandboring. Danmarks Geologiske Undersøgelse, III raekke, No. 31, 1954.

IV. FRANCE

Area: 549,183 km^2 Population: 56,100,000

A. General

The French Republic covers a little more than a twentieth of the area of Europe. It is the largest country of the continent except for the European part of the USSR. Only with Spain does it share Atlantic and Mediterranean coastlines. No point in France is more than 450 km from the sea. Its land and sea frontiers are about equal in length.

1. Relief

The great divergences of geologic structure is reflected in sharp topographic contrasts. In France, the balance of plains, plateaus, and young and old mountains make it a veritable summary of the European continent.

The young mountains take the form of rectilinear Alpine chains (Pyrenees) or arched chains (Alps, Jura). These form the highest relief and include the highest point in Europe (Mont Blanc, 4,807 m), the largest part of European territory above 1,000 m (37,255 km^2) and all European territory above 2,000 m (8,425 km^2). Formed by erosion, especially in the Alps, they consist of deep valleys and steep hills. They clearly reflect the diversity of their components, a mixture (except for the Jura) of crystalline and sedimentary rocks of all kinds, among which the limestones give rise to jagged relief.

The ancient mountains, namely, the Armorican Massif, 65,000 km^2; the Massif Central, 86,000 km^2; and the Vosges, Ardennes, and Provençal massifs compose the backbone of the territory and enclose flat sedimentary areas. They have been smoothed by prolonged erosion to form ancient peneplains. Their relief, however, is accentuated by the fact of rejuvenation caused by the after-effects of the Alpine phase. Uplifts have led to superimposed valleys and additional volcanic relief, especially in the Massif Central and the Vosges. These mountains nowhere exceed 2,000 m: the Massif Central rises to 1,885 m, the Armorican Massif to 417 m, and the Vosges to 1,424 m.

The plains of the great sedimentary basins (Paris and Aquitaine basins) are in few places as flat as the great plains of the north and east of Europe. Their structural surfaces rise towards the periphery of the basins, forming with their most resistant components, the limestone "cuestas" overlooking the underlying geologic formations and facing the borders of the basins. In only a few places does their altitude exceed 200 m but may be 500 m on the plateaus which prolong them and the thresholds which separate the border massifs (east of the Paris basin, especially; Lorraine, Burgundy). A whole system of fluvial valleys develops here and gives rise to hills, terraces and alluvial plains.

There are also subsident plains that are less extensive (Alsace, Limagnes in the Massif Central, Bresse and the Rhodanian corridor) associated with the major river valleys (Rhine, Loire and Allier, Saône and Rhône), and the coastal plains of variable size (Flanders, lower Languedoc).

The present coastlines (more than 3,000 km) are the result of the late Quaternary Flandrian transgression. The alternation of rocky hills, steep and jagged formations, and low sandy and rectilinear hills, some of which are lagoonal, reflects the variety of relief and the geological components of the coastal regions. The mouths of the rivers are also varied, the estuaries of the Atlantic coast (Seine, Loire, Garonne) contrast with the Mediterranean delta of the Rhône.

Modelled on the relief structure, the hydrographic system is organized into four major river basins - Seine, Loire, Garonne and Rhône. Flowing towards the three main coastlines (English Channel, Atlantic Ocean and Mediterranean Sea) these rivers drain 63 per cent of the territory. In addition, a part of the Rhine basin, which forms a river frontier of 170 km between France and the Federal Republic of Germany, drains 4.3 per cent of the territory (13 per cent of this great European basin). Only the Rhône has an exterior upper basin, in Switzerland (7,443 km^2), and only the north-east frontier is crossed by rivers, carrying water to Belgium, Luxembourg and the Federal Republic of Germany, which drain in France, together with the Rhine, an area of 38,664 km^2 (table 9, figure 13). The total length of French waterways has been calculated at 270,000 km.

2. Climate

Situated between latitudes 42° and 51° north the territory is entirely included within the temperate zone, but its relief, shape and the situation of its coastline produce a climatic variety which may be summarized under three main types of climate, oceanic, continental and Mediterranean (table 10).

A temperate oceanic climate dominates most of the country, from north to south-west, where its penetration is favoured by the relief and by the predominance of west winds, with transitional forms of climate in the interior Paris and Aquitaine plains that are less humid and have more sharply contrasting seasonal thermal differences.

A drier and harsher continental climate prevails over the mountainsides and massifs and in the sheltered depressions of the east and centre of France, giving way to a mountainous climate at higher altitudes. The Western slopes of the mountain barriers which restrict the oceanic influences are the regions with the most precipitation (Vosges, Jura, Massif Central, northern Pre-Alps and Western Pyrenees).

Mediterranean climate conditions prevail from the shores of the Mediterranean, over a distance of 100 to 150 km inland at most, up to the eastern Pyrenees, the southern Massif Central and the southern Alps, which from a protective screen for the coastal plains against the influences coming from the north and west. It is characterized by high summer temperatures and relatively abundant precipitation (often higher than that in the centres of the Aquitaine and Paris basins) but irregular and fairly concentrated in time.

The 800-mm average annual precipitation is unequally distributed in both space and time (figure 14). One third of the country receives half the precipitation. It ranges in an average year from 500 to 2,800 mm (extreme values) depending on locality: maximum readings in the northern Alps (including 0.4-9 m of snow), the west and the south of the Massif Central, western Pyrenees; minimum readings in the Mediterranean coastal plains. The minimum assured precipitation for nine years out of 10 is of the order of 50 to 75 per cent of the annual average.

By contrast, real evapotranspiration is less diversified and variable. It is generally within the range of 500 to 600 mm per year, decreasing to less than 500 mm in regions with less precipitation (Mediterranean coast) and in dry years, but it exceeds 600 mm in regions and years of heavy precipitation (figure 15).

Table 9. France: Principal river basins

River	Area of basin in France (km^2)		Length of main river in France (km)	Average flow in France (km^3/year)	Average flow in France (m^3/s)	Average runoff (mm/year)	Regular flow (m^3/s)
Seine	78 600		776	14	450	180	210
Loire	118 000		1 010	27	850	230	250
Garonne	81 000		647	31.5	1 000	390	300
Rhône	88 157		522	46	1 460	50	1 000
Rhine	23 700		313	7.5	240	320	120
Coastal rivers: Atlantic, Channel	102 793	Adour	335	23	730	220	310
		Charente	360				
		Vilaine	229				
		Somme	245				
Coastal rivers: Mediterranean	40 697	Aude	223	15	475	370	90
		Herault	160				
		Var	120				
Rivers flowing into neighbouring countries	16 236	Meuse	489	8	255	490	120
		Escaut	180				
France as a whole	549 183		-	180	5 700	300	2 400

Figure 13. France: Main river basins and water agencies

3593.9x

The boundaries and names shown on this map do not imply official endorsement or acceptance by the United Nations.

Table 10. France: Climate, temperature and precipitation

Aspect	Climate			
	Oceanic			
	Armorican and Aquitaine	Sequanian (Paris)	Continental	Mediterranean
Average seasonal temperatures	Mild winter, cool summer	Colder winter, warm summer	Cold winter, hot summer	Warm winter, hot summer
January/July° C	2-6/15-22	0-4/15-20	Minus 5-2/15-20 (in the mountains: minus 10-0/ 10-15)	4-6/20-22
Average annual number of days of frost	20-60	60-80	80 (in the mountains: 100)	20-40
Annual average precipitation	750-800 mm in 150-160 days	600-750 mm in 160 days	750-1,000 mm, 1-2 m in high mountains in 150-180 days (except Southern Alps and Massif Central: 100-120 days)	600-800 mm in 70-80 days
	3-10 days of snow	5-20 days of snow	20-50 days or more of snow	3 days or more of snow
Rainy season	Armorican: September-January Aquitaine: April-June, October-December	May-November	May-November	October-December
Dry season	Armorican: February-August Aquitaine: January-March, July-September	December-April	North: January-April South: December-March	June-August

Figure 14. France: Annual precipitation

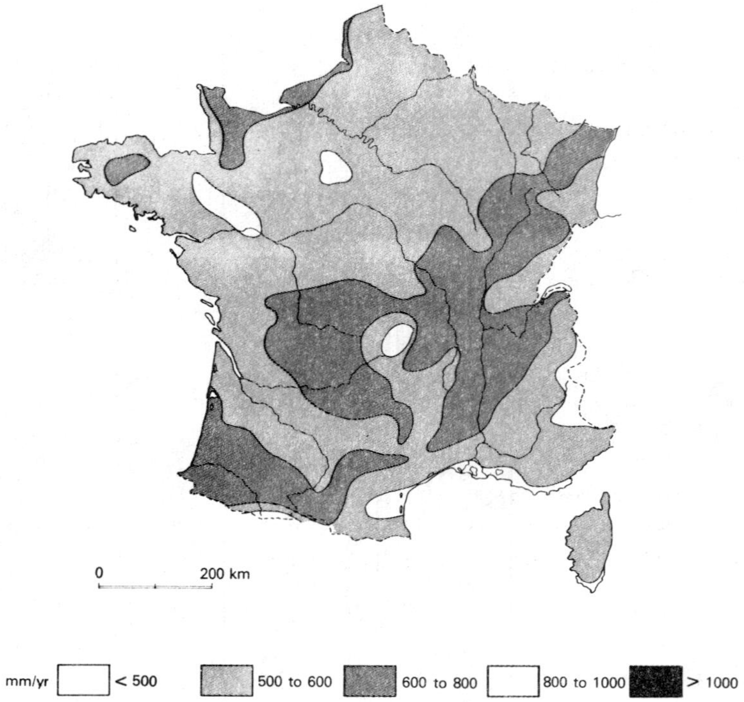

Figure 15. France: Annual evapotranspiration

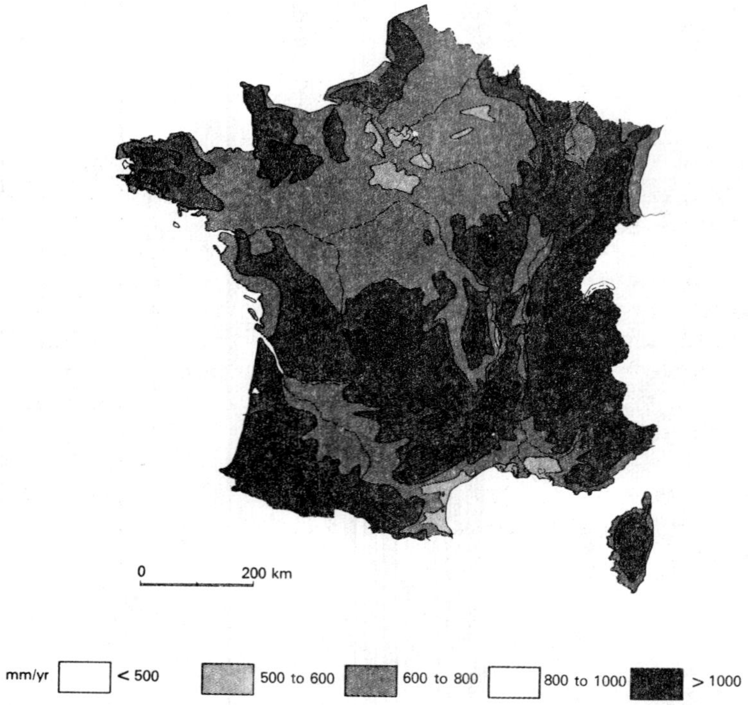

The boundaries and names on these maps do not imply official endorsement or acceptance by the United Nations.

3. Surface water

Geomorphologic and climatic features give way to three major surface water systems, Atlantic, mountain and Mediterranean.

The Atlantic or oceanic region, the most extensive (more than 50 per cent of the territory), includes all the lower areas of the south-west, the west, the centre and the north. The water system is essentially pluvial: high water in winter, low water in summer. Summer is the least rainy season and the one with the highest potential evaporation - often alleviated in the sedimentary areas with controlling aquifers. The average runoff coefficients vary between 30 and 50 per cent.

A mountain region includes all the fairly high mountains in which the lower temperatures reduce the evaporation. The altitude increases the precipitation, the mountain slopes accentuate the runoff so that runoff coefficients reach 60-80 per cent. The water system ranges from pluvio-nival to glacial (Alps) with high water in spring and sometimes including summer (snowmelt), low water at the end of the summer and in autumn, and also relative low water in winter (low liquid precipitation). Most of the large river systems emerge from this area.

A Mediterranean region predominates in the south-east, with highly irregular water systems. There are two seasons of high water (spring and autumn) and a marked low water in summer, alleviated sometimes in certain basins by controlling karstic aquifers. The runoff coefficients are reduced to less than 30 per cent, on the average, by heavy evaporation. Intense and violent floods occur, mostly as a result of autumn storms.

The regimes of the major rivers result from the more or less complex divisions of their basins among these three regions. The Seine, with an oceanic pluvial regime, is the most regular of the French rivers; a mountain tributary, the Yonne, may nevertheless give rise to devastating floods. The Loire, which has an upper mountain basin with a more irregular regime and very low water in summer, is sustained in its upper reaches by tributaries with an oceanic regime. The Garonne has high mountainous basins with the most irregular regime, with high spring water swollen by snowmelt and sometimes violent floods. It also has very low water because of the lack of regulators and the heavy summer evaporation. The Rhône has the most complex regime because of its mainly mountainous basin with important glacial and lacustrine regulators (Lake Geneva) and also because of its aquifers (Jura) and the oceanic influence of the Saône basin. There is, therefore, a link between the high water of winter supplied by the Saône, the high water of spring and summer from the Alpine tributaries, and the autumn floods from the tributaries of the Jura and the Massif Central.

The global water budget in France and in the six basins administered by the six water agencies (figure 13) which contribute financially to water management is summed up in table 11.

Table 11. France: Average annual water budget

Basin (Water agency)	Area (km^2)	Entering flow: (1) Precipitation	Entering flow: (2) Flow from neighbouring countries	(3) Evapo-transpiration (deficit in internal runoff)	Departing flow: (4) Natural runoff of internal origin (1) minus (3)	Departing flow: (5) Total departing natural runoff (2) plus (4)	Share of (5) runoff in neighbouring countries and reimported with (2)
		(in cubic kilometres per year)					
Adour-Garonne	116.9	92	1	53	39	40	--
Artois-Picardy	19.9	14	-	10	4	4	-
Loire-Brittany	155.9	105	--	70	35	35	--
Rhine-Meuse	31.5	29	2 a/	17	12	14	1
Rhône-Mediterranean-Corsica	130.1	125	12	63	62	74	4 b/
Seine-Normandy	94.9	75	-	57	18	18	--
France	549.2	440	15	270	170	185	5
Per cent	-	100	-	61	39	-	-

a/ Rhine frontier not included.

b/ Tributaries of upper Rhône in Switzerland.

4. Geology

The geologic features of France, at the heart of the western edge of the Eurasian plaque formed by western Europe, display both a great litho-stratigraphic diversity and a complex structure. These are the results of a varied and active geologic history. A first glance at the geologic map confirms and completes the relief analysis (figure 16). France contains a juxtaposition of seven main well-contrasted formations. First, there are two ancient massifs (Armorican, Massif Central) and several secondary massifs (Ardennes and Vosges to the north-east, Maures to the south-east). These are the visible parts of the Hercynian basement including itself folded smoothed chains and pre-Hercynian sedimentary basins. There are two chains of more recent uplift: the Alps, with the Jura and Corsica to the south-east and the Pyrenees to the south-west. Finally, there are three post-Hercynian sedimentary basins: Paris, Aquitaine and the sub-Alpine basin squeezed between the Massif Central, the Provençal basement and the Alps which have to a large extent incorporated it in their folded structures. France thus reflects the two main structural divisions of middle Europe: Hercynian and Alpine.

Nearly all the stages of the stratigraphic scale from Precambrian to Quaternary are represented. Precambrian and Paleozoic sediments (schist, sandstone, conglomerate, limestone) are partly metamorphosed and conserved in a number of trenches encrusted in the ancient massifs or covered by more recent formations and known mainly as a result of mining research and exploration (coal basins) or oil prospecting. Secondary sediments (sand, clay, salt-bearing and carbonate rocks) are the basic components of the three major basins and their submarine extensions. Here are alternating facies of numerous sedimentary sequences especially of marine origin (table 12). Tertiary and Quaternary sediments (sand, clay, molasse, limestone, gypsum, alluvium) are partly continental with lacustrine, fluviatile formations and locally glacial or eolian (loess). They occupy mainly the centre of the Paris basin and the south and west Aquitaine basin, and several tectonic troughs in the centre and east of the country (Auvergne, Burgundy, Alsace). All have been involved in the complex construction of the Alpine chains. Eruptive, intrusive or extrusive rocks also belong to different geologic ages from the Precambrian granite of the Armorican massif to the Quaternary lavas of the Auvergne.

Two major orogenic phases, Hercynian and Alpine, are responsible for the structural features found in France. The Hercynian orogeny affected the territory as a whole: a phase of foldings, Variscan directions and intrusions, sometimes metamorphism, followed by a general episode of late Hercynian fractures and folds which often influenced subsequent sedimentation and foldings. The Alpine orogeny affected only the south and south-east of France. It gave rise at the end of the Tertiary to the Pyrenees-Provençal and later the Alpine chains partly modelled on the Hercynian formations. Reworked and raised higher, cores of the old basement form the basic structure of axial chains of the Pyrenees, the Alps and Corsica, whereas the Mesozoic and Tertiary sediments have been strongly folded and in some cases thrust down into alluvial drifts. Scattered and collapsed rocks are abundant in the outer zones and in the revivals of ancient structures, namely, Corsica, Maures, south and east of the Massif Central and the rift of the Rhine trench. The Quaternary volcanism of the Auvergne in Languedoc, the frequent signs of still active neotectonics and appreciable seismicity in various regions are the symptoms that this phase is continuing.

Figure 16. France: Geologic structure

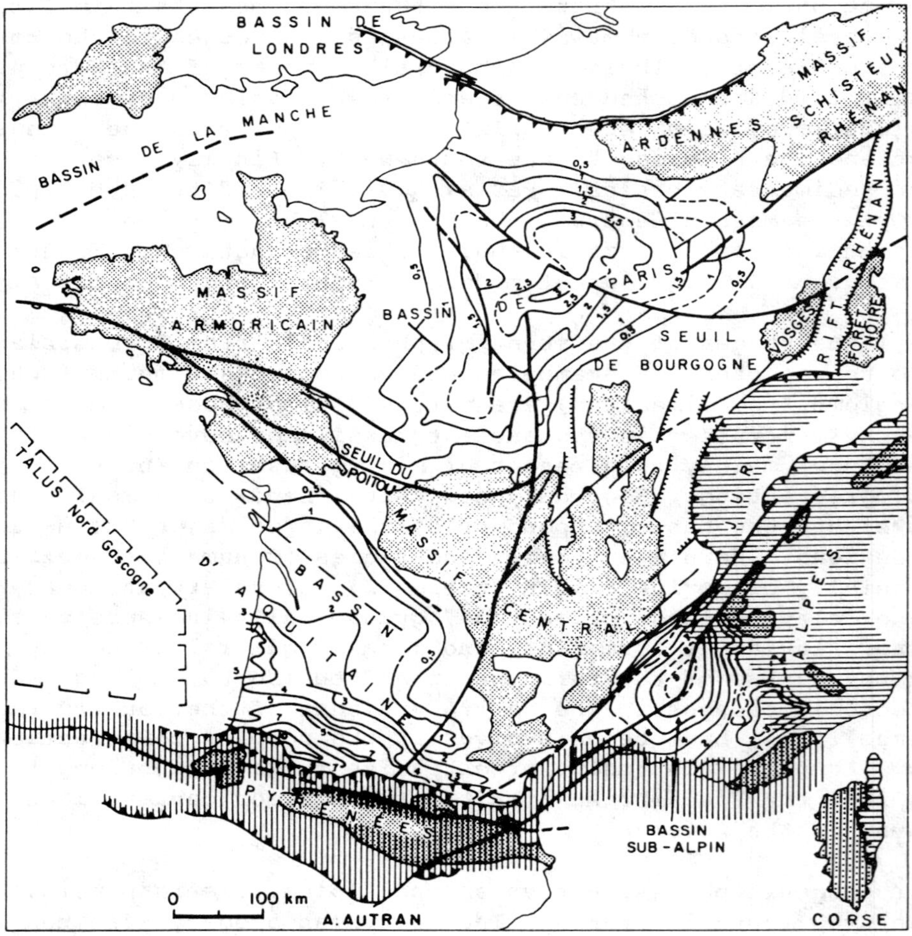

Post-Hercynian (Mesozoic, Tertiary, Quaternary) sedimentary cover and equally deep lines of the Hercynian basement (in km).

Outcropping Hercynian basement. 3 and 4: Alpine orogenic zones:

Alpine region:

Pyreneo-Provençal region.

Main tardi-Hercynian Mesozoic and Tertiary repeated faults.

External North-Hercynian overthrust.

Alpine and Pyrenean external overthrusts.

The boundaries and names shown on this map do not imply official endorsement or acceptance by the United Nations.

3593.12x

Table 12. France: The three main sedimentary basins

Basin	Area (km^2)	Coastline	Maximum thickness of sedimentary cover (m)
Paris	160 000	Channel	3 200
Aquitaine	60 000	Atlantic	10 000
Sub-Alpine	25 000	Mediterranean	9 000

B. Hydrogeology

1. Background

From the work of the precursors of the seventeenth century (P. Perrault, 1674; E. Mariotte, 1686) to the contributions of physicians, engineers and geologists of the nineteenth century (F. Arago, 1834; E. Belgrand, 1846; J. Dupuit, 1848; H. Darcy, 1856; J. Gosselet, 1886; A. Daubree, 1887; E. A. Martel, 1894) the progress of knowledge on groundwater in France has been inseparable from development of the science of hydrogeology itself. This progress has also been directly linked to the emergence of problems caused by urban and industrial growth, which have underlined the importance of groundwater, its analysis and the search for solutions.

The discovery and development of underground artesian layers (in Artois, province in northern France), with the first flowing boreholes being drilled at Tours (1830) and Paris (1840); the need to forecast and calculate the dewatering of coal and iron mines; the desire to understand the origin of thermal and mineral waters and to develop their rational use; the first speleologic explorations; the awareness of the risks of contaminating springs and wells and the search for means of protecting them; all these developments led to the progress of hydrogeology in France. Until the beginning of the twentieth century, its advancement, nevertheless, remained empirical and sectoral while concentrating on certain areas only.

Since the middle of the twentieth century, the growth in demand for community water supply, for industrial use and for irrigated agriculture caused a sharp increase in the use of groundwater. Its sometimes excessive and intensified exploitation gave rise to efforts to evaluate supply and required more complete, pertinent and more quantified knowledge. There grew up a generation of professional hydrogeologists, many working in Africa, especially the Maghreb. J. Archambault, H. Schoeller, G. Castany were pioneers, assisted by the Universities and the Grandes Ecoles, the Bureau de Recherches Géologiques et Minières (BRGM) and engineering firms. After a stimulating analysis by P. Lafitte and J. Ricour in 1956, methodical inventories, documentary research and prospecting and modelling operations developed. They made it possible to complete monographs and cartographic work which today reflect a fairly advanced state of hydrogeologic knowledge for most of France. The main lines of this research are summarized here. Contemporary efforts are concerned with everything to do with the quality of groundwater, its sources and its protection.

Finally, the applications for hydrogeology have been extended to purposes other than prospecting and exploiting groundwater as a raw material. After the contributions of petroleum research (mainly the Paris and Aquitaine basins), researchers studied the feasibility of underground storage of fluids (especially gas) in aquifers. They sought to evaluate and exploit geothermal resources and to analyse and guarantee the safety of surface or underground deposits of waste including atomic waste. This has recently helped to advance the knowledge of deep hydrogeological conditions, especially those of areas with low permeability, both in the ancient basement and in the depths of sedimentary basins.

The great variety of geologic features and climatic systems give way to a great diversity of hydrogeological conditions and groundwater systems. Hydrogeologic structures of different sizes and complexities and aquifer systems of differing behaviour form part of the geologic and hydrographic structures of France. The hydrogeologic structure map (aquifer systems) of France on a small scale, published in 1980 (BRGM) gives an overall picture of these conditions (figure 17).

A schematic hydrogeologic description can be made on the basis of the major geologic areas.

2. Ancient massifs

The crystalline and schistose areas which dominate these massifs do not contain extensive aquifers. But Brittany is far from being as impervious as a "roof of slate" and in the heart of Limousin, in the Massif Central, the name of the plateau of Millevaches (a thousand springs) is in itself significant. The continuity of runoff of the dense river system in these regions is not due only to the high frequency of the rains. Many of the ancient massifs, especially of granite and similar rocks, have a considerable capacity because of their weathered zones, which can be several tens of metres thick. Thus they help considerably to control the runoff of the rivers (few less than 3 to 5 $l/s/km^2$ at low water). The springs are fairly weak but are numerous and regular. Moreover, all this land is fissured and fractured to various extent and water may circulate up to several hundreds of metres in depth and over quite varied distances. In some cases, these distances are great, a fact which explains the origin of certain mineral springs. An essential feature of this domain is the multitude of elementary microsystems, with changing and unstable limits depending on the connections of the fissured systems. It also explains the importance of the vertical components of the underground formations which are generally three-dimensional.

The volcanic lands of the Auvergne, with its mixture of highly porous rocks and unequally fissured lavas, form complex aquifers. Their situation resembling that of a water tower, ensures an abundant supply from rain and snow of 400 mm per year in the chain of Puys whose aquifer discharges on the average of 3.5 m^3/s.

Figure 17. France: Hydrogeology

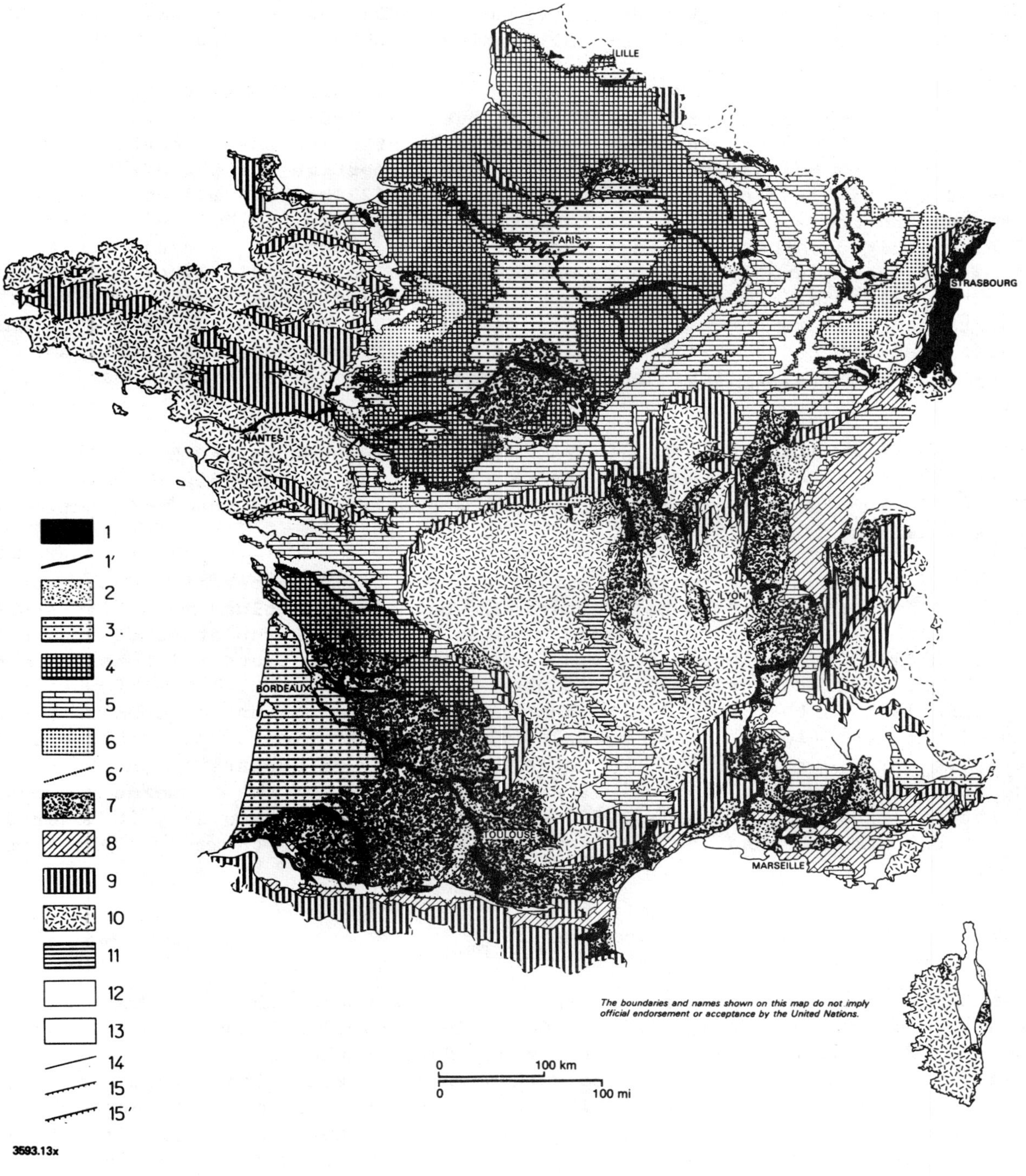

3. Alpine mountains

The Alpine mountains are made up of blocks of different rocks juxtaposed or superposed, dislocated and fractured in depth. Therefore, the Alps, Pre-Alps and Pyrenees have aquifers of only medium extent, divided into compartments, but very numerous. They are mainly karstic limestone massifs many of which form dominant relief: Chablais, Bauges, Chartreuse, Vercors and Devoluy in the Pre-Alps; the massifs of Arbailles, Pierre-St-Martin, Comminges and Pays-de-Sault in the Pyrenees.

In these limestone massifs, with their surface riddled with sinkholes, geologists have discovered the most remarkable karstic systems of conduits: the Vernaison cave (20 km) and the Berger gulf, the deepest in the world (1,122 m, discovered in 1956), in the Vercors; the Glaz hole (23 km, 603 m deep) in Chartreuse; and the gulf of Pierre-St-Martin (1,110 m) in the Pyrenees. Because of their high altitude, these aquifers receive a heavy charge of rainfall (up to 1,000 mm per year) largely due to snowmelt which is more regularly occurring than infiltrated rainwater. The autumn flow from the collecting springs and rivers is more sustained than in the plain. The permanent water reserves contained in the permeated areas of these massifs is difficult to evaluate but is probably not enormous and is certainly less than the reserves of the karstic aquifers lower down.

Besides these limestone reservoirs, the overburden deposited by glaciers and torrents (moraines, dejection cones and terraces) play an appreciable role as regulators of runoff, for these are the best accumulators and they form small but numerous aquifers.

The Jura holds a vast aquifer complex where a small number of independent underground runoff basins can be identified. Both in the folded mountains and in the high plateaus, this is the classic area in which the circulation of groundwater is not controlled by the boundary of surface watersheds: there are considerable losses from rivers and springs and enclosed basins develop. The most famous underground link is that which joins the discharge from the upper Doubs to the spring of the Loue (10 to 15 km) revealed for the first time by the transport of absinthe released during a fire at the Pernod factory at Pontarlier on 11 August 1901. The subsoil receives and collects most of the abundant regional water intake (from 350 to nearly 1,500 mm per year) before returning it lower down in the Doubs, the Loue and the Ain, mainly: about 250 m^3/s (8 km^3 per year) on the average.

4. Plateaus

On the edge of the ancient massifs or mountain chains there are vast plateaus of karstic limestone (from Jurassic to Lower Cretaceous) or sandstone (mainly Triassic). They form a ring of extensive aquifers on the periphery of the great sedimentary basins: from Lorraine to Burgundy, Berry and Poitou; from Quercy to the Causses to the south of the Massif Central, to the garrigues of Languedoc and upper Provence (plateaus of Vaucluse and Verdon). In the limestone, many of the unconfined surfaces of the submerged areas are deep, more than 100 m, and even to 300 m.

These aquifers are well supplied (150-600 mm per year) and of great capacity. In Lorraine, for example, the limestone reservoir of Côtes de Meuse holds 8 km^3, the limestone reservoir of Côtes de Moselle 50 km^3, and the unconfined reservoir in the lower Triassic sandstone about 100 km^3. They are good water regulators and they help feed the rivers in the basins of the Seine, Loire, Saône, Durance, Charente, Dordogne and Garonne (right bank). The underground circulations, many of the karstic type, especially in the southern half of France, predominate over surface flows with loose or non-existent water systems and they can often short-circuit them. For example, the Serein loses water for the benefit of the Cure in the basin of the Yonne. The total losses from the Bandiat and the Tardoire (tributaries of the Charente) are recovered by the spring of the Touvre near Angoulême and the waters of the Nesque find their way to the Fontaine de Vaucluse. Examples of such underground intakes are many. It is in these karstic aquifers that the longest circulation distances have been discovered by tracing; the record is that achieved in 1968 in the Vaucluse aquifer between the Belette sinkhole and the Fontaine de Vaucluse; 46 km in a straight line covered in 25 days.

Some of these aquifers, of medium size, are on the Mediterranean coast (Massif des Calanques to the east of Marseille, mainly). They flow directly into the sea at the rate of 6 m^3/s, on the average, through coastal or submarine springs whose water is more or less brackish. Their currents carry along part of the sea water that invades these deep reservoirs and this is mixed with the fresh water. However, most of the karstic aquifers in French territory are discharged by continental springs collected by rivers. These springs have an abundant discharge which for many exceeds 1 m^3/s at low water and some tens of cubic metres per second at flood time, but irregularly, especially in the south. They include the most important springs in France: Fontaine de Vaucluse (average flow: 25 m^3/s; up to 150 m^3/s in flood period; secular minimum; 4 m^3/s. Touvre de Charente, Fontaine-Levêque du Verdon, and others.

5. Sedimentary basins

The superposition of layers with sharply contrasting permeability characteristics gives rise to alternating aquifers, aquicludes and aquitards. The emergent bowl-shaped structures form a series of rings, especially in the Paris basin, in the northern and north-west edges of the Aquitaine basin and on the western edge of the sub-Alpine basin. These basins therefore abound in extensive aquifers. They consist both of directly fed unconfined aquifers and deep confined aquifers which prolong them and are more or less intercommunicating. According to their structure and runoff, one may distinguish three types as follows:

(a) The major unconfined single-layer aquifers are formed of mostly carbonate rocks (chalk, Jurassic, Cretaceous or Tertiary limestone) but also siliceous (Cretaceous or Tertiary sand and limestone). With an average recharge of 150 to 300 mm per year, they help appreciably to maintain the flow of the rivers which drain them and which divide them into aquifer systems. These are more or less independent and of extent variety from several hundreds to several thousands of square kilometres.

The most important layers are those of the chalk aquifers in the north of France and a large part of the Paris basin (Normandy, Picardy, Champagne, Sénonais), the layer of La Beauce (7,000 km^2 with an average total flow of 600 million m^3 per year and a capacity estimated at 20 km^3), the layer in the sands of Les Landes (13,000 km^2, 14 km^3 per year on the average), the main supplier of the multilayered system of western Aquitaine. Although it is situated above the system, it is difficult to exploit because of its mediocre productivity.

(b) The complex multilayered aquifers consisting mainly of varied and heterogeneous Tertiary sediments (alternating layers of sand, limestone and clay superposed, juxtaposed and many bevelled). These form the central areas of the Paris and Aquitaine basins - Ile-de-France (layer of Soissonnais) and Bordeaux - and several other small areas (basin of Orchies in the north; basin of Aix-en-Provence). These aquifer systems have an unconfined upper level fed by rainwater infiltration and several deep confined layers which all communicate with each other more or less easily, depending on the heads. Some of them locally have a surface pressure. For example, in western Aquitaine the layer of Eocene sand and limestone is extended by the "infra-molassic" layer in the centre of the Aquitaine basin, the main aquifer (50,000 km^2) of the multilayered system in this basin. It is also the most utilized, receiving more than a half of its intake and releasing more than a third of its discharge by exchanges with continuous aquifers. There is a calculated total flow of 130 million m^3 per year.

(c) The large confined aquifers (several hundreds and up to 1,000 m thick), mainly sand or sandstone (Triassic, Jurassic, Cretaceous) or limestone (Carboniferous) of northern France (figure 18). Their total intake and flow are limited but their capacity is enormous and may be reckoned at about 100 km^3. However, these reserves are of little practical use because the layer is confined. Nevertheless, their slow renewal ensures the presence of fresh water in many of the reservoirs but these also contain brackish or salt water. These reservoirs are the classic artesian layers recognized since the nineteenth century, such as the layer of greensand (Albian) in the centre of the Paris basin, the layer of Lower Triassic sandstone of Lorraine, and the layer of Carboniferous limestone in the region of Lille. The deeper aquifers contain only brackish or salt water but are now of geothermal interest (Middle Jurassic limestone in the Paris basin).

6. Rift valleys

The Rhine valley, the corridor of the Saône and the valley of Bresse, the Limagnes of the Massif Central, the Geneva valley between the Alps and Jura, lower Dauphiné, between the Alps and the Massif Central, all have deep structures similar to those of the basins. They have the same alternation of aquifers with confined layers or semi-permeable layers. The confined layers give rise to little or no outcroppings, being limited mainly by faults. Their intake is low and the water is already salt at medium depth. The semi-permeable layers consist of sandy and argillaceous molasse of unequal water-bearing capacity and predominate on the surface. The exceptions are the extensive molasse areas of lower Dauphiné, in southern Lyon (up to 600 m), giving rise to high transmissivity and a large storage capacity estimated at more than 100 km^3 and, above all, the valley of the Rhine. This valley has a thick layer (in places exceeding 200 m) of permeable and productive alluvium with abundant recharge, especially by infiltration from rivers of the Vosges. The average total flow is estimated at 1 km^3 per year. It has a large capacity, a reserve of about 35 km^3 at levels sustained by the Rhine to the east, which provides the plain of Alsace with a good supply of groundwater.

Figure 18. France: Principal artesian aquifers

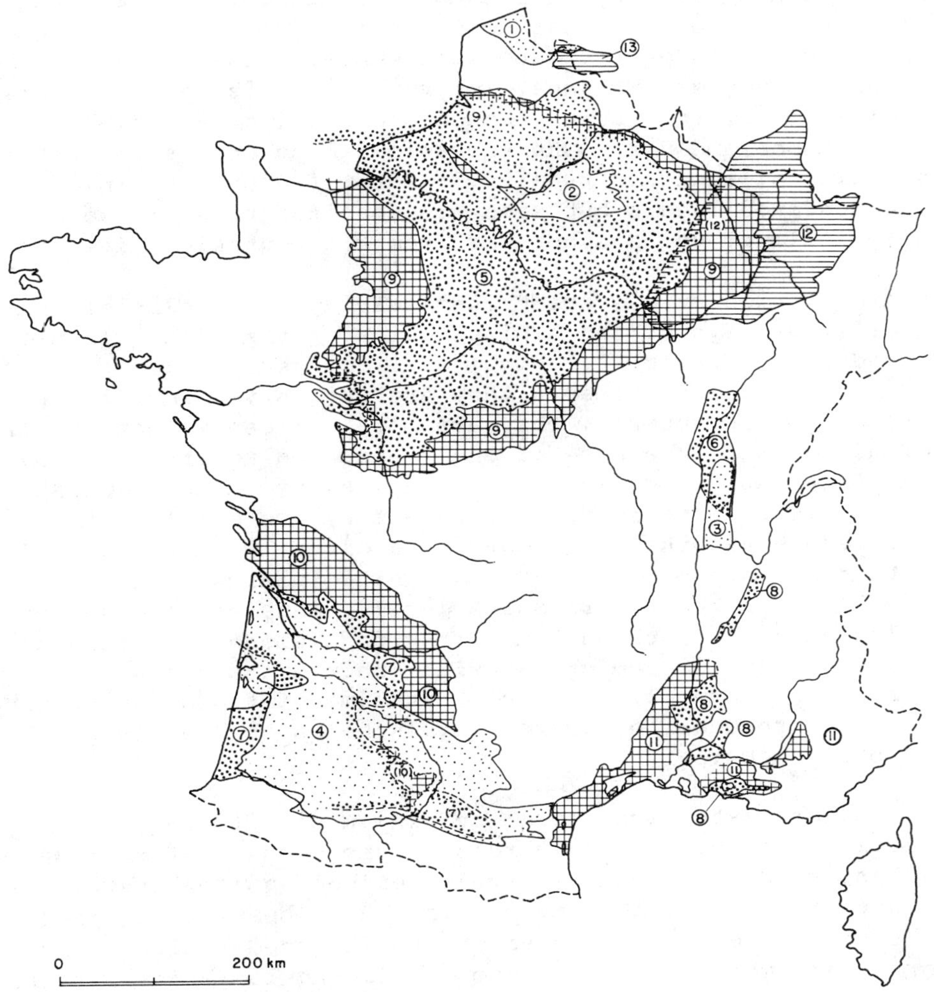

Levels from top to bottom:

Tertiary
1-Landean sands of Flanders.
2-Brach sands and Soissons chalk.
3-Miocene sands of Bresse.
4-Lower Eocene sands of Aquitaine.

Cretaceous
5-Green sands of the Paris Basin and Cenomanian sands of Touraine.
6-Albian sands of Burgundy.
7-Cenomanian sands and cretaceous limestone of Aquitaine.
8-Cretaceous limestone of the Rhodanian corridor and Provence.

Jurassic
9-Upper and middle Jurassic limestone of the Paris Basin.
10-Upper and middle Jurassic limestone of Poitou and the Aquitaine Basin.
11-Jurassic limestone of Languedoc and Provence (Aix Basin, Valensole).

Trias and Palaeozoic
12-Sandstone of lower Trias ("Vosgian Sandstone") of Lorraine.
13-Carboniferous limestone of the North.

The boundaries and names shown on this map do not imply official endorsement or acceptance by the United Nations.

3593.14x

7. Alluvium

The sand and gravel which cover the bottom of the valleys form a whole system of water-bearing channels which are more or less continuous. They duplicate the river system with which most of the layers are closely linked. The exceptions are several extensive alluvial aquifers similar to those of the sedimentary layers described above. These include a part of the Alsace plain, the valley of the Rhône and the plain of Crau in Provençe. The most important alluvial aquifers are found in the valleys of the large rivers: Seine, Meuse and Moselle, Rhine, Loire and Allier, Saône, Rhône, Durance and Var. The terracing which occurs in certain valleys isolates the alluvial aquifers from river water (Garonne). In most cases these aquifers have unconfined layers but some are confined under mud and clay coverings, principally in lower valleys not far from the mouths of rivers.

These aquifers are of high or average transmissivity. The glaciofluvial alluvions of the interior valleys of the Alpine chains or the foothills are the most permeable and productive (table 13). They are, however, of low capacity, except that on the plain of Alsace, because they are not very thick, for example: $3 \times 10^5\ m^3/km^2$ in the Crau. In general, they are more permeable than the land in which they are embedded, even in the case of chalk and sand aquifers. Many of the alluvions serve as an intermediary drain between the layers on the plateaus and the rivers. This drainage role is also important where they are embedded in formations which are nearly impermeable and therefore cannot be directly exploited. These aquifers include certain molasse beds of lower Dauphiné (plains of Bièvres) or southern Aquitaine (valleys of the Adour, and Ariège). Thus, the actual water recharge of these alluvial layers, through their surface alone, is often secondary to that from the hills and also in comparison with their exchanges with the rivers. These exchanges are natural, especially in time of flood, or caused by man-made withdrawals which reduce the aquifer levels below the levels of the surface water, which often occurs.

Nearly all the aquifers extensive enough to play a significant role in the natural system of total runoff or to play a role in the economy of regional water have been identified and studied. About 200 well-defined aquifer systems of regional importance with an area of 100 km^2 and more have been inventoried in France, including 175 unconfined single-layer or multilayered aquifers, many limited by the major rivers which traverse them, including 15 large alluvial aquifers, 40 karstic aquifers, 20 multilayer aquifers (at least in part), and 25 deep confined aquifers linked for the most part with one of the preceding aquifers (figure 18).

Most of these aquifers are included entirely within one or other of the great river basins of which they form the natural regulating reservoirs. The best endowed is the basin of the Seine, with 40 unconfined regional aquifers covering most of its extent. The least endowed is the basin of the Garonne with about 20 aquifers of average extent. Some aquifers are shared between two great basins (La Beauce between the Seine and Loire, the sandstone aquifer of the Vosges between the Moselle and Saône, Causse du Larzac between Garonne and Hérault) or are shared between adjacent coastal basins (aquifers of Craie d'Artois and of Picardy). At the national level, the dividing lines between surface and groundwater differ little. In the regions formed of karstic limestones the predominance of ground circulation may give rise to notable differences, in particular in the thresholds between sedimentary basins (Burgundy, Poitou, Grands Causses to the south of the Massif Central).

Table 13. France: Characteristics of the principal aquifers

Aquifer formation	Position	Situation	Type of conductivity P - porous F - fissured M - mixed	Porosity in unconfined layers	Thickness (m)	Transmissivity (m^2/s)
Gravel alluvium	Pliocene, Quaternary	River valleys, terraces	P	0.03-0.2	5-40	10^{-3} to 5 X 10^{-1}
		Plain of Alsace	..	0.05-0.15	10-250	1 to 5 X 10^{-1}
		Plain of Roussillon	..	0.01-0.2	10-150	5 X 10^{-4} to 3 X 10^{-2}
Sands	Quaternary	Coastal dunes, Les Landes (W Aquitaine)	P	0.05-0.1	10-150	10^{-3} to 2 X 10^{-2}
	Pliocene	Languedoc (deep)	P	..	10-30	10^{-4} to 10^{-3}
	Miocene	Bresse (deep)	P	..	80	10^{-3}
Sands of Ostricourt	Eocene	North, Flanders (Belgium)	P	0.01-0.05	15-45	10^{-6} to 2 X 10^{-3}
	Cenomanian	SW Paris basin, N Aquitaine basin	P	0.05-0.01	30-90	10^{-4} to 10^{-2}
Greensand	Albian	Paris basin, Burgundy	P	0.05-0.2	30-200	10^{-4} to 4 X 10^{3}
Sands, sandstone and limestones mixed	Oligocene-Miocene	Centre, Paris basin, S Ile-de-France, Beauce, W Aquitaine basin	M	0.01-0.15	50-130	10^{-3} to 10^{-1}
Molasse	Miocene	N Sub-Alpine basin, Lower Dauphiné	M	0.1-0.05	100-600	10^{-3} to 10^{-2}
Coarse limestone, Soissonais sand, inferior sand	Eocene	Centre Paris basin, N Ile-de-France	M	0.1-0.0005	70-120	10^{-3} to 10^{-2}
	Lias	Centre and W Aquitaine basin	M	..	50-300	10^{-3} to 10^{-2}
Luxembourg sandstone	Upper Triassic (Rhetian)	NE Paris basin, N Lorraine	M	..	50-100	5 X 10^{-4} to 1 X 10^{-3}
		E Paris basin, Lorraine	M	..	10-30	5 X 10^{-5} to 2 X 10^{-4}
Vosges sandstone	Permian, Triassic	E Paris basin, Vosges, Lorraine	M	0.10-0.15	50-500	5 X 10^{-4} to 2 X 10^{-2}
Chalk	Upper Cretaceous	Centre N and W Paris basin	M (local karst)	0.0001-0.05	100-500	10^{-5} to 5 X 10^{2}
Limestone, dolomites Limestone of Brie, Champagny and St. Ouen	Eocene Oligocene	Centre Paris basin, S and SE Ile-de-France NW Aquitaine basin	F (local karst)	0.001-0.01	40-100	2 X 10^{-4} to 10^{-1}
	Upper Cretaceous	N Aquitaine basin, Charentes (deep)	F - karst	0.01-0.02	300-350	10^{-4} to 2 X 10^{-2}
	Lower Cretaceous	Sub-Alpine basin, Alps	F - karst	0.01	200-400	10^{-1} (Vaucluse)
	Urgonian	Provençe				10^{-2}

Table 13 (continued)

Aquifer formation	Position	Situation	Type of conductivity P - porous F - fissured M - mixed	Porosity in unconfined layers	Thickness (m)	Transmissivity (m^2/s)
	Upper and Middle Jurassic (several layers)	Paris basin (periphery, deep), Lorraine, Champagne, Burgundy, Berry, Poitou, Normandy	..	..	100-800	10^{-6} to 10^{-2}
		N and NE Aquitaine basin, Quercy Charentes	F, M (dolo-mites) karst	0.01	100-200	3×10^{-4} to 4×10^{-2}
		Jura, Alps, pre-Alps, S Massif Central, Causses	..	..	600-500	..
	Lias	S Paris basin, Poitou	F, M - karst	0.001-0.01	5-80	10^{-5} to 5×10^{-4}
	Triassic (Muschelkalk)	E Paris basin, Lorraine, Alps, Provençe	F - karst	0.005-0.1	30-100	5×10^{-5} to 5×10^{-3}
	Devonian Carboniferous	North Ardennes, S Massif Central (Montagne Noire), Pyrenees	F - karst	0.01-0.2	30-50 fissured zones	5×10^{-5} to 10^{-2} 10^{-1} to fissured zones
Volcanic rocks, lavas	Quaternary	Centre Massif Central, Auvergne	P.F.	0.01-0.1	10-200	5×10^{-5} to 10^{-2}
Crystalline and metamorphic rocks	Precambrian	Armorican Massif, Massif Central, Vosges, Alps Pyrenees, Maures, Corsica	F	..	fissured zone 100-200	..

C. Aquifer characteristics

1. Lithology

The different geologic formations, sedimentary for the most part, which make up the aquifer systems have characteristics that are both geometric (thickness) and petro-physical (storage and conductivity). These vary widely and are uneven in homogeneity and continuity. Table 13 recapitulates for the most widespread aquifer formations in France the lithologic types, their dominant characteristics and the state of knowledge obtained from studies and model operations. The range of parameters relating to each formation is wide in most cases: the porosity and permeability vary considerably depending on the local facies (alluvions, more or less argillaceous sand or the degree of fissuring) and their vertical distribution (carbonate rocks, eruptive rocks). For example, chalk is much more fissured and hence more transmissive in the valleys than in the plateaus and a hundred to a thousand times that of the deeper aquifers, under cover.

The productivity of the waterworks allowed by local transmissivity is therefore highly diverse and widely dispersed: table 14 shows the distribution of the commonest aquifer formations in accordance with their range of productivity and figure 19 shows their relative distribution as regards the first accessible aquifers (usually unconfined layers).

Table 14. France: Productivity of the aquifers

Level of productivity* (m^3 per hour)	Aquifer systems
High: 100 or more (exceptionally, to 1,000) in carbonate rocks	Thick, non-argillaceous alluvium of wide valleys, especially among mountains and foothills (glaciofluvial rocks, chalk, limestone, dolomite), which are highly productive, fissured and partly porous, layers of sand and sandstone of high productivity (deep aquifers)
Average: 10-100	Most alluvial aquifers, most sedimentary aquifers with unconfined layers of average productivity: certain molasse deposits, sand and Tertiary limestone, chalk and fairly well fissured limestone, sandstone, volcanic rocks
Low: 1-10	Alluvial aquifers (especially in valleys of ancient massifs), argillaceous sand or thin beds of sand. Gravel, molasse, carbonate rocks with few fissures (especially at depth) or marl, altered schist, fractured crystalline rocks

* Production of a well or borehole which is complete and correctly built and equipped.

Figure 19. France: Average yield of tubewells

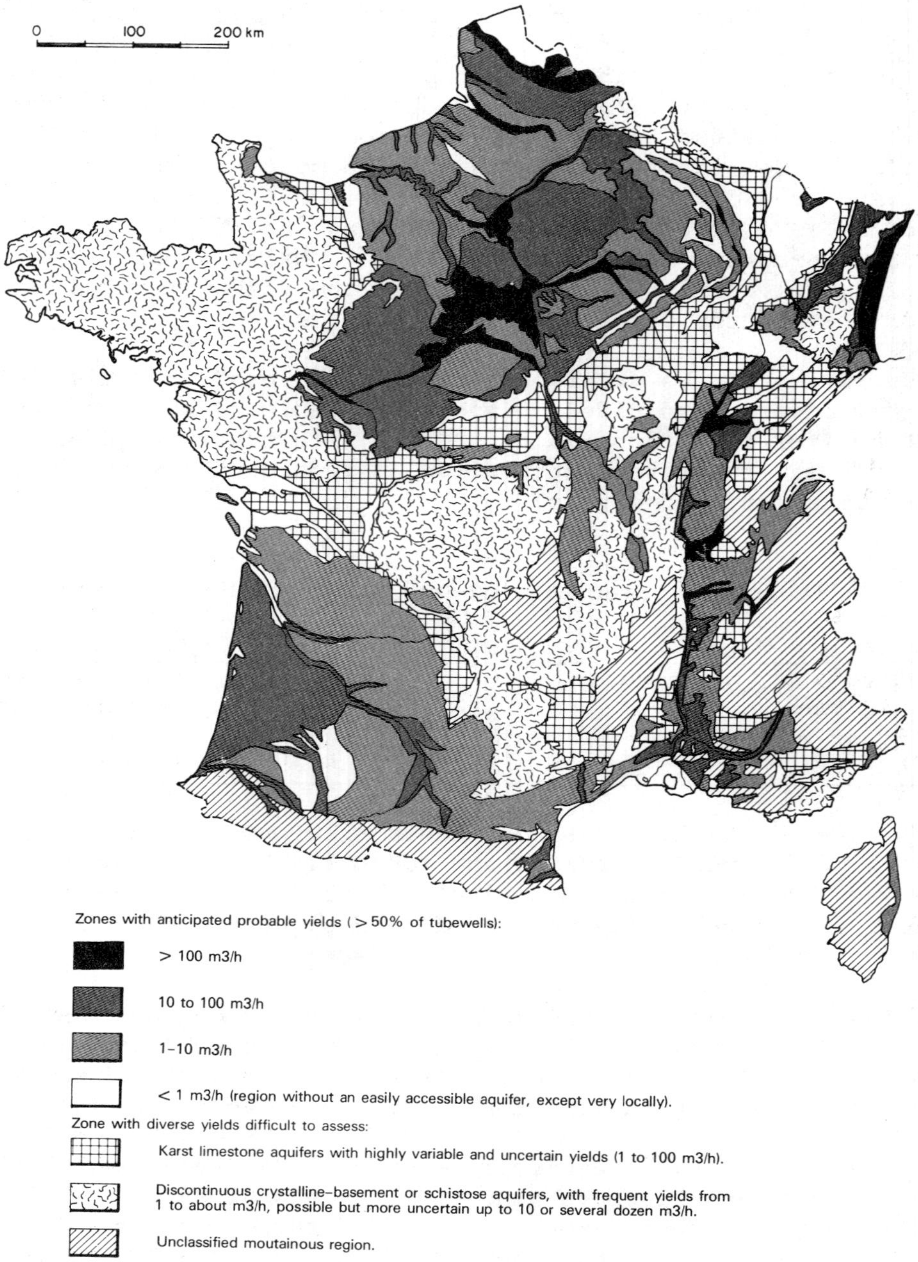

3593.15x

The boundaries and names shown on this map do not imply official endorsement or acceptance by the United Nations.

2. Hydrology

The availability of records of piezometric levels varies from region to region both as regards duration and number of observations. We know, however, roughly the basic pattern of variation in the levels of unconfined layers of the main continuous aquifers, which are much better known than the karstic aquifers. The variation reflects the regulating roles of the runoff of these irregularly fed aquifers. The differences from year to year also reflect the differences in efficient precipitation. The widest variations may attain one or several tens of metres in the centre of the most extensive and compact aquifers, not dissected to any large extent by the collecting rivers. These include the chalk aquifers of La Beauce and the Jurassic limestone plateaus. Here the variation over a number of years may substantially exceed the seasonal and annual variation. They range from one to several metres in the less extensive aquifers of which no sector is far from the limits of water potential.

The regular and annual systems of the underground layers that are well supplied at their limits - Alsace for example - or influenced by irrigation - such as the alluvial plains of Provence or Languedoc - contrast with the irregular pluri-annual systems of most unconfined layers fed by rainfall, which are of low or medium capacity and discharge through springs whose flows are themselves highly variable.

The intakes of meteoric origin, concentrated in France over two to four months in winter and the beginning of spring, vary considerably from year to year. The annual recharge is less than half the recharge of the average year, less than once every three to five years. Recharge is negligible 1 year out of 10 generally, but 1 year out of 5 in the Mediterranean region. Years of high recharge occur twice as often as do years of negligible recharge. Variation coefficients (ratio of standard variation to the average) of annual intakes ranges from 50 to 100 per cent. The yield of springs discharging from porous aquifers is fairly regular. Conversely, the yield of springs discharging from fractured carbonate aquifers varies considerably through the season and from year to year, especially when these aquifers are in karst. They range from 1 to 100 and sometimes 1 to 1,000 in the case of Vauclusian springs which emerge or re-emerge from true underground rivers.

As a contribution to the evaluation and management of groundwater in France, a number of researchers and study groups have assessed water budgets for about 100 aquifer systems of different types which include unconfined and confined aquifers, aquifers in different states, whether natural or exploitated, in state of equilibrium or not. Table 15 gives numerical data relating to a selection of aquifers given in figure 20. The data are important because of their extension or utilization and representation in terms of area, total average input flow (equivalent in the natural stage to the average output flow) and expression in equivalent height of water (mm per year) as to facilitate comparisons or proportions of input according to origin and withdrawals.

Table 15. France: Data on water budgets of aquifer systems

Type of aquifer formation	Situation (numbers refer to figure 20)	Aquifer system	Runoff regions L=unconfined C=confined	Area (km^2)	Average global flow (input = natural output) (Mm^3/year)	(mm/year)	Origin of input a: precipitation, infltn. b: surface water, infltn. c: affluents, drainage *: induced water supply	Withdrawal total (m^3/year, date)
	Alsace (1)	Quaternary alluvion, Rhine trench	L	2 600	= 1 000 (700-1 200)	385	a: 200 Mm^3/year (20%) b: 500-1 000 (30%)	480 (1976)
	Rhône Valley (2)	Quaternary alluvion	L	480	initial: 140 As of 1975: 275	290 (573)	1975: a: 140 Mm^3/year b + : 135	275 (1975)
		Plain of Lyon	L					
Gravelly alluvion	(3)	Quaternary alluvion, Comtat, Plain Aigues-Sorgues	L	530	165	310	a: 25% b: 75% (including irrigation and canals: 20%)	50 (1975)
	Provence (4)	Quaternary alluvion, plain of Crau	L	520	200	385	a: 19% b: (irrigation: 69%) c: 12%	35 (1975)
	Roussillon (5)	Plio-Quaternary alluvion, multilayer	L, C	800	58 Including confined Pliocene (with 18)	72	a: 19% b: 4% (Pliocene: c = 50%)	40 (1975)
Sand	Aquitaine (6)	Quaternary sand, Les Landes	L	13 000	1 400	110	a: 100%	60 (1975)
	Ile-de-France, Paris (7)	Sands of Soissonnais, coarse limestone (Eocene, multilayer)	C	700	95	135	c: for the most part (including flow from the Seine)	190 (1976)
	Aquitaine (8)	Lower Eocene sands and "infra-molassic aquifers"	C	50 000	As of 1965: 130	3	c + 100% (including: drainage 70 Mm^3/year)	36 (1965) 55 (1975)
	SW Paris basin, Touraine, Anjou (9)	Cenomanian sand	C	25 000	..	..	..	21 (1976)
	Paris basin (10)	Albian greensand	C	75 000	25	03	c: 100%	34 (1936) 16 (1977)
Limestone	East Paris basin, Lorraine (11)	Lower Triassic sand-sandstone	C	20 000	50	25	c unconfined 100%	164 (1978)
	Paris basin, Ile-de-France, Orléanais (12)	Limestone of Beauce (Oligocene)	L	6 000	465	78	a: 100%	112 (1976)
	Ile-de-France, Brie (13)	Limestone of Champigny	C	3 500	630	180	b + c 100% loss of surface water	65 (1976)

Table 15 (continued)

Type of aquifer formation	Situation (numbers refer to figure 20)	Aquifer system	Runoff regions L=unconfined C=confined	Area (km^2)	Average global flow (input = natural output) (Mm^3/year)	(mm/year)	Origin of input a: precipitation, infltn. b: surface water, infltn. c: affluents, drainage *: induced water supply	Withdrawal total (m^3/year, date)
	Flanders (14)	Chalk, Dunkirk region	L	630	100	160	a: artificial supply	44 (1975)
			C	1 105				
	(15)	Chalk, Deule-Scarpe region	L	306	96	87	a: 50% b: 12% (losses, canals) c: 38% drainage	93 (1975)
	Picardy (16)	Chalk, basin of Somme	L	5 670	1 030	182	a: 100%	90 (1973)
Carbonate rocks, chalk, limestone, dolomite	Provence (17)	Limestone, Cretaceous, plateaus of Vaucluse, karst	L	1 230	600	488	a + and b: 100%	..
	East Paris basin, Lorraine (18)	Jurassic limestone, Côtes de Meuse	L	1 500	300	200	a: 100%	12 (1976)
	(19)	Jurassic limestone, Côtes de Moselle	L	3 300	600	180	a: 100%	240 (1975)
	Jura (20)	Limestone, basins of the Deubs, and Loue	L	6 350	4 200	660	a: 100%	..
	S Massif Central (Grand Causses between Lot/Herault) (21)	Jurassic limestone, karst	L	4 500	2 000	445	a + b: 100%	..
	Flanders, Belgium (22)	Carboniferous limestone	L, C	1 582	initial 1973: 125	56	Initial state: a: 90% c: 10% 1973: c - 30 Mm^3/year c - 29%	90 (1973) 73 (1977) 3C and F
Volcanic rocks	Massif Central, Auvergne	Volcanic ash, Chain of Puys	L	260	110	425	a: 100%	40

Note: M = million.

Figure 20. France: Aquifer systems

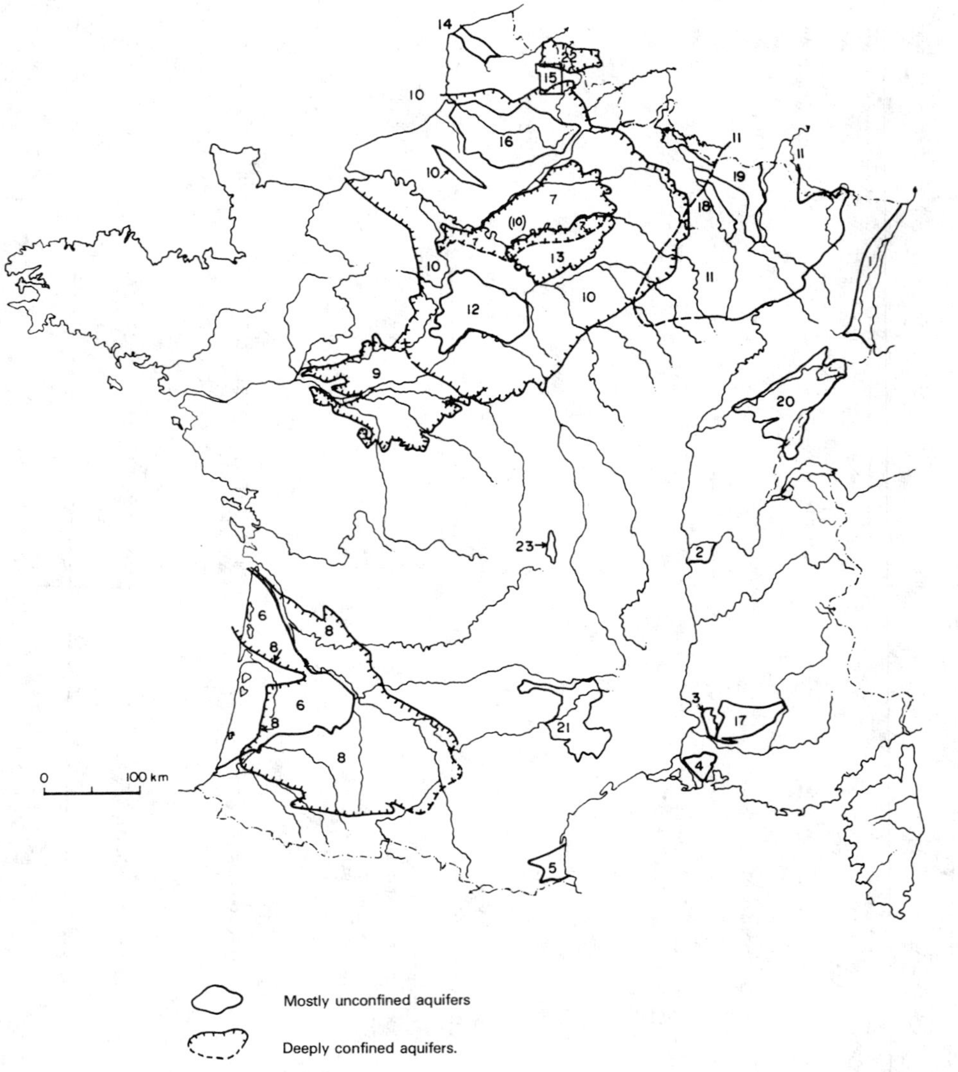

The boundaries and names shown on this map do not imply official endorsement or acceptance by the United Nations.

3593.16x

The analysis of the complete range of the budgets established leads to the following conclusions. The average unit input flow in the unconfined layers varies from 50 to more than 1,000 mm per year, the highest figures corresponding to the karstic aquifers in the mountains (Alps, Jura). The input for confined aquifers, usually by drainage, is by comparison minimal; from several tens to less than one millimetre per year also because of their large areal extent and in absolute terms generally lower than 100 mm per year in their natural state. The distribution of flows both by origin of input (infiltration of precipitation, surface water or irrigation; natural flows or drainage and by type of discharge, namely, emergence or drainage by rivers; lateral flows or drainage; withdrawals) is highly diverse. Input by precipitation predominates normally for the unconfined aquifers but can be minimal, e.g., Crau. Captive aquifers are recharged and discharged essentially by drainage in the natural state.

Appreciable discharges which result in reduction in reserves must and can be taken into account in certain cases since their order of magnitude may exceed that of natural input, for example, in the Triassic sandstone of Lorraine. An attempt to make general estimates of average input flows based on climatic and geologic data has been carried out and is expressed in figure 21.

Figure 21. France: Recharge of phreatic aquifers

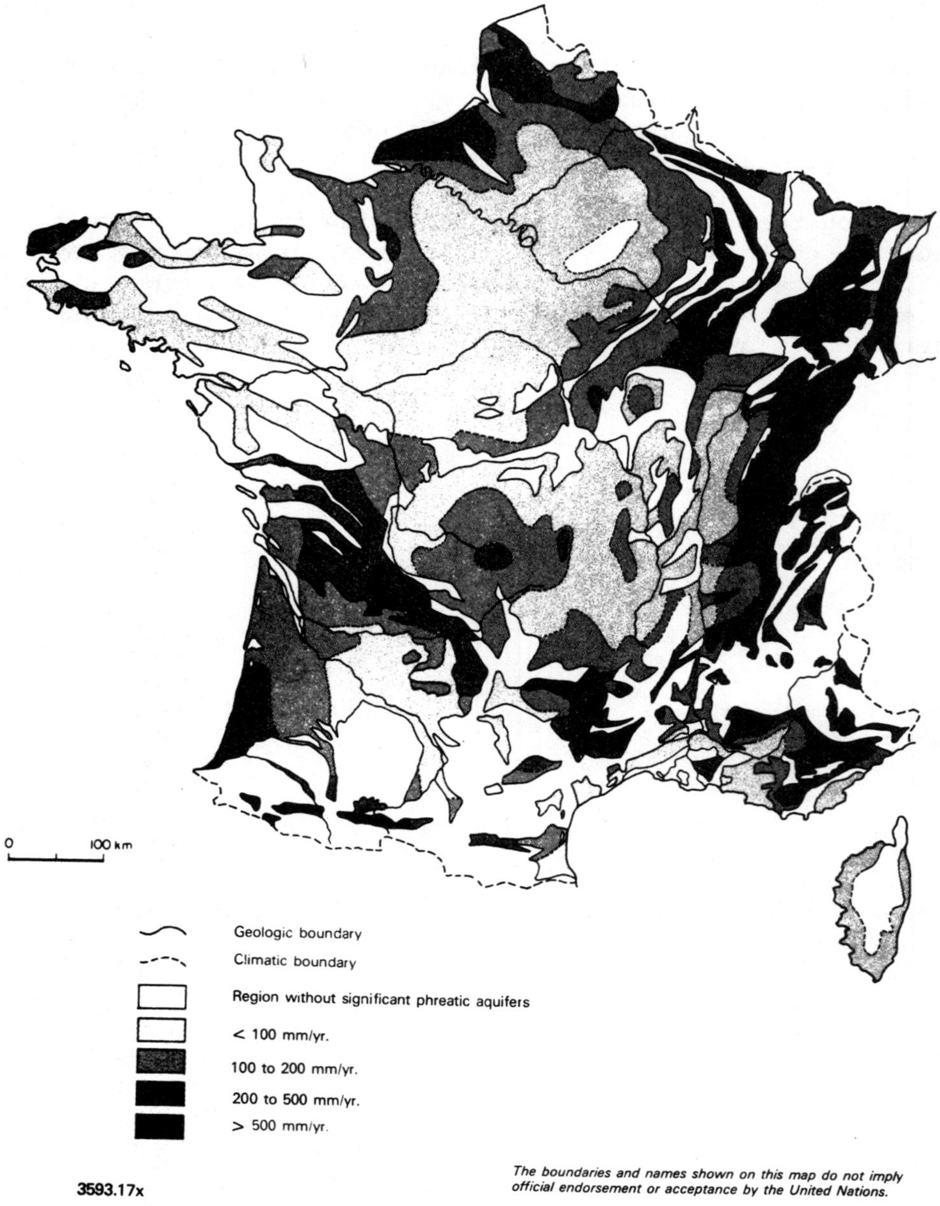

The boundaries and names shown on this map do not imply official endorsement or acceptance by the United Nations.

Finally, on a more macroscopic and global scale, one may combine the total water budgets of the basins and of France as a whole as sketched above. The distribution of ground flow may be estimated by distinguishing the total underground flow, the main components of the regular surface runoff and the discharges from the underground layers of the main regional aquifer systems which form the natural base for renewable groundwater resources and which can be compared with the withdrawals (table 16).

Table 16. France: Underground flows

Basin	(1) Approximate underground flow	(2) Average local discharge of main regional aquifers, included in (1)	(3) Hidden underground runoff to sea (approximate)
	(in cubic kilometres per year)		
Adour-Garonne	22*	9*	0.6
Artois-Picardy	3.5	3	0.02
Loire-Picardy	20	6	0.02
Rhine-Meuse	8.5	7	--
Rhône-Mediterranean and Corsica	31*	15*	0.2*
Seine-Normandy	15	10	0.25
Total France	100	50	1.1

* Including high proportion of unsteady discharge from karst.

Data on reserves of the aquifers, however, are less up to date and much more approximate. Expressed in terms of average water height, these reserves would range in France between 0.1 m and more than 10 m depending on the aquifer. For France as a whole the reserves of groundwater, up to 1,000 metres in depth, are probably about 2,000 km^3 of which a tenth could represent the variable part (regulating reserves). The estimate for the regions would range from 1 to 10 million m^3 per km^2.

3. Water quality

Although the quality of groundwater varies from region to region and from aquifer to aquifer, it is fairly well known. This is because about 63 per cent of the groundwater is extracted to provide drinking water. Figure 22 is a hydrochemical map showing the differing quality of groundwater in the various regions in 1977.

Figure 22. France: Chemical characteristics of groundwater

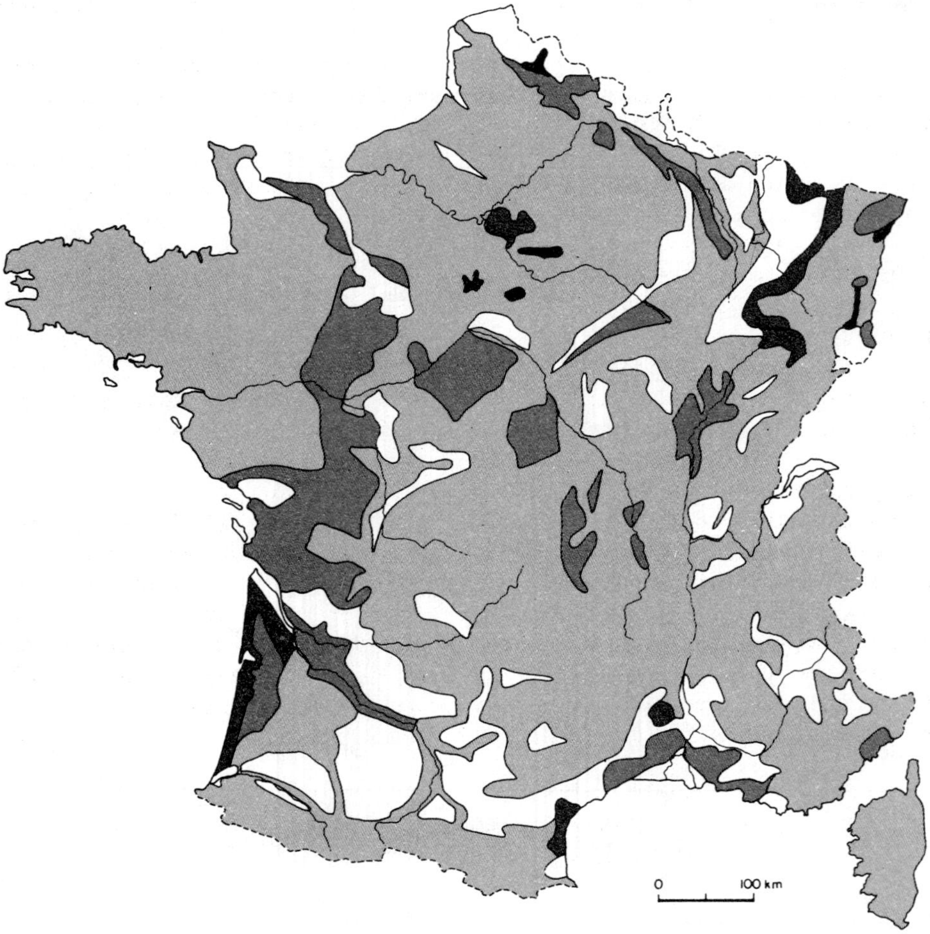

Graded by quality:

- Water of good chemical quality, suitable for multiple uses when treated (primarily for drinking-water purposes).
- Water of a chemical composition, acceptable either without treatment for various industries or, after simple treatment, for more stringent uses (grade 1).
- Water of poor quality, generally caused by high TDS (total dissolved solids) or iron content, unsuitable for most uses unless thoroughly treated.
- Water of poor quality, resulting from human activities.
- Uncharted areas (no data, aquifers too scattered or not exploited, no phreatic aquifer).

3593.18x

The boundaries and names shown on this map do not imply official endorsement or acceptance by the United Nations.

Good quality water with low mineral content (less than 100 mg/l), satisfactory for almost all uses, predominates in the mountains and in the ancient massifs. Here, however, much of the water is hard and some is loaded with iron in the crystalline and siliceous rocks and in the fluviatile or glaciofluvial alluvions. Water of medium quality, drinkable only after treatment but suitable for industrial use, is the most widespread in the sedimentary basins and valley alluvium. Water from carbonate aquifers often requires softening, while that from the alluvions and certain sands require the removal of iron; such water can be used everywhere for irrigation. Water of bad quality (containing relatively high concentrations of sodium chloride and/or calcium sulphate) is localized in certain regions of these basins: Eocene aquifers of the Paris region (north and east) high in gypsum and sulphates, Eocene carbonate aquifers containing fluorine in Ile-de-France and western Aquitaine, the Muschelkalk limestone aquifer in Lorraine in contact with salt-bearing land, and the coastal aquifers of Aquitaine and Languedoc affected by sea-water intrusion. Deep aquifers contain fresh water with less than 1,000 mg/l down to fairly great depths, e.g., 1,000 m in the Paris basin and 1,500 m in the basin of Aquitaine. Other aquifers with low renewal and in contact with salt-bearing formations, contain salt water at lower depths, e.g., the Lower Triassic sandstone aquifer in Lorraine west of Metz-Nancy, the layer of Albian greensand in the north of the Paris basin, the deep layers of Alsace and Brest or the coastal areas, such as Languedoc and the Rhône delta.

Applying the simplified multi-parametric classification that is compatible with the classification recommended by EEC but adapted more specifically to groundwater as presented in table 17, table 18 summarizes the predominant quality of the water of the main types of French aquifers, on the same lines as table 13.

In addition to the natural causes of poor-quality groundwater, there are man-made causes: local chronic or accidental pollution - but with persisting effects - of mainly industrial origin and more diffuse and extensive pollution of agricultural origin (increase in nitrate content).

Table 17. France: Hydrochemical classification of groundwater quality

Quality (in decreasing order)	Parameter (maximum values)					
	$CaCO_3$ hardness	Dry residue	Cl	SO_4	NO_3	Others
	(in milligrams per litre)					
1	50	100	20	20	10	-
2A	250	850	20	20	15	-
2B	300	2 000	20	250	50	-
3	300	2 000	250	250	50	-
4	2 500	850	-	-	-	Fe, 0.3; Mn, 0.2
5	180	2 000	-	-	-	Fe, 1; Mn, 1
6	500	2 000	-	-	-	Fe, 2; Mn, 1
7	500	2 000	-	-	50	Fe, 2; Mn, 1

Table 18. France: Chemical quality of water in the principal aquifers

Aquifer lithology	Stratigraphic positions and locations	Water quality class (table 17)	Water quality parameters (TDS = Total Dissolved Solids)	Factors: Natural	Factors: Artificial
Gravel, alluvium	Quaternary, Pliocene	2 to 7	TDS, TH, Cl	Nature of soil and reservoir	Urban and industrial effluents, leaching of wastes, sea-water intrusion
			Fe	Surface-water input, sea-water intrusion induced	
Sandy	Quaternary (dunes and coastal bars)	2A, 3, 7	Cl	Sea-water intrusion, rainfall in coastal areas	..
	Plio-quaternary (Les Landes)	2 to 7	Fe	Nature of soil and reservoir	..
	Pliocene (Astian, Languedoc)	2A, 4, 7	Cl, Fe	Nature of reservoir, sea-water intrusion	..
	Miocene (Aquitaine, Sologne)	2B	Cl, SO_4	Surface water input	..
		3	Cl, TH	Ferruginous facies	
		5, 7	TH, Fe		
	Eocene (northern France)	2A, 7	TH, TDS, Cl	Nature of overburden and reservoir, sea-water intrusion	..
	Cenomanian, Albian (Paris basin)	2A	TH	Nature of covering	
		2B	Cl, SO_4, TDS	Leaching of Pyrites	
		6	TH, Fe	Limonites, glauconite	
		7	Fe	Presence of lignite	
Molasse	Miocene (Lower Dauphiné, Rhône valley, Aquitaine)	2A, 2B	TH, Cl	Nature of reservoir	..
		3	Cl		
Sand and limestone complexes	Eocene (Paris basin, Soissonnais)	4, 6	TH	Nature of reservoir	Urban effluents
		7	TH, SO_4, Fe, TDS, F	Gypsum facies	
	Aquitaine (west)	1, 2A, 2B	Cl	Nature of reservoir	..
		3	Cl, TDS	Marine invasion	
		4, 5, 6	TH, Fe	Ferruginous facies	
		7	Cl, TH, SO_4, F	Gypsum facies	
Sandstone	Permian-Triassic (Lorraine, Lower Normandy)	1, 2A	TDS, TH, SO_4	Nature of reservoir, gypsum facies, ferruginous facies, leaching of pyrites	Dewatering of iron mines
		6	TDS, TH, SO_4, Fe		
		7	Fe		
	Paleozoic "Armorican sandstone"	1, 4	Fe	Nature of reservoir	..

Table 18 (continued)

Aquifer lithology	Stratigraphic positions and locations	Water quality class (table 17)	Water quality parameters (TDS = Total Dissolved Solids)	Factors	
				Natural	Artificial
Chalk limestone	Upper Cretaceous (Paris basin, Artois, Charentes)	2A	TH, Cl	Nature of reservoir, marly, Glauconitic facies, "siderolitic" argillaceous cover, sea-water intrusion	Nitrate pollution (fertilizers)
		2B	SO_4, NO^3, Mg		
		3	Cl		
		6	TH, Fe		
		7	Cl, SO_4, Fe		
Limestone, dolomite	Oligocene (Paris basin, Brie, La Beauce)	2A	TH	Nature of reservoir	
		2B	SO_4, NO_3		
		3	Cl	Marly facies	Nitrate pollution (fertilizers)
		4, 5	Fe	Pyritic	
		6	TH, Fe	Ferruginous cover	
		7	NO_3, Fe		
	Eocene (Paris basin, Brie)	2A, 3	Cl	Nature of reservoir	..
		6	TH		
	Lower Cretaceous (South Alpine basin, Provence, Languedoc)	2A, 2B	TH, TDS	Nature of reservoir	..
	Middle and Upper Jurassic (Paris basin, Jura, Provence, north and north-east Aquitaine)	1, 2A	Cl	Nature of reservoir	..
		2B	TH, SO_4, Mg	Marly facies	
		3	Cl	Ferruginous substratum	..
		6	TH, Fe		
	Triassic (Lorraine, Provence)	7	TH, SO_4	Gypsum facies	..
	Carboniferous (North, Armorican Massif)	3, 6	TH, Fe	Nature of reservoir, artesianism, leaching of pyrites	..
		7	SO_4		
Schist, crystalline rocks	Precambrian (Armorican Massif)	1, 2A	TDS, TH	Nature of reservoir	..
		2B	TS, TH	Influence of sedimentary cover	
		4, 5	SO_4, Mg, Fe, NO_3		
Volcanic rocks	Quaternary (Auvergne)	1, 2A	TDS, TH	Nature of reservoir	..

<u>Note:</u> Table is for water quality characteristics as related to lithology of aquifers.

Finally, France is rich in thermo-mineral waters of all kinds: carbonated water, water with sulphur or chlorine, spread among different hydromineral and hydrothermal provinces. It is present in the sedimentary basins and in the Alpine chains as well as in the basement areas with new structures and Recent volcanism: 1,200 springs have been given the "mineral water" label assigned by medical and health authorities.

D. Groundwater development

1. Exploration technology

(a) Prospecting and inventories

All known and proven technologies for hydrogeologic investigations are applied in France. They are based largely on an advanced state of geologic knowledge of the surface and underground. They include surveys of water points, construction of piezometric surfaces, hydrometry of springs and waterways at low water, analyses of fluctuations in natural levels, active hydrodynamic tests, tracings, hydrochemical and isotopic analyses, exploration by drilling and geophysical prospecting mainly by electrical resistivity methods.

The basic inventory undertaken since the 1950s in the north, the Paris basin, western Aquitaine and Alsace was followed by a more extensive inventory in other regions (west, south-east) without attempting to cover the whole territory. There was also a systematic collection of data obtained from study projects for specific purposes. In karstic areas, efforts were concentrated on the study of springs from all aspects, hydraulic, chemical, thermic and biological. Tracers were used especially by speleological groups whose exploration results were methodically preserved. In basement areas, especially in crystalline and metamorphic terranes and in karstic areas, the methods of analysing fractures, based on photogeologic materials and remote-sensing pictures have become much more advanced and are currently applied.

Besides hydrochemical studies, mainly descriptive and concerned directly with the evaluation of groundwater quality, isotopic hydrology has been promoted by researchers of the University of Paris. This has contributed, through an independent approach coordinated with hydrodynamic approaches, to the genetic and cinematic study of groundwater in a certain number of, mainly deep, aquifers.

Exploratory drilling programmes usually involve rigs of the type also involved in engineering, groundwater development and studies at shallow depths and geothermic or petroleum prospecting at great depths. Thus, the number of rigs involved in the prospecting for groundwater are difficult to itemize (probably some hundreds of drills). A basic role is played by test pumpings to identify the parameters of aquifers: data have been collected and statistically analysed in great numbers, sometimes linked with tracer studies to estimate flow dispersal. Recently, following the increased use of geothermic resources of medium or low enthalpy, including the storage of heated or cooled water in aquifers, techniques of identification *in situ* of the thermic parameters of aquifers have been developed.

(b) Observation networks

In certain rare cases, the observation of hydrogeologic variables dates back a long time. There exist century-old records of a small number of aquifers, in natural or artificial systems - or origins of springs used for the supply of towns. However, methodical networks of piezometric observations have been developed only since the 1950s. The intensity and frequency of these observations have varied because the objectives have been varied. They have been concerned not so much with basic knowledge as with the impact of water development. As a result, the number of water points observed has tended to decrease during the 1970s and 1980s in spite of efforts of the public services to consolidate and establish on a permanent basis a minimal general network which will serve as an instrument of management control. There are currently about 8,000 piezometric observation stations, as opposed to 12,000 in 1967 and 9,000 in 1975, in service in France. They are unevenly distributed and are to be found mainly in the different alluvial aquifers of the Rhône basin. They are used mainly by BRGM, the regional services of the Ministry of Agriculture and various regional development organizations, including the National Rhône Company.

The regular testing of the chemical properties of groundwater, besides the testing required at water points used for drinking-water supplies, has developed only recently. But here again progress is uneven and selective: it mainly applies to the testing of the quality of groundwater which has deteriorated or is threatened. About 2,300 water points are checked and exploited by various organizations.

(c) Data banks

The "data bank of the underground" established and maintained by BRGM contains hydrogeologic data relating to about 200,000 water points, mainly boreholes, springs and underground works. It contains information on the nature and parameters of aquifers, on water levels, on natural or artificial discharges, and on the chemical characteristics of the water. There are special files with records of the data obtained by the exploitation of piezometric networks. From the 3,886 piezometers handled by BRGM in 1985, about 80,000 measurements are recorded every year. In certain cases, for example springs and smaller rivers, hydrometric records are kept together with data on deep aquifers of geothermic interest. More specialized data banks have been built up on a regional basis. They are nevertheless connected with the general system and also with the data banks instituted since 1981 in each great basin. They contain data on drinking-water supply systems and on the chemical characteristics of groundwater and its development. Thus they constitute a record of the quality of groundwater.

These data banks may be linked to computers that can handle the appropriate statistical and geographic data and which can be consulted by certain users via telephone networks. In addition to these factual data bases used for specific references, there are bibliographic data banks, partly analytic, incorporated in the general bank. They cover the whole field of earth sciences (Pascal-Geode and Geode) in which BRGM and CNRS are major producers of data. In 1985 the French bibliography contained 79,000 references. Moreover, this bibliographic data bank is being developed at the regional level (south-east).

(d) Cartography

As a special means of representing and disseminating knowledge, groundwater cartography has gone hand in hand with the early advances in hydrogeologic research. A good example is the hydrogeologic map of Paris and its surroundings on a scale of 1:25,000 compiled by A. Delesse in 1862. This was probably the first such map in history. However, the hydrogeologic mapping of French territory did not really begin until a few decades ago.

Regular coverage on a large scale (1:50,000, 1:100,000) by topographic sheets following the international legend, was begun in 1964 by BRGM in several regions (north, Paris basin, south-east): some 20 or 30 sheets were supplemented by regional or departmental maps at still greater scales (1:100,000) or medium scales. They included a map of the Paris basin on a scale of 1:1,500,000 compiled in 1967. There were also regional atlases which reduced the need for structural and parametrical maps of individual aquifers (Paris region, La Beauce, Normandy, Bordeaux region). As part of the French contribution to the International Hydrogeological Map of Europe on the scale of 1:1,500,000, small-scale national maps and then an atlas of groundwater in France were compiled on a scale of 1:1,000,000 in 1970. The continuation of a general hydrogeologic map, considered to be of low priority in many regions, was then stopped as an autonomous project (although still used to illustrate regional monographs) and was replaced by a series of thematic maps with more specialized objectives. These objectives included the productivity of waterworks, their vulnerability to pollution, the quality of groundwater and the possibilities of its use with heat pumps. These maps were compiled on different scales and with different methods of representation as part of regional programmes. Sometimes they were gathered together in regional atlases (Alsace, Languedoc-Roussillon). At the same time small-scale national maps were compiled. Figures 19 and 22 are based on certain of these maps and the most recent one shows the nitrate content of groundwater in France (1985). A general hydro-geodynamic map of France, based on more scientific conceptions, was compiled by BRGM in 1980. It shows aquifer systems on a scale of 1:1,500,000. Finally, monographs were compiled on the groundwater resources of the countries of the European Economic Community. These included maps on the scale of 1:1,500,000 covering the whole of French territory. They show the hydrogeological conditions of the territory, the progress of waterworks, and the quality and vulnerability of groundwater. These maps were an extension of a number of previous maps on the same subject. Figure 23 shows the present state of coverage of French territory by maps of different types.

(e) Models

Hydrodynamic simulation models began to be used in France in the early 1960s. Electric analog techniques involving resistors and capacitors were first developed by the Geological Service of Alsace, and by petroleum study groups. They were rapidly superseded by numerical techniques backed up by computers in the 1970s. Many computer programs were worked out by different study groups and research centres, including the geological computer centre of the School of Mines of Paris and BRGM. These made it possible to simulate increasingly complex structures and phenomena. At the same time, progress was made in the analysis of aquifer systems and the use of computers became more advanced and simplified. Their use has become common in practical hydrogeology with the expansion of microcomputers. More than a hundred models of regional aquifers of different size and complexity have been built. This has helped in the study of many problems and in the adoption of

Figure 23. France: Groundwater resources map coverage

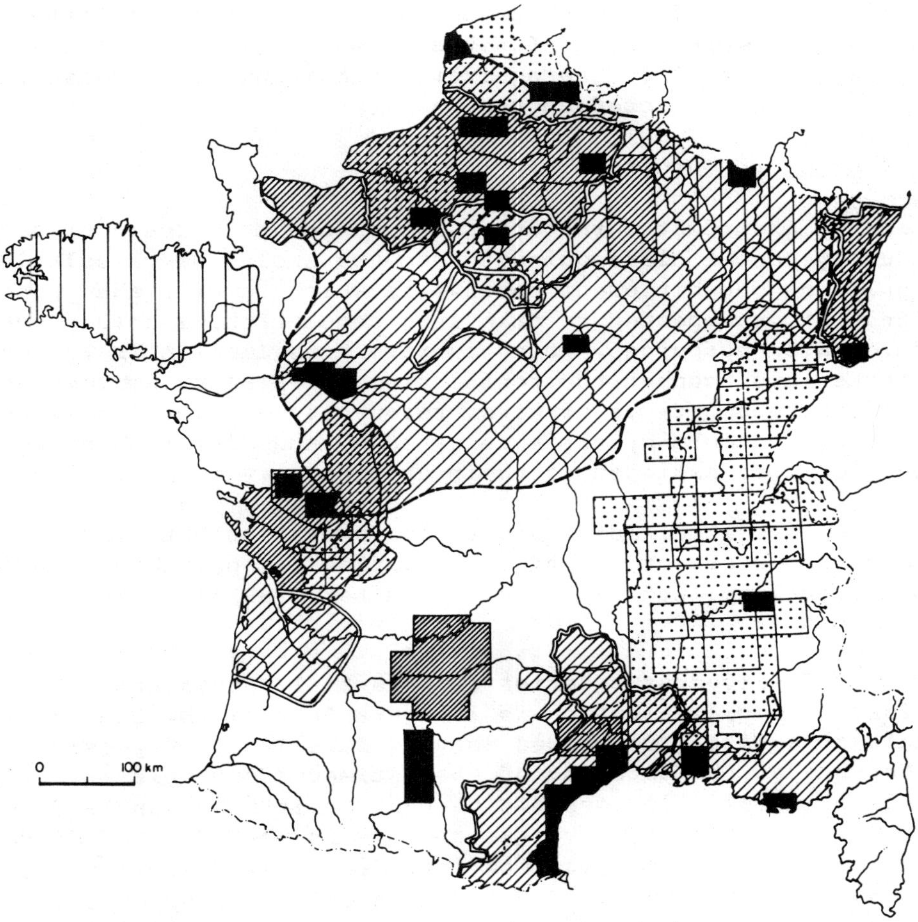

The boundaries and names shown on this map do not imply official endorsement or acceptance by the United Nations.

1/50,000

1/100,000

1/200,000 to 1/500,000

Hydrogeologic atlases

Yield maps.*

Water quality maps and/or maps showing level of exposure of aquifer to contamination
* Special maps showing ground-water potential for heat pumps are not included.

3593.19x

decisions concerning the management of certain underground layers whose exploitation is controlled. Following the general use of hydrodynamic models and based on them, hydrocinematic, hydrogeochemical and hydrothermic models began to be devised to deal with the problems of protecting and conserving the quality of groundwater. Moreover, global models showing reservoirs and transfer functions can now be linked with hydrologic models of similar basins. They are being applied to the solution of complex problems of a quantitative nature including the behaviour of karstic aquifers, the flow in non-saturated areas, or of a qualitative nature including development and forecasting of nitrate content. The progress of models of meteoric water infiltrated into the aquifers is helping to increase the accuracy of the hydrodynamic models mentioned above.

(f) Evaluation of water resources

The evaluation of groundwater resources, at an adequate scale of aquifer systems, has gradually become distinguished from the exclusive calculation of input flows and the regulating capacities of reservoirs which remains the physical basis of such evaluation. It is expected to take into account the interdependence of ground- and surface-water resources, which is rarely negligible in France, and at the same time to take into account the conflicting constraints imposed upon the respective users of these two types of water resources. It is also expected to relate the exercise to economic criteria selected by the different categories of users and to the priorities assigned to their objectives.

The exercise involves a multidimensional confrontation between the characteristics and parameters of demand such as flows, conditions of access and productivity, costs, quality, and the spatial and temporal distribution of these variables.

Water resources conceived as a series of potential values cannot be reduced to a single expression. Today, evaluation is inseparable from the construction of the hydrodynamic models of aquifers described above. This makes it possible, firstly, to verify the coherence of the data and of the estimates of their structure and operation. Secondly, it makes it possible to test their behaviour and measure their reactions - including qualitative reactions - in the light of different projects or scenarios of their exploitation or indirect influence. Hence their viability can be tested on the basis of the chosen criteria, especially in cases where exploitation may significantly transform the initial conditions and lead to oversupply (table 15). This procedure has been applied in France for the last 15 years to a certain number of intensely used aquifer systems in order to optimize their management. They include the large alluvial aquifers with unconfined layers in Alsace, the plain of Lyon, and Crau; several chalk aquifers, especially in northern France and that of La Beauce; the multilayered aquifers of Aquitaine, the Lille region (Carboniferous chalk or limestone) and Roussillon; the deep greensand aquifers in the Paris basin and the Lower Triassic sandstone of Lorraine. These are some of the most noteworthy examples; most of the aquifers are shown on figure 20 accompanying table 15.

2. Methods of implementation

In France, the conditions of occurrence of groundwater vary widely: in terms of yields, depth, probable rate of success of drilling, operation, availability. Consequently, there are great differences in the competitiveness of groundwater as a source of supply, as compared with surface water which itself is irregularly distributed in space and even more in time. This diversity is accompanied by a great diversity in water demand as regards quantity, locality and quality; hence, the diversity of the methods and intensity of the exploitation of groundwater in France. A rapid survey of the natural regions will illustrate the situation.

In the ancient massifs, groundwater resources remained undeveloped for a long time because development was considered impractical. However, local exploitation of groundwater has become possible usually through boreholes of average depth (40-60 m) rather than by wells built traditionally over springs or drainage galleries. This can satisfy a demand which is dispersed, limited and mainly individual. Flows of some 10 m^3 per hour are feasible if the projects are adequately designed and carried out. The multiplication of such drillings in the Armorican massif after the drought of 1976 has proved this point. Most boreholes built without study on the utilization sites can provide several cubic metres per hour. Statistics gathered on more than 4,500 boreholes have shown that 50 per cent were producing more than 0.1 m^3/h per drilled metre or more than 5 m^3/h in general. Depending on the areas, productivity was higher than 20 m^3/h in 5 to 20 per cent of the cases. However, drillings made on the advice of a geologist gave better results. For example, out of 70 boreholes drilled in the region of St. Malo, almost 50 per cent yielded more than 10 m^3/h and more than 25 per cent yielded more than 20 m^3/h. In the course of the last 10 years the technique of percussion drilling with compressed air has been considerably developed (down-the-hole hammer). As a result, the exploitation of groundwater in basement areas has expanded especially in the Armorican Massif (Brittany). There, more than 6,000 productive boreholes have been drilled in a few years, in both granitic rock and schist.

The exploitation of the volcanic aquifers of the Massif Central consists mainly in using the abundant and pure spring waters. In some cases, however, long gravitational channels are built as far as the base of the reservoirs (chain of Puys in Auvergne).

In the Alpine mountains, groundwater is rarely exploited directly. This is because of the difficulties of access, the complex structures of the aquifers and the fact that there are plenty of springs and surface water of good quality. The most notable aquifers in the karstic limestone massifs are commonly the deepest and give rise to random drillings. The most usual method of exploiting them is to build wells over their springs and adapt to their minimum discharge. Nevertheless, the morainal aquifers and, naturally, the highly productive aquifers in the glaciofluvial alluvions may be exploited to supply urban agglomerations, especially those with a large winter population, at a time when surface water is low or, in the mountains, nonexistent.

In the vast limestone plateaus, which lie between the mountains and the sedimentary plains, the partial or total withdrawal from springs is the main method of using the karstic aquifers. Direct exploitation remains difficult because of the great depths involved and the uncertainty of drilling results. This method is little developed without being excluded in certain cases. The main concern is to protect these reservoirs which are vulnerable to surface pollution. Three quarters of a century after the campaigns of the speleologist Martel and in spite of an array of well-enforced water regulations, the many sinkholes and fissures in these plateaus are too often considered as sewers in which it is tempting to get rid of a variety of wastes. Yet these aquifers have enormous capacities. They are admittedly difficult to quantify because the development of fractures and cavities in the rock mass varies greatly in depth. However, they could be put to better use. The simplest method is to act upon the level of the springs: to lower them in dry periods, i.e., periods of the greatest use, in order to increase their discharge and to reduce the discharge in subsequent floods. This method was used already in 1965 by pumping in the outlet of a karstic aquifer in Languedoc, the source of the Lez (0.2 to 20 m^3/s, 2 to 3 m^3/s on an annual average, in a natural system). This was replaced in 1982 by the construction of wells dug directly into the deep karstic channel in order to supply Montpellier (11.7 m^3/s). This could be done elsewhere; for example, a project has been studied in Vaucluse to provide a secondary source of supply in dry years. The particular case of coastal aquifers with karstic channels submerged below sealevel, thus favouring mixtures of fresh and salt water, has provided an opportunity to experiment with a special technique: the underground and submerged dam of Port-Miou east of Marseille which has raised the fresh-water level by several metres.

Most groundwater use in France takes place in the large sedimentary basins which are rich in extended aquifers, both accessible and productive, as we mentioned above.

Firstly, there are the extended water-table aquifers present in most of the Paris basin and its northern extension in Artois. Also the aquifers in the north and east of the Aquitaine basin, mainly the chalk aquifers and several Tertiary carbonate aquifers in La Beauce, Brie and elsewhere. Their widespread and direct exploitation is facilitated by several factors: First, the fact that the layers are unconfined means that the reservoirs have high storage capacity, which reduces the effects of withdrawal from springs and rivers. Also, the extent of their recharge areas is equivalent to most of their area and they are easily accessible owing to moderately deep water levels. Finally, nearly all have average to high productivity, yielding several tens of cubic metres per hour, many with more than 100 m^3/h and several with much more than 100 m^3/h. In the Picardy chalk area, for example, statistics taken from more than 300 tubewells have shown that their specific discharges exceeded 10 m^3/h for 56 per cent, 50 m^3/h for 15 per cent and even 100 m^3/h for some.

The exploitation of these aquifers, often withdrawing 5 to 10 per cent of their global average discharge, does not affect greatly their general system. In a few limited areas, however, the intensity and concentration of withdrawals has resulted in considerable reduction in the water levels. These have become more or less stabilized today, for example, in the chalk aquifer in the Lille region. It has been suggested that the spontaneous renewal of low-water levels by these aquifers should be increased by well-regulated seasonal pumpings. This idea has been put into practice in the chalk aquifer in the valley of the Lys, the river used to supply the town of Lille.

The multilayered systems mainly in the Cenozoic series of the Ile-de-France and in western Aquitaine contain deep aquifers with confined layers that are more permeable and productive than the upper unconfined layers. Thus in these regions they are the most exploitable and the most exploited. They serve as drains in this stratified system. Their resources are renewed when they are exploited by drainage from upper aquifers which are better supplied but less productive, or by lower reservoirs which produce additional capacity. Their exploitation can thus achieve a balanced system if the withdrawals are not constantly increased.

The intensive and steadily increasing exploitation of these aquifers in some sectors such as the northern Paris region: 6 m^3/s for the multilayered Eocene system of the Soissonnais; Bordeaux region: 1 m^3/s on the average between 1900 and 1960, 1.5 m^3/s in 1975 for the Lower Eocene sand aquifer has caused a gradually increasing depression of the piezometric level. The concentration of boreholes reduces their individual production. For example, in the case of the exploitation of the Lower Eocene sand aquifer in the Bordeaux region (Gironde) 25 flowing boreholes produced together 1.2 m^3/s in 1890, whereas 85 produced, in 1960, 1 m^3/s, including 40 per cent still flowing - and 129 produced, in 1975, 1.5 m^3/s entirely by pumping.

Finally the exploitation of the deep aquifers, with artesian layers, has been in many cases facilitated by the surge of initial flows which have offset the need to drill to great depths. However, as was to be expected, this flow has rapidly decreased and generally disappeared as a result of boreholes proliferations (regions of Paris, Lille, Tours).

Also, the withdrawals have been increased from 1.5 m^3/s with 180 productive boreholes in 1900, to 10.5 m^3/s with 1,170 productive boreholes in 1975, for the four most exploited confined aquifers in France (in addition to the multilayered aquifers previously mentioned). These aquifers are by order of present importance:

a. "Vosgian sandstone" (Lower Triassic of Lorraine);

b. Carboniferous limestone of the North;

c. Greensand (Albian) of the Paris basin;

d. Cenomanian sand of Touraine.

Large volumes of water have thus been withdrawn from these aquifers for over a century. For example, more than 6 km^3 of water has been taken from the confined Lower Triassic sandstone aquifer in Lorraine between 1900 and 1975, including water pumped out of the mines. This is more than double its natural supply (40-50 million m^3 per year) during the same period. From the greensand aquifer in the Paris basin about 1.2 km^3 have been withdrawn from the first - the famous artesian well of Grenelle dug in 1841 - until 1975. This withdrawal is lower than the natural global input (about 20 million m^3 per year) but it is concentrated exclusively in the Paris region.

There were inevitable losses of pressure brought about locally by the concentrated exploitation of these aquifers. The lowering of the piezometric surface reached about 100 m over 100 years. This gave rise to the first measures of conservation. An authorization system, from 1935 onwards, was applied initially in the Paris region to protect the greensand aquifer to maintain yield. The strict application of these regulations sometimes exceeded objectives. In Paris, for example, the water levels of the boreholes in the greensand aquifer not only stopped falling but have even gone up since 1971. Admittedly, various causes have contributed to the reduction in withdrawals.

Finally, there are the alluvial aquifers, which are particularly well maintained by the rivers to which they are subordinated and by the aquifers in the plateaus consisting of chalk and other materials. They possess all of the conditions favourable for the exploitation of groundwater: ease of access because of shallow depths, high productivity (often 100 to 1,000 m^3/h), dynamic and rapid renewal by recharge from the rivers, and proximity to the places of use. They also lend themselves to the construction of radial wells with high yields.

Their exploitation is intensive in many valleys: from 15 m^3/s in the plain of Alsace, to more than 20 m^3/s in the valley of the Rhône from Lyon to the sea. The discharges in the rivers affected by the 70 m^3/s withdrawn in the alluvial aquifers of France are usually strong enough to sustain these indirect withdrawals; this does not exclude, however, a few local problems especially in dry periods (valleys of the Loire and Garonne).

The growth of these uses is nevertheless constrained by the risk of silting up of certain river banks, by the difficulties experienced in land use and management in certain areas due to conflicts with urbanization and mining (e.g., in the valley of the Var) and by the difficulties of protecting the quality of the water including the quality of the recharging rivers.

On the whole, among the types of waterworks constructed at the present time, boreholes predominate. Most are not deep and are carried out mainly by small or mid-sized enterprises. About 3,000 boreholes are drilled in France every year to exploit groundwater. Of these fewer than 5 per cent exceed 100 metres and only 25 per cent exceed 50 metres in depth; the total length drilled is 100,000-150,000 per year.

3. Costs

The costs of production vary, depending on such factors as the difficulties of access (depth of boreholes, hazards); the methods utilized in groundwater withdrawal which determines the level of investment funds needed; the period chosen for amortization; and operating expenses, which depend on the depth of water levels and the cost of energy. Over the last decade, the use of compressed-air drills has considerably reduced, in constant terms, drilling costs for average depths in hard rocks or difficult terrain.

A summary inquiry was carried out in 1971 and was based on a standard water project and uniform rules of amortization. The results showed that the costs of production of raw groundwater, excluding treatment and transport, then varied in France between minimums of 0.02 to 0.04 frames per cubic metre (F/m^3) in the most favourable areas such as large alluvial aquifers, like those of Alsace and the Rhône valley; and the "Northern Chalk" of Artois and 0.5 F/m^3 in the deeper aquifers of moderate productivity, requiring pumping. In 1983 francs, this would correspond to a range of 0.06 to 1.5 F/m^3. Previously, I. Cheret (l'Eau 1972) had estimated that the average costs of harnessing groundwater and surface water, especially for the production of drinking water, varied from 0.38 to 0.55 F/m^3 (at 1983 prices). More recently, an inquiry by the Ministry of Agriculture showed that the average price of drinking water distributed from underground sources was 12 per cent cheaper than that of surface water. These were 1980 statistics relating to domestic consumption, the average global price being 3.80 F/m^3. This difference is attributable mainly to the lower costs of treatment and transport of groundwater in spite of the economies of scale achieved by the use of surface water for extensive urban water distribution networks.

Finally, these costs depend too much on local variables to perceive any dominant regional differences. Analyses of an individual aquifer have shown variations in costs similar to that indicated for a national inquiry: 0.035-0.40 $F(1983)m^3$ in an alluvial aquifer of Languedoc, for example. Costs do not vary much at production discharges above a certain threshold, in a fairly wide range (75-350 m^3/h in the case mentioned).

4. Withdrawals and reserves

(a) Statistics

For the last 15 years (1965-1980) or so, users of both ground and surface water have been required to pay fees to the water agencies. In the absence of accurate and regular surveys, these have been the main source of statistics on withdrawals but they have been completed and refined in only a number of sectors. The last complete statistics relating to 1981 are summarized in table 19. They cover the distribution of groundwater withdrawals (including that of springs) for each basin and economic sector. Their relative importance can therefore be evaluated in total withdrawals:

a. Most of the withdrawals are divided among three main sectors of exploitation, namely, 50 per cent for production of drinking water, 25 per cent for industry unconnected to water supply systems and 18.5 per cent for agriculture (irrigation);

b. The greatest volumes of water are withdrawn from two basins: Rhône-Mediterranean-Corsica (37 per cent of the total) and Seine-Normandy (22 per cent);

c. The proportion of water demand covered by groundwater is highest in the drinking-water sector (63 per cent, up to 93 per cent in Artois-Picardy). It is appreciable for unconnected industries (33 per cent) and for irrigation (about 27 per cent). Additional irrigation, which has developed considerably in France outside the traditionally irrigated areas of the Mediterranean - especially in the Paris basin and in Aquitaine and Alsace - usually makes use of groundwater;

d. The pumping out for the drainage of mines is relatively unimportant (6 per cent) but is noteworthy in the Rhine-Meuse basin (30 per cent).

A total of about 7 km^3 of groundwater is reckoned to be withdrawn annually at present (1980) in France. This is one fifth of all withdrawals of raw water, except that from the navigation canals and for hydro-electricity, in this particular case with marine discharge. The proportion rises to 40 per cent if surface-water withdrawals that are nearly all renewed by thermal power stations for cooling purposes are not accounted for.

Nearly half of this volume (40 per cent) is withdrawn from alluvial aquifers and hence, to a large extent, from rivers. More than one third, however, is withdrawn from the unconfined aquifers of the sedimentary basins. Of this amount at least 1.5 km^3 per year, or 20 per cent of the total, are taken exclusively from the chalk aquifers and nearly one fifth from the confined aquifers of the multi-layer systems and from the deep aquifers.

Table 19. France: Groundwater withdrawals, 1981

Basin	Community water supply 1/	%	Industries having independent sources of water	%	Thermoelectric power plants	%	Agriculture (irrigation) 2/	%	Mines (dewatering)	Miscellaneous (heat pumps, etc.)	Total	Per cent of all water withdrawals: Of all sectors 3/	Per cent of all water withdrawals: Thermoelectric plants excluded
Adour-Garonne	230	35	80	8	--	--	100	20	--	--	410	20	31
Artois-Picardy	310	93	200	45	0	0	60	60	5	10	525	73.5	74
Loire-Brittany	530	55	80	25	--	--	100	15	10	5	725	33.5	60
Rhine-Meuse	260	78	440	35	20	0.4	40	100	370	25	1 125	17	49
Rhône-Mediterranean-Corsica	1 200	67	580	48	5	0.1	800	25	5	15	2 605	15	27
Seine-Normandy	970	62	320	32	--	--	200	70	0	30	1 520	23	49
Total France	3 500 4/	63	1 700	33	25	0.15	1 300	27	390	85	7 000	20	39

Withdrawals in million cubic metres per year and percentage of all water withdrawals (figures rounded); columns from Community water supply to Miscellaneous: Withdrawals by use sector.

1/ For domestic industrial, commercial and government services connected to community water supply network systems (includes withdrawals from autonomous domestic systems).

2/ Figures corrected, from underestimated statements of dues.

3/ Not including surface-water withdrawals for navigation canals and hydropower (cases of discharges to the sea).

4/ Including 900 spring-tapping systems.

A large proportion of withdrawals is concentrated in a small number of aquifers and of extensively exploited well fields. These are shown, for the most part, in table 15. The withdrawals exceed 100 m^3 per year in 10 aquifers which alone provide more than a third of the production of groundwater in France. They are listed below in order of decreasing withdrawals estimated for the years 1975-1980.

	Aquifer	Withdrawal (million m^3 per year)
1.	Alsace, alluvial aquifer of Rhine valley	480
2.	Lower Seine valley, chalk plus alluvions	400
3.	Total of interlinked aquifer systems in chalk of the region north of the Artois threshold (sectors, Calais to Lille, Cambrai, Valencienne)	380
4.	Lyon plain, alluvial aquifer (Rhône valley)	275
5.	Côte-de-Moselle, Jurassic limestone aquifer	240
6.	Soissonnais sands plus coarse limestone, multilayered Eocene aquifer, Paris region and north of Ile-de-France	190
7.	Isere valley, Grenoble sector, alluvial aquifer	180
8.	Lower Triassic sandstone in Lorraine (confined aquifer)	164
9.	Beauce limestone aquifer and sands of Fontainebleau	110
10.	Rhône, valley sector from Vienne to St. Rambert d'Albon, alluvial aquifer	100

A considerable proportion of the remaining withdrawals is distributed between some tens of aquifers where they often exceed 30 million m^3 per year or about 1 m^3/s (figure 24). The rest is much more scattered.

(b) History

The history of groundwater withdrawals is not well known. Their amount has probably doubled three times since the beginning of the twentieth century. The exploitation of confined aquifers - the best known because of the smaller number of drillings and their gradual regulation since 1935 - grew rapidly in two phases (1920-1940, 1945-1970) and has stabilized since the 1970s.

There has been a fairly rapid growth in withdrawals caused by the rapid growth of water demand during the 1950s and 1960s. Different trends were subsequently observed when the system of payments made it possible to prepare statistics during the early 1970s. These statistics show a marked global reduction and different sectoral trends. In the production of drinking water there was a slow, linear or decelerated growth in withdrawal. In industrial withdrawals there was a regular decrease, the joint result of deterrent measures by water agencies and the slowing down in economic activities, except in the Rhine-Meuse area. Withdrawals for

Figure 24. France: Main areas of groundwater development

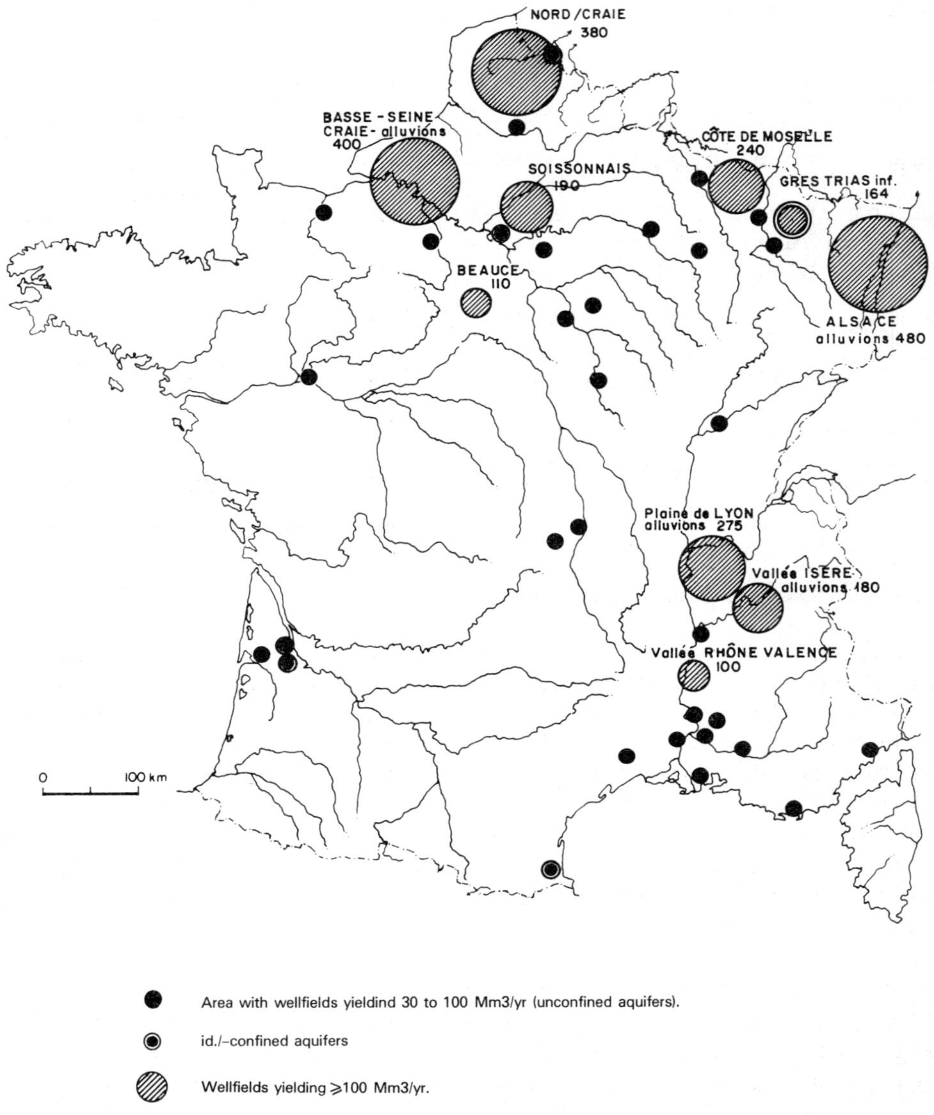

3593.20x

The boundaries and names shown on this map do not imply official endorsement or acceptance by the United Nations.

agricultural purposes showed an irregular and poorly identified decrease. Table 20 shows the trends from 1970 to 1981 for the withdrawals from the two best-known sectors (production of drinking water, unconnected industries) in the three best controlled basins. On the less global scale of specific aquifers, the trends may differ from these general trends.

Table 20. France: Trends in groundwater withdrawals, 1970-1981

River basin water agency	Production of drinking water			Unconnected industries		
	1970	1975	1981	1970	1975	1981
	(in million cubic metres)					
Artois-Picardy	234	257	293	326	248	201
Rhine-Meuse	198	246	251	557	624	709
Seine-Normandy	721	853	917	545	417	322
Total	1 153	1 356	1 461	1 428	1 289	1 232
Percentage	100	118	127	100	90	86

(c) Ratio of withdrawals to reserves

The numerical comparison of withdrawals with the natural average flow of the aquifers, confined to this single, physical and quantitative component of groundwater resources, brings out varied rates of exploitation. For each aquifer the rate is still very low for many of the extended unconfined layers and is not more than a few percentage points. But it can be fairly high in some cases which are brought out by table 15: for the Beauce aquifer, 24 per cent; for the Côtes de Moselle aquifer, 40 per cent; for the Alsace alluvial aquifer, 48 per cent; for chalk aquifers in the north, 44 per cent between La Lys and Dunkerque taking account of artificial recharge 97 per cent between La Deule and La Scarpe taking into account leakage-induced inputs).

In some alluvial aquifers high rates of exploitation are observed for systems that are largely artificial with induced recharge caused by withdrawals. Thus all the input is withdrawn in the plain of Lyon. This is also the case of the multilayered aquifers where intensive exploitation has caused drainage leakage-induced flows: a rate of 42 per cent in the aquifer of lower Eocene sands of Soissonnais where this imbalance reflects a decrease in storage. Rates higher than 100 per cent also reflect decreases in the reserves of deep confined aquifers: the most notable case is that of the Lower Triassic sandstone in Lorraine, 328 per cent (all figures for 1975-1980).

On a more global but less significant scale, this comparison has been attempted for the large basins by comparing the figures in tables 16 and 19. The following tabulation shows average rates of exploitation higher than 15 per cent in four basins out of six and 14 per cent for France as a whole, which is not negligible at this level of integration.

Basin	Ratio of withdrawals to average natural flows of regional aquifers (in per cent)
Adour-Garonne	4.5
Artois-Picardy	17.5
Loire-Brittany	12
Rhine-Meuse	16
Rhône-Mediterranean-Corsica	17.4
Seine-Normandy	15.2
France as a whole	14

5. Problems

Various problems related to water availability from the quantitative point of may occur as a result of exploitation itself, rendering it more difficult or affecting other activities. Firstly there is the case of the exploitation of a modest number of aquifers or parts of aquifers of moderate extent which play an important role in the economy of regional water. In practice a ceiling is reached for these withdrawals and it becomes difficult to increase them, or to maintain them, even by multiplying the water points or increasing input artificially, a practice still confined in France to some 10 local operations. Moreover, a pertinent modulation of a seasonal or indeed over several years of the withdrawals appears to be desirable for certain aquifers with low or poorly defined capacity whose exploitation is increasing, for example in areas of crystalline rocks. Methods of regulation and forecasting are being designed to meet this problem. However, situations of overexploitation, with an imbalance unrelated to the growth of withdrawals, are rare in France. The problem of stabilization of withdrawals arises in some cases, not so much as to restore a satisfactory balance in the aquifer itself but rather as to keep the dynamic water levels within adequate limits allowing for permanent pumping.

Second, considerable lowerings of the piezometric surface have occurred locally, sometimes on a regional scale, due to extensive exploitation. In some cases levels have not been stabilized. They affect certain unconfined aquifers (chalk in the north) and more generally the confined aquifers such as the Carboniferous limestone in the region of Lille, Soissonnais to the north of the Paris region, the greensand in the centre of the Paris basin and the Lower Triassic limestone in Lorraine. They have in fact reduced productivity and added to the cost of production. The artesian outflow has in most cases disappeared. These reductions, however, have not had any appreciable effect on the stability of the soil. Related cases of subsidence have not been observed in France.

Third, the concentration of well fields in several zones, especially urban areas, has generated or may generate conflicts over use, especially if thermal applications are being developed.

Fourth, in some regions an assessment of the condition of aging deep wells appear necessary. Rehabilitation or replacement may be necessary. Last, there have been lowering of water levels especially in industrial areas in aquifers that have been exploited for a long time. They include the aquifers in the Paris region. On the other hand, mines have been closed and pumping stopped. This has caused for several years a rise of levels in the unconfined aquifers which interferes with the occupation of the subsoil and gives rise to new kinds of conflict.

The problems related to groundwater quality are wider and more general. They are due more to the impact of activity of the occupiers of the land than to the effects of exploitation. The most serious problem is the spread of pollution in the unconfined aquifers. This occurs particularly in areas of intensive agriculture in which fertilizers, especially nitrates, are used. The nitrate content of the water already exceeds 50 mg/l in various localities and may come close to 100 mg/l. In 1985, 1.2 million inhabitants were supplied in France with water having a nitrate content higher than 50 mg/l. Of this water, two thirds came from underground sources. Furthermore, a growth in nitrate content is likely up to 1990 and 2000, even if fairly substantial reductions are achieved in the input.

A number of other types of chemical pollution of ground and subsurface water occur every year in a more irregular and fortuitous fashion. They may be caused by an accident in the transport of harmful materials, by a weakness in the industrial process or by fire or other disaster. They may also occur as a result of a failure to confine the position of surface or underground wastes or they may result from a deficiency in the system of purifying the used water. The duration of these kinds of pollution are variable; some date back a long time and are overcome only very slowly. An example is the salt pollution of the alluvial aquifer in the plain of Alsace which began in 1910 from the potassium mines and which extends over 70 km. Further, groundwater used for drinking purposes is often exposed to pollution risk due to delays encountered in adoption of protective measures. Such measures have been defined and fully legalized in France in only 10 per cent of the cases.

Changes in water quality caused by exploitation include saline intrusion in coastal aquifers in contact with sea water. They also occur in the case of certain deep aquifers. Such cases, however, are rare and localized. Nevertheless, the risk of causing further salinization restricts the possibility of exploiting certain aquifers.

E. Administration and management

1. Legislation and users

As a general principle, the right of accession and appropriation with regard to the groundwater is linked to the property of the land (Civil Code). Groundwater is neither domanial, except in the overseas departments (since a 1973 Law), nor *res communis* like non-domanial surface water.

However, French legislation has instituted several specific groundwater regulations which partly restrict action affecting it, with the aim of conserving and protecting water resources. Compulsory reporting of withdrawals (exceeding 8 m^3/h) by operators - and termination of projects - similar to a registry, was instituted by the Water Law of 1964 (article 40). A system of authorization for drillings was instituted by a 1935 decree and subsequent extensions (most recent: 1985) in several departments, above a fixed discharge and below a depth varying according to area (2-80 m). The right of access was restricted for conservation purposes and for the quantitative protection of groundwater. Any diversion and dumping of used water likely to harm the quality of groundwater is prohibited (Water Law of 1964, article 40) together with the regulation of classified establishments (decree of 1977). By law, operators must pay charges fixed and collected by regional water agencies. Special regimes similar to mining concessions are applied to thermal and mineral water and to geothermy.

Many economic agents or users in different categories and with different objectives have the legal capacity and the economic power to exploit groundwater directly. There are more than 14,000 management units concerned with the production and distribution of drinking water. They supply some 36,000 local communities (communes). Most exploit groundwater for this purpose: 85 per cent of the communes, comprising 60 per cent of the French population, are supplied by groundwater from wells, boreholes, or springs.

More than 30 per cent of the 40,000 farmers practising irrigation utilize groundwater on about one third of the entire irrigated land (according to the 1970 agricultural census). Thousands of industrial or tertiary enterprises, not connected to the water supply, have their own supplies taken directly from groundwater (table 19). Extractive industries such as mines and quarries also make substantial withdrawals. For several years the users of geothermal energy of medium or low enthalpy have exploited hot water at great depths without encroaching on the fields of action of the other operators. Also, some 25,000 heat pumps, working in shallow "ordinary" groundwater, have been installed since 1970 for heating purposes. This development indicates a new category of operators, both individuals and groups.

There are other equally numerous agents who are not operators but whose activities have a chronic or incidental impact on groundwater, on its availability, its regime and, above all, its quality. These include farmers who irrigate and who use fertilizers, managers of surface or underground stocks of useful but harmful materials, various occupiers of the land and transport enterprises.

These users are in great numbers. The effects of their actions may be harmful and sometimes conflicting. The conciliation of these conflicting interests and the recognition that government interest is permanent cannot occur automatically. Nor can it be the result of market forces, which are not particularly efficient in this field. Intervention in the management of groundwater by the public authorities is usually considered necessary and is actually practised in France, particularly to solve or remedy the problems described above.

2. Groundwater management

(a) Objectives

The management objectives differ from region to region. They depend on the quantity and quality of the resources, the socio-economic aims of groundwater use, and the modalities of its relationship with surface water. The most common objectives, which themselves must be conciliated, are either of general interest to the users or of more general public interest. The aims are:

1. To satisfy the water demand of those using the water - both as regards quality and quantity - by protecting them both against their own overexploitation

and against third persons who also cause harm. In other words, to solve the problems mentioned above;

2. To assign groundwater for priority uses from the point of view of the general interest, mainly the supply of drinking water to communities, and in part to secure supplies permanently;

3. In some cases to integrate the utilization of acquifers within water management systems as to utilize their regulative capacity;

4. To conciliate the use of groundwater and various forms of development of the underground;

5. To preserve certain natural roles of groundwater, especially as a factor in the system of surface water (dry periods) or as a permanent component of water resources development plans.

In fact, the management of groundwater resources cannot in most cases therefore be done independently from that of surface water and that of underground space.

(b) <u>Authorities and machinery</u>

Two types of water management authorities coexist in France. Each of them has one of two categories of instrument for affecting the behaviour of the agents mentioned above. Government authorities have the power and responsibility to enforce the existing regulations. These include reporting and authorization systems, administrative monitoring and penalties for offenders. Since 1976, the main competent government agency has been the Ministry of the Environment; however, this Ministry delegates its monitoring responsibilities to the external services, both departmental and regional, of the Ministries of Industry, Agriculture and Public Works with a fairly complex share of responsibilities. These Ministries with sectoral competence, together with the Ministry of Public Health, have also special responsibilities. They must monitor the classified establishments which possibly cause pollution; they must conserve and protect the resources used to supply public drinking water; and they must regulate withdrawal of mineral or thermal water. Special regulations are aimed at protecting the quality of groundwater used for the production of drinking water. They include rules for the occupants of the land within an appropriate area in each case. Advice is given by an official hydrogeologist and the process is initiated by a declaration of public utility.

The six financial agencies instituted by the 1964 Water Law under the jurisdiction of the Ministry of the Environment enjoy a broad autonomy. They use financial instruments to help protect and assign water resources, particularly groundwater. The forms of economic influence are practised: deterrents by payments or stimulation by credit assistance.

Water agencies charge fees based on the volumes of groundwater withdrawal above a certain hourly or annual threshold. In certain cases they depend on the net volume consumed (not returned). The payments vary according to basin, zone and sometimes season. They become an indirect means of allocating resources because the amount of payment depends on surface-water charges and also on the uses (descending scales). The amount of these payments remains low (a range of 0.008 to

0.35 F/m^3 for withdrawals, 0 to 0.17 F/m^3 for consumption in 1985). This has little effect on water production whose cost is increased only by a few percentage points.

Financial incentives granted by the water agencies and by certain administrations are aimed at stimulating a greater development of groundwater in zones where its exploitation is neglected or where water is abundant. They compensate for the risk of drilling failures in geologically difficult terrane or they encourage certain agents to use an alternative source of supply, such as surface water, in order to keep groundwater for higher priority use. This assistance is also given for studies and health protection measures for drinking water.

Thus, the national or regional authorities involved in the management of the large basins have a number of shared tasks. Firstly, there is the management of the different sectors (rural or urban water supply, agriculture, industries and mines, transport) and, secondly, there is the management of different areas, particularly groundwater and partly subsoil resources. There is also a horizontal division of competence between these authorities, each one acting in its own capacity. This horizontal division is preferred to the vertical division between specific management authorities in each territorial field taking action in all sectors of utilization and equipped with all the management instruments.

(c) Assistance in management decisions

In order to help them take decisions and use the management tools described, administrations and organizations involved in the management of groundwater may call upon the knowledge and forecasts available from general studies, the follow-up of developing situations, and calculations.

Methodical inventories, accumulation of basic statistics in computerized information banks and their regular use for cartographic purposes, and the follow-up of variables by piezometric and hydrochemical observation networks help to maintain and bring up to date the state of hydrogeologic knowledge which facilitates the pertinent analysis of actual situations. Nevertheless, in certain cases, special studies are undertaken at the request of management authorities to deal with specific problems, especially as regards forecasting.

It has been pointed out that numerical simulation models of many aquifer systems are currently used in France. They may be hydrodynamic, global or random. One of their aims is to forecast the impact of additional withdrawals that may be the subject of authorization requests for development projects; another is to facilitate short-term forecasts of trends in natural variables. These include the level of the aquifer, the flow of the source, the content of substances which determine water quality (for example, nitrates) or temperature. These forecasting tools have become important instruments in management decisions.

The Public Groundwater Service responsible for much of the research, the general studies, the operation of observation networks, the establishment and maintenance of data banks and the elaboration and use of simulation models for management purposes is the Bureau de Recherches Géologiques et Minières, a public establishment which comes under the jurisdiction of the Ministries of Industry and the Environment. It provides advisory services for the administrations and organizations participating in the management of groundwater in France.

Management of groundwater in France means that the public authorities, the administration and its regional bodies control and guide the acts of the many agents which use or have an impact on groundwater. It is based on the recognition of a basic reality: the unity of water resources. The management of groundwater is not separate from the management of water in general.

It is inspired by a principle to protect the special rights and interests of users while subordinating them, nevertheless, to the general interest. It seeks to reconcile several objectives, namely, to satisfy as far as possible the different demands by reducing conflicts, to develop priority uses, and to protect water resources in both quantity and quality so that it may be used indefinitely in the future. It is adapted to a need to ensure that management decisions are in line with natural regional contexts, which cover local fields of action, sometimes conflicting, of many individual users whose coexistence must be monitored.

It proceeds from a desire to decentralize management activities and to avoid imposing on users such administrative constraints or controls as are premature or useless; to make the users aware that they are sharing a resource and that they are all subject to natural conditions; and to make them adopt a responsible attitude and to accept a common discipline. It cultivates an ambition to coordinate all the different activities that may be taken by the different authorities. It adopts the rule of basing management decisions on scientific procedures and on up-to-date knowledge of the different situations.

3. Training

Professional hydrogeologists are trained in France mainly at the Ecole nationale supérieure de géologie appliquée at Nancy and at several universities: Besançon, Bordeaux, Montpellier, Paris - at several centres - and others which teach mainly students who have first received general training in geology. The school at Nancy offers a third-year option in hydrogeology and has trained more than 200 engineer-geologists in this speciality during the last 40 years.

University teaching is partly included in broader training programmes either in the field of applied geology or in the field of water science and technology. In many cases it is linked with research activities leading to a doctorate. A recent (1985) reform was designed to provide a single diploma for advanced study in hydrology. It consolidated several of these university centres and the Geological Computing Centre of the Ecole nationale supérieure des Mines in Paris. They are thus able to link their complementary specialities.

Several other important engineering schools include in their programmes a number of subjects relating to groundwater. These are adapted to specific professions but they may also help in the specialization of hydrogeologists. The subjects include hydrodynamics, computer applications, knowledge and management of water resources and exploitation techniques. These schools are the Ecole nationale supérieure du Génie rural, des eaux et des forêts, the Ecole nationale des Ponts et Chaussées, the Ecole nationale supérieure d'électrotechnique, informatique et hydraulique (Toulouse), the Ecole nationale supérieure d'hydraulique (Grenoble), the Institut de mécanique des fluides de l'Université Louis Pasteur (Strasbourg) and the Conservatoire national des Arts et Métiers.

Finally, continuing or refresher training is organized by several of these higher educational establishments - in particular the Ecole des Mines of Paris, the Ecole des Ponts and Chaussées, and by specialist bodies such as the International Training Centre for the Management of Water Resources (Sophia-Antipolis), or the Fondation de l'eau at Limoges, as part of diversified programmes and by establishments dealing with special studies and research, such as BRGM.

All these courses are open to foreigners, especially those from French-speaking and developing countries.

4. International cooperation

French cooperation in the field of hydrogeology is not limited to the Francophone developing countries with which France cooperates actively for various reasons. The scientific and technical research organizations of France have developed closer relations than those which existed officially between the former members of the Community by opening their laboratories and their missions to African participants in development research. The traditional field of action and cooperation is to a large extent in the arid and semi-arid zones: French hydrogeologic research has developed in the Maghreb, the African Sahel, in Djibouti, in all regions where groundwater is of vital importance and where the search for it gives rise to difficult problems. The French idea of the use of groundwater is solidly based on Pasteurian health principles, on concerns for appropriate technology the furthest removed from traditional methods, and on dynamic and innovative drilling enterprises.

French cooperation takes the form of support, by France itself or in cooperation with other international organizations, of large-scale programmes of assistance for the establishment of water points: many teams of hydrogeologists (public, private and non-governmental organizations) are engaged in water development at the village level in West Africa. Technical advisers and experts are outposted to the Water Ministries in most Francophone States and to such regional organizations as the Inter-African Committee for Hydraulic Studies. Research, closely linked with training, is carried on to develop programmes on such major subjects as the recharge of groundwater under arid climates and the hydrogeology of fractured rocks.

Scientific cooperation with the developed countries facing the constraints and problems of advanced technologies is also actively supported by the Ministry of Foreign Affairs. The aims of this research are management of water quality, the prevention of pollution, the handling of high-risk wastes, and so on. It involves cooperation with the United States of America, Canada, most of the EEC countries, Switzerland and several countries of eastern Europe. Among the main subjects are models, movement of solids, hydraulic reactions to thermal operations, phenomena of exchanges between water and rocks, and denitrification.

References

1. Background - geography, climatology, general hydrology, geology

Arlery, R. Les climats de la France. La Météorologie, No. 16, March, p. 121-1,256, Soc. Météor. Fr., Boulogne, 1979.

Beaujeu-Garnier. Le relief de la France. Paris, 1953.

Castany, G. Bilan d'eau de la France. Evaluation des ressources en eau. Evolution des prélèvements et prévisions pour l'an 2000. UNESCO-WMO, 2nd meeting water problems Europe, Brussels, September 1977.

Debelmas, J. Géologie de la France. Doin, 2 vols., Paris, 1974.

Dottin, O., J. Goguel and others. Carte géologique de la France, et de la marge continentale à 1:1,500,000 et notice explicative, 102 p. (Bur. Rech. Géol. Min., Orléans), 1980.

Martonne, (De) E. Géographie physique de la France (A. Colin, 2e éd. Paris), 1947.

__________. France physique. Chap. XVI - L'Hydrographie, généralités (in Géographie Universelle, VI. La France, p. 358-390. A. Colin, 3e éd. Paris), 1955.

Pomerol C. and others. Géologie de la France, grands itinéraires (and Geology of France), Masson, Paris, 1980.

2. Hydrogeology

(a) On the whole of France

Albinet, M. Carte de vulnérabilité à la pollution des nappes d'eau souterraine de la France à 1:1,000,000 (Ed. DATAR-BRGM. Paris, Orléans), 1970.

Anonymous/BRGM. Atlas des eaux souterraines de la France (Ed. DATAR-BRGM, Orléans), 1970.

ARLAB. Bilan des ressources en eau souterraine de la France (Comm. Commun. Eur. - EUR 7943 FR., 268 p. - annexes, Publ. Th. Shäter Druckere: GmbH, Hannover), 1982.

Bodelle, J. and J. Margat. L'eau souterraine en France (Masson, 208 p., Paris), 1980.

Delaroziere-Bouillin, O., J. Forkasiewicz and J. Margat. Variations de l'alimentation des nappes en France: une donné importante pour la gestion des nappes d'eau souterraine (Coll. nat. "Eaux souterraines et approvisionnement en eau de la France", BRGM, Nice, T.I., p. 203-22, Orléans), October 1977.

Lafitte, P. and J. Ricour. La recherche minérale la plus importante de la France métropolitaine: l'eau. Role de l'hydrogéologue (Rev. industrie minérale, Paris), January 1956.

Landreau, A. and B. Lemoine. Carte de la qualité chimique des eaux souterraines de la France (à 1:1,000,000) et notice explicative (Minist. Envir. Cadre de vie/ BRGM, doc. 77 SGN 606 HYD, Orléans), 1977.

Margat, J. Définition des principaux systèmes aquifères du territoire français (Coll. nat. eaux sout. approv. eau Fr. Nice, T. II, p. 371-378, BRGM, Orléans), 1977.

__________. Carte hydrogéologique de la France à 1:1,500,000 ét Notice (éd. BRGM, Orléans), 1980.

__________. Bilans des nappes d'eau souterraine en France (Coll. intern. "Calcul du bilan des eaux souterraines". Com. bulg PHI/UNESCO, Varna, Sept.-Oct./et note BRGM, 11 p.), 1982.

Margat, J. and Y. Emsellem. Collecte et stockage des informations sur les caractéristiques hydrodynamiques des réservoirs aquifères et l'état des réserves d'eau souterraine (Soc. hydrot. Fr., Paris, Oct. 1970, C.R. XIe journée hydraulique. BRGM, Bull., 2e série, III, No. 3, 1971, No. 3, p. 3-10).

Pointet, T. Les réseaux piézométriques et la situation des ressources en eau souterraine en France (C.E.M.P.E.), 4e conf. intern. planif. et gest. eaux, Marseille (Doc. BRGM 43, I, p. 133-140, Orléans), May 1982.

Risler, J.-J. Carte des eaux minérales et thermales de la France, à 1:1,000,000 (BRGM, Orléans), 1973.

(b) Basic regional bibliography

Albinet, M. and others. Carte hydrogéolique du Bassin de Paris à 1:500,000 (BRGM, Orléans), 1967.

Astie, H. and others. Synthése des études sur le système multicouche des nappes tertiaires du Nord de l'Aquitaine (La Houille Blanche, p. 907-918, Grenoble), 1969.

Astie, H., J. Chamayou and others. Les eaux souterraines en Gironde. Cartes à 1:500,000 (Doc. BRGM, 26 maps, 55 p., Pessac, Orléans), 1977.

Babot, Y., Cl. Malaux and J. Ricour. Le réservoir des grès du Permo-Trias en Lorraine. Utilisation actuelle et perspectives de valorisation (Tec. Sc. municip. - L'eau, No. 3, p. 123-140, Paris), March 1962.

Besbes, M., G. De Marsily and M. Plaud. Bilan des eaux souterraines dans le bassin aquitain (Mém. Assoc. intern. hydrogéol., vol. 11, Budapest Conference, p. 294-303, Budapest), 1978.

Caous, J. Y., D. Common, J. C. Roux and others. Atlas hydrogéologiques départmentaux des régions de Picardie et de haute-Normandie, Aisne, Eure, Oise, Seine-Maritime, Somme. Avec cartes hydrogéologiques à 1:100,000 et notices (BRGM, Amiens, Rouen), 1976-1980.

Caous, J. Y. and J. C. Roux. Ressources en eau souterraine de la Picardie (Bull. BRGM(2). III, Hydrogéologie-Géol. de l'ing. 1, p. 19-52, Orléans), 1980-1981.

Castany, G., Cl. Megnien and others. Les bassins de la Seine et des cours d'eau normands. T. 1 - Ressources d'eau et données hydrologiques. Fasc. 4 - Eaux souterraines (Miss. dél. bass. Seine-Normandie/Ag. fin. bass. Seine-Normandie. Paris), 1974.

Cazal, A., M. Plaud, P. Pouchan and J. Vouve. Esquisse hydrogéologique des nappes profondes de l'Aquitaine orientale. (Mém. BRGM, No. 76, p. 795-819, Orléans), 1969.

Desprez, N., Cl. Megnien and others. Atlas hydrogéologique de la Beauce. Cartes (à 1:250,000) et Notice explicative. (Doc. BRGM, 2 cartes, 135 p., Orléans), 1975.

Diffre, P. Hydrogéologie de Paris et de sa banlieue (La Houille Blanche, No. 8, p. 665-671, Grenoble), 1972.

Drogue, C., A. M. Laty and H. Paloc. Les eaux souterraines des karsts méditerranéens. Exemple de la région pyrénéo-provençale (France méridionale) (Hydrogéologie-Géol. de l'ing. 4, p. 293-312, Ed. BRGM, Orléans), 1983.

Duprat, A., L. Simler and J. Valentin. La nappe phréatique de la plaine du Rhin en Alsace (Thèse Et. Univ. Louis Pasteur, Strasbourg et Doc. Cons. Europe, Groupe trav. Nappe phréatique rhénane), 1979.

Flandrin, J. and H. Paloc. Contribution à la connaissance du réservoir aquifère de la Fontaine de Vaucluse (Mém. BRGM, No. 76, p. 611-642, Orléans), 1969.

Mania, J. Gestion des systèmes aquifères. Application au Nord de la France (Mém, Soc. Géol. Nord., T. 15 - 228 p., Villeneuve d'Ascq), 1978.

Marchal, J. P. and others. Atlas des eaux souterraines départementaux de la Région Languedoc-Roussillon. Aude, Gard, Hérault, Lozère, Pyrénées-Orientales (Minist. Industrie, BRGM, Serv. Géol. rég. Languedoc-Roussillon, Montpellier), 1977-1983.

Megnien, Cl. and others. Atlas des nappes aquifères de la région parisienne (cartes à 1:200,000) et Notice explicative (Minist. dév. indust. scient. Préfect. et District. rég. Paris. éd. BRGM Paris), 1970.

__________. Hydrogéologie du Centre du bassin de Paris (Thèse état. Univ. Paris VI, Mém. BRGM No. 98, 1979, 532 p., Orléans), 1976.

Paloc, H. Carte hydrogéologique de la France. Région karstique nord-montpelliéraine. Notice explicative (Mém. BRGM No. 50, 205 p., planches et carte à 1:80,000), 1967.

Paloc, H. and others. Carte hydrogéolique de la région des Grands Causses à 1:200,000 et Notice explicative (CERGA-BRGM. Montpellier-Orléans), 1972.

Vouve, J., A. Cazal, M. Plaud and P. Pouchan. Introduction à l'étude hydrogéolique du bassin d'Aquitaine (Centre d'hydrogéol. Fac., Sc. Univ. Bordeaux, A.F.B. Adour-Garonne, 54 p., annexes, Talence), 1969.

Anonymous. Les ressources en eaux souterraines du bassin Rhin-Meuse. Carte hydrogéologique à 1:500,000 et Notice explicative (Miss. dél. bass. Rhin-Meuse, Moulins-les-Metz), September 1975.

Collectif. Congrès national d'hydrogéologie. Journées Henri Schoeller (Bordeaux, mém. BRGM, No. 76, 896 p., Orléans), March-April 1969.

__________. Hydrologie en pays calcaire (1er colloque Besançon, Oct. 1971, 2e colloque, Besançon, Oct. 1976, 435 p., Besançon).

__________. Les eaux souterraines et l'approvisionnement en eau de la France (Colloque national, Nice, Oct., 2 vols., 698 p., BRGM, Orléans), 1977.

__________. Hydrogéologie de la craie du Bassin de Paris (Rouen, May, doc. BRGM 1 et 3, 2 vols. 627 et 155 p., BRGM, Orléans), 1978.

__________. Le karst de France ("Karstologia", No. sp., Assoc., fr. karstol. - Fédér. spéléo.), 1986.

3. Exploitation and management

Albinet, M. and M. Tirat. Premier essai de recensement des prélévements d'eau souterraine dans les divers systèmes aquifères du territoire français (Doc. Ministr. Envir. Cadre de vie, BRGM 78 SGN HYD, Orléans), 1978.

Babot, Y. and J. P. Vancon. Gestion des ressources en eau souterraine par modèles hydrodynamiques, hydrochimiques et hydrothermiques. Exemple de la plaine d'Alsace (C.E.M.F.E., 4e Conf. intern. planif. et gest. eaux, Marseille, May, Doc. BRGM 43-1, p. 361-366, Orléans), 1982.

Beckelynck, J. and others. Gestion intégrée des ressources en eau souterraine et superficielle. Le cas de l'alimentation en eau potable de la Métropole lilloise (Proc. intern. Symp. on Ground Water in Resources Planning. UNESCO, IAH, IASH, Coblenz, August-Sept. II, p. 685-696), 1983.

Bodelle, J. and M. Tirat. Utilisation des eaux souterraines en France. (Coll. Nat. Eaux sout. approv. eau France - BRGM, Nice, Oct. 1977 - rap. général - thème 2, Orléans), 1977.

Brachet, J. L'action des Agences financières de bassin dans la protection des eaux souterraines. (Association nat. prot. eaux, journées Protection nappes souterraines, Paris, p. 28-37), 1975.

Bourguet, L., O. Delaoziere-Bouillin and A. Dreyfus. Etude et cartographie du cout de l'eau souterraine en France (Minist. Protect. Nature Environment, BRGM-BURGEAP, avec carte à 1:1,000,000, Orléans, Paris), 1971.

Cheret, I. Le système français de tarification pour la gestion de l'eau (in Gestion de l'eau, O.C.D.E., p. 435-442, Paris), 1972.

Crampon, N. and D. Olivier. Controle de gestion des eaux souterraines en France. Coordination administrative et para-adminstrative. Exemple de la Moselle (Comm. intern. irrig. drain., 93 congr. Moscou, R9, question 31, p. 106-134), 1975.

Emsellem, Y. Modèle mathématique et gestion des ressources en eau souterraine des nappe tertiaires du nord de l'Aquitaine occidentale (Symp. Sci., Intern. Informatique et problèmes de l'eau, Montpellier, I.B.M., p. 149-164), 1971.

Erhard-Cassegrain, A. Comportements socio-économiques et utilisations de la ressource en eau. Cas pratique de l'agglomération tourangelle (Doc. BRGM, 79 SGN 506 HYD, Orléans), 1979.

Gerente, M. Réflexions sur l'action de l'Administration dans le domaine de la gestion et de la police des eaux souterraines (Colloq. nat. Eaux sout. approv. eau Fr., BRGM, Nice, Oct. 1977, T. 1, p. 299-306, Orléans), 1977.

Levy-Lambert, H. L'utilisation optimale des nappes souterraines (La Houille Blanche, 3, p. 231-237, Grenoble), 1965.

Margat, J. La gestion des eaux souterraines en France (Doc. - BRGM, 86 SGN 169 EAU, Orléans), 1986.

Margat, J. and J. C. Roux. Interaction des impacts des aménagements et des exploitations sur les eaux de surface et les nappes souterraines (cas historiques en France) (Soc. hydrot. Fr., XIX journée Hydr., Paris, Sept. III.14, p. 1-10), 1986.

Nicolazo-Crach, J. L. and Cl. Lefrou. Les Agences financières de bassin (Ed. P. Johannet, 211 p., Paris), 1977.

Ollagnon, H. Propositions pour une gestion patrimoniale des eaux souterraines: l'expérience de la nappe phréatique d'Alsace (Minist. Agric., RCB No. 36, March - Paris), 1979.

Panier, J. La police des eaux souterraines ("Géologues", Un. fr. géol., No. 41, p. 9-15, Paris), 1977.

Ricour, J. and others. Intégration de la mise en valeur du réservoir des grés du Trias inférieur dans l'aménagement de la Lorraine (Proc., Internat. Symp. on Ground Water in Water Resources Planning. UNESCO, IAH, IASH. Coblenz, August-Sept. I, p. 171-188), 1983.

Roux, J. C. Les hydrogéologues agréés en matière d'hygiene publique ("Géologues", Un. fr. géol., No. 74-75, p. 33-42, Paris), 1985.

Collectif. Protection des eaux souterraines captées pour l'alimentation humaine (Colloq. nat., BRGM, 4 vols., Orléans), March 1977.

Collectif/PIREN/EAU (CNRS). Aspects économiques et sociaux liés à la gestion des eaux souterraines en millieu alluvial. Le cas de la nappe phréatique d'Alsace (Proc. Internat. Symp. on Ground Water in Water Resources Planning. UNESCO, IAH, IASH, Coblenz, August-Sept. I, p. 223-232), 1983.

V. THE GERMAN DEMOCRATIC REPUBLIC

Area: 108,333 km^2 Population: 16,640,000

A. Physiography and climate

The German Democratic Republic has some 350 km of shores on the Baltic Sea in the north, for some 500 km to the south it extends to the central massifs of the Erzgebirge and Thüringerwald. From south to north (general downslope direction) it includes highlands 500 to 800 metres high and a few peaks over 1,000 metres; a region of foothills ranging from 200 to 500 metres in altitude; then lowlands (moraine landscapes with lakes, sandy areas and valleys) having an average height of 50 metres.

Offshore to the north, the Baltic Sea is shallow with many open bays and islands, of which Rügen is the largest with 926 km^2.

In accordance with this regional division there is in the south and partly also in the west, a solid rock area. In the central and northern areas, that is, over 80 per cent of the territory, unconsolidated formations predominate.

The country has five principal river basins: the Elbe, about 78,900 km^2; the Oder, about 5,600 km^2; the Obere Werra and Obere Aller, 5,100 km^2; and the Rodach and Itz, 800 km^2. In addition, 18,000 km^2 of land in the north drain directly into the Baltic Sea.

Reflecting the surface configuration and the geologic formations, the network of watercourses in the highlands is dense and evenly distributed with gradients averaging 1 to 2.5 per cent) while in the lowlands (river gradients 0.001 per cent) areas vary in density of rivers and lakes. In the lowlands are artificial waterway systems for drainage and shipping canals. In the highlands there are numerous reservoirs with a total capacity of about 1.5 km^3.

The German Democratic Republic is in the temperate zone and has a moist, moderate, variable climate transitional between oceanic and continental. High-pressure weather systems result in warm, dry summers and cold, dry winters. These alternate with marine low-pressure systems with considerable precipitation and which in winter give rise to mild weather and in summer to cool weather. Temperatures in January vary from -2° to 0° C; in July the average temperature ranges from 17° to 19° C. The average annual temperature fluctuates from 8.5° to 9.5° C.

The results of research into the regional distribution of rainfall are expressed in the precipitation maps of the Climatic Atlas and in the Hydrographic Series of Maps for the German Democratic Republic (scale 1:1,000,000). A schematic map of the annual precipitation, reflecting the records from 1901 to 1950, is presented in figure 25.

It shows that annual precipitation in some areas is as low as 500 mm, locally even less than 450 mm. In most of the other lowlands precipitation ranges from 500 to 600 mm and in the coastal region it is 600 to 650 mm. The greatest precipitation is recorded in the Harz Mountains, namely some 1,400 mm. Precipitation between 1,000 and 1,200 mm occurs in the Thuringer Wald and the Ore Mountains.

Figure 25. German Democratic Republic: Annual precipitation

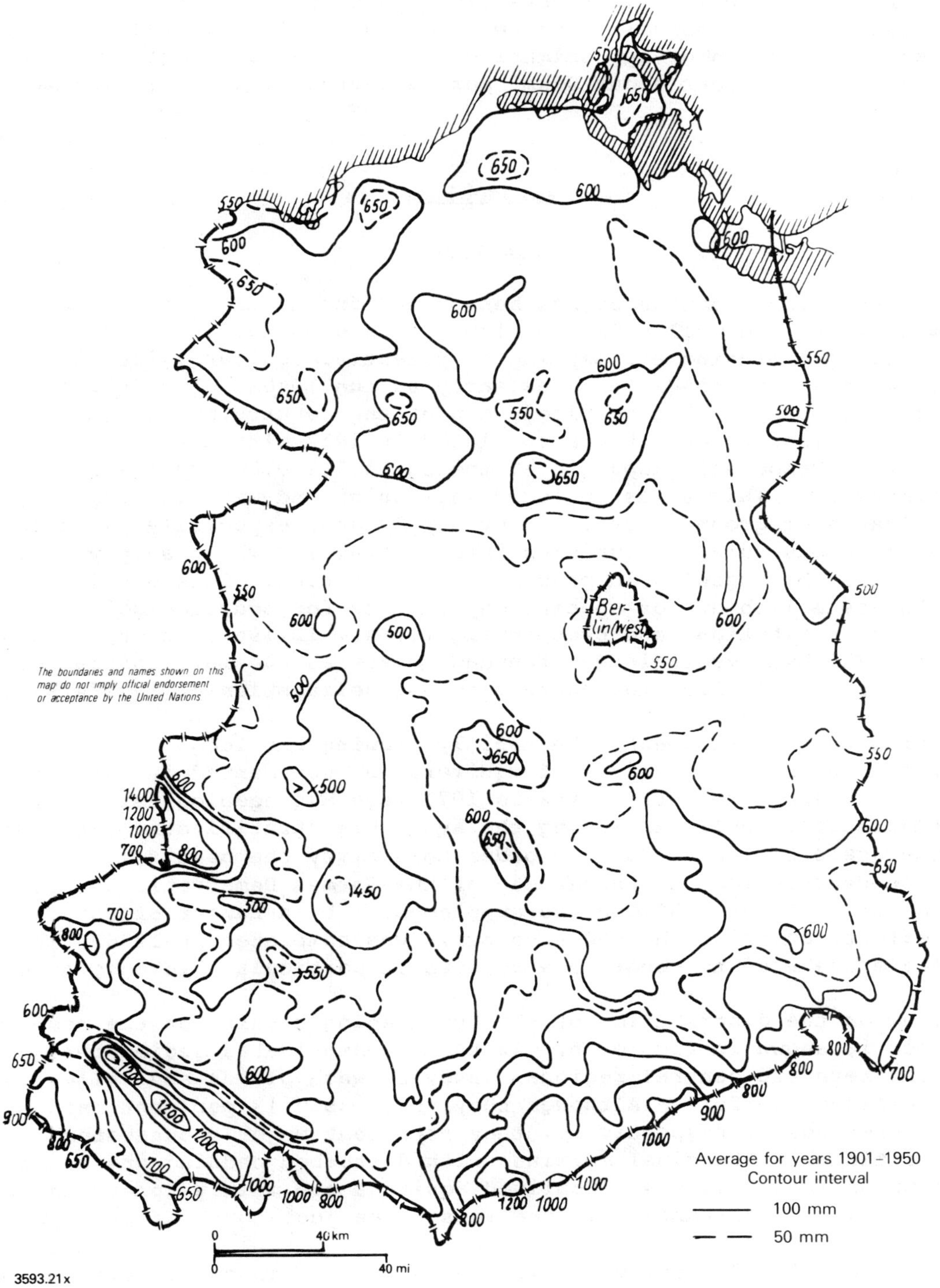

Records for the years 1901-1970 give an average precipitation for the entire country of 607 mm per year. With an evapotranspiration of 447 mm per year, a total runoff rate of 160 mm per year remains. The available part of the subterranean runoff of the total runoff varies in the different river basins from 0.65, in unconsolidated rock areas with a high retention capacity, to 0.13 in solid rock regions where the retention capacity is low. For the whole area one can assume a ratio of runoff to retention of approximately 0.5, i.e., from a total runoff amounting to 17.7 km^3 per year about 9 km^3 per year are available as stable water resource.

B. Hydrogeology

1. Background

The exploration for groundwater has been based for decades on the experiences gathered by waterworks and well drilling firms. The growing demands for water gave rise to systematic prospecting and mapping of potential aquifers, including an assessment of the chemical conditions, as early as the 1940s. In 1947, the Geological Survey produced a hydrogeologic map of the country at the scale of 1:100,000. Another mapping effort was completed in 1952 with the Hydrogeological Atlas of the German Democratic Republic at the scale 1:200,000 by the Governmental Geological Commission. This atlas included maps on groundwater quality. With the growing significance of groundwater in water supply and, especially, with the integration of numerous rural communities into centralized water supply schemes, there was a considerable upswing in prospecting activities. This development was reflected by the establishment of special departments for hydrogeologic tasks in geologic research institutions and prospecting firms. In 1968, the nationally owned enterprise VEB Hydrogeologie was founded to act as the main contractor for hydrogeologic prospecting, from forecasting to the delineation of aquifers.

In line with this development, the rapidly growing knowledge about distribution, thickness and structure of aquifers was documented in form of maps covering the whole territory. From 1962 to 1971, the Hydrogeological Survey Map of the German Democratic Republic including the available stratification records, was compiled at the scale of 1:200,000. In order to satisfy the need for detailed information, the Hydrogeological Survey Map of the German Democratic Republic was transferred to the scale of 1:50,000 and supplemented by a number of special maps. This mapping effort, covering the whole country, was completed in 1986. Maps of even larger scale are used to report the results on particular hydrogeologic units.

An overview on the distribution of the hydrogeologic units on the basis of lithologic types is shown in figure 26. In the unconsolidated sediment region, covering the northern and central regions, sandy outwash plains and valley moraines form thick unconfined aquifers. Although they cover only 13 per cent and 20 per cent of the region, respectively, they represent major groundwater resources. Additionally, terminal moraines with disturbed bedding are to be considered as unconfined aquifers. Confined aquifers are mainly spread in ground moraines and - of less importance - in the loess area just north of the mountains.

The description of hydrogeologic units which follows is based primarily on lithologic-hydrogeologic characteristics, taking into account the geologic structure and geomorphology of the area.

Figure 26. German Democratic Republic: Hydrogeologic units

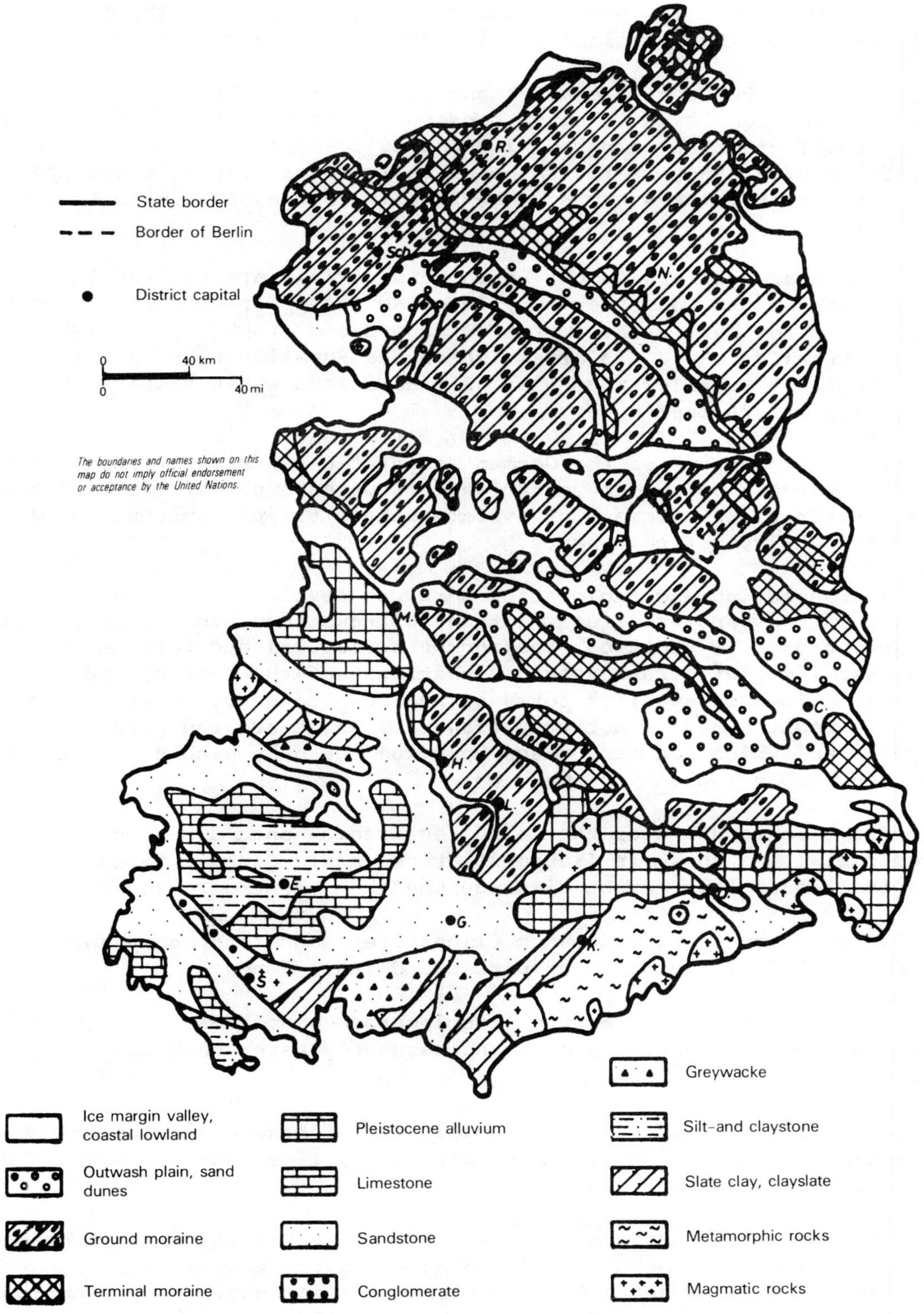

3593.22x

A rough division into aquifer complexes and aquiclude complexes is given by stratigraphic correlation schemes for Pleistocene and Tertiary regions. On a regional scale, it is possible to distinguish eight aquifers and aquicludes in Pleistocene rock which, however, rarely occur all together at one location. However, owing to lithofacies peculiarities, regional aquifers may split up locally into subaquifers and within regional aquicludes local aquifers may occur.

In the case of the Neogene aquifer and aquiclude complexes, the delimitation of regional aquifer complexes is not a consequence of the significantly higher permeability coefficient as compared with the adjacent rocks, but is based rather on the existence of a sandy facies in potentially water-bearing stratigraphic complexes. Aquifer complexes may also be locally interspersed with beds of lignite or silt and clay.

All fresh-water aquifers in the North German Plain are related to sediments from the Pleistocene and lower Tertiary (Miocene and upper Oligocene) periods and are protected against salt-water intrusion from deeper rocks by widely occurring mid-Oligocene Rupelian clay. In areas, where the Rupelian clay is absent due to erosion or diapirism, shallow anomalies of mineralized water testify to the uprising of deep mineralized water.

The solid-rock regions of the German Democratic Republic are classified into 12 units with hydrogeological characterizations of each class. It includes lithofacies distinctions, seepage of surface water, the predominant forms of runoff and their different runoff components.

As less prospecting and exploration work was devoted to the solid rock regions as compared with the sandy aquifers, little is known about the hydrogeologic and chemical parameters. In addition, the common horizontal and vertical alterations of lithofacies, i.e., of stratigraphic conditions, of the fabric, and texture of the rocks result in very specific forms of hydrogeologic processes. Of special importance are fault zones (fractures, fissures) which are the preferred flow paths of groundwater and which, in many cases, provide the only usable groundwater in otherwise impermeable rock.

The hydrogeologic units of the solid rock region should be understood as lithofacies units. As shown in figure 26, three basic genetic types of hydrogeological conditions may be distinguished:

a. Consolidated table sediments (limestone, sandstone, claystone) which are aquifers in certain stratigraphic complexes and fault zones;

b. Sediments in molasse basins (conglomerate, sandstone, claystone, clay slate) which are more compacted and are consequently water-bearing mainly in fault zones;

c. Basement rock massifs with magmatic rock (granite, granodiorite, volcanic rock) and metamorphic rock (gneiss, mica slate, phyllite) where groundwater is limited exclusively to fault zones.

In addition to these basic types, usable groundwater is found in channels and depressions filled with unconsolidated sediments in the transitional range between unconsolidated and solid-rock regions (Marginal Pleistocene) where lithologic conditions are quite variable and stratigraphic conditions are complex.

Under the loose, weathered cover of the basement rock, groundwater occurs in fault zones alone. There is no interconnected groundwater zone and there is clear correlation between tectonics, depth and degree of weathering and groundwater content.

2. Groundwater quality

The groundwater classification established by the Ministry of Geology, in cooperation with authorities for water management and the public health service, allows a classification by chemical, physical, biological, and hydrogeological criteria. Depending on the specific case, classification is based on groundwater type or utilization class.

Groundwater types serve as the basis for registration and assessment of the quality of water, the determination of its genesis, representation of the relation between water and aquifer, the regional delimitation of fresh and mineral waters, and the hydrogeologic mapping and delimiting of regions. Salinity and mineral content also provide the bases for classification.

Classification by main constituent is based on the assumption that ions contained in groundwater in concentration of more than 25 mmol-equivalent-per cent shall be recorded as dominating ions, whereby the sum of anions or cations shall be equated with 100 mmol-equivalent-per cent.

Such concentrations are attained only by Na, Ca, Mg and Cl, SO_4, HCO_3, in various combination, in natural waters unaffected by human activity. Ca-HCO_3 and Ca-HCO_3/SO_4 waters, e.g., appear to be the predominant types for the fresh water of the country.

Atlases with an areal representation of such groundwater types are available.

In order to classify groundwater, quality is evaluated with the view to use as drinking water, by the following groups of characteristics (see table 21):

a. Toxic constituents such as noxious heavy metals, organic compounds and certain other secondary constituents and trace substances;

b. Criteria of toxicity-indicating character referring to physiologically critical substances in connection with a generally increased organic load of the water, e.g., chemical oxygen demand, ultraviolet light absorption, extractable substances;

c. Other criteria, notably primary and secondary constituents, and properties which - if certain limits are exceeded - might cause restrictions on water use.

Table 21. German Democratic Republic: Chemical constituents of groundwater

Concentration	Molecular dispersions (true solution) $<10^{-6}$ mm				Colloid-dispersions (colloidal solution) 10^{-6} to 10^{-3} mm	Coarse dispersions (suspension) $>10^{-3}$ mm
	Electrolytes		Non-electrolytes			
	Cations	Anions	Gases	Compounds		
Primary constituents (>10 mg/l)	Na K Mg^{2} Ca^{2}	Cl (NO_3) HCO_3 SO_4	O_2 N_2 CO_2	SiO_2 $.nH_2O$	$SiO_2.nH_2O$ Clay minerals	Clay minerals, fine sand, organic solids
Secondary constituents (0.1-10 mg/l)	Fe^{2} Mn^{2} NH_4 Sr^{2} (Cu^{2}) (Zn^{2}) (Pb^{2})	F Br J NO_3 NO_2 HPO_4	H_2S NH_3 CH_4	Organic compounds	Hydrated metal oxides, silicates, humates	Hydrated metal oxides, oils, fats
Trace substances (<0.1 mg/l)	Li Rb Ba^{2} Cu^{2} Zn^{2} Pb^{2}	HS S^{2} (Br) (J)	Rn	Organic compounds	-	-

Altogether, 44 criteria or constituents are considered for assessing water quality. This number may be reduced, but this would also reduce the level of precision of the quality classes determined. Table 22 gives some examples for the assessment of the quality class with regard to potability.

The use of groundwater for drinking purposes depends essentially on whether it is endangered by existing or potential anthropogenic contamination.

Nitrate is one of the most important pollutants today. Nitrates in drinking water not only threaten the health of infants but also raise the carcinogenic risk of the population. Furthermore, nitrate may accelerate the internal corrosion of pipes. The main cause of nitrate contamination of groundwater is the extensive use of nitrate as fertilizer in agriculture.

In general, it is relatively easy to combat point or linear contamination sources through measures at the respective objects. Consequently, it is the scattered introduction of nitrates which pose the greatest problems in groundwater quality. But steps are being taken to reduce nitrate contamination so that ample potable water can be assured at reasonable cost.

Table 22. German Democratic Republic: Classification of groundwater quality

Assessment criteria	Groundwater quality class				
	1	2	2	4	5
Toxic constituents:					
As (mg/l)	--	0.05	1.0	1.0	>1.0
Pb (mg/l)	--	0.05	0.1	0.1	>0.1
Cd (mg/l)	--	0.005	0.01	0.01	>0.01
Criteria of a toxicity-indicating character:					
COD-Cr mg/l O_2	$\leqq$ 6.0	10.0	15.0	25.0	>25.0
Other criteria:					
Mg (mg/l)	$\leqq$ 70	125	150	200	>200
Cu (mg/l)	$\leqq$ 0.005	1.0	2.0	4.0	>4.0
Fe (mg/l)	$\leqq$ 0.1	20	100	100	>100

3. Exploration technology

In the German Democratic Republic, groundwater investigations seek to determine the geometry and geologic features of the aquifers, the genesis of the water, its quality, variations, flow and limits, the extent of reserves, the safe yield, and the most appropriate means of development.

The process of investigations moves, in general, from regional scope to the local level in various steps during which hydrogeologists use different instruments and methods. The process of presentation comprises the processing of acquired or available information (parameters, conditions) and reports of results.

Hydrogeologic prospecting entails much rock and groundwater sampling. Here, the usual techniques are employed: digging, rod sounding, and auger drilling for investigations in shallow strata. A wide range of deep drilling techniques are used for prospecting profiles deeper than 5 m (dry drilling, rotary drilling, counterflush drilling, core-drill method, air lift, cable-tool drilling).

As an integrated process, groundwater exploration and monitoring comprises also geophysics and borehole logging (diameter, conductivity, temperature, resistivity, and measurement of flow).

Inferences of many hydrogeologic parameters can be drawn from pumping tests, water-level gauging, and tracer experiments.

Hydrogeologic isotope investigations are routinely used today. Tritium, carbon-14, oxygen-18, denterium and other environmental isotopes are used to study the genesis, age or residence time of water or the degree of protection needed against human activities.

The processing and storage of the great amounts of data generated develop along two main lines of work: first, information is compiled and issued in the form of continuously updated mapping; second, information of hydrogeologic relevance is added to computerized data banks, including such designated files as basic data, stratification record, technical data, pumping tests and water analyses.

In the German Democratic Republic, the exploration, classification and certification of groundwater reserves are regulated by law. Groundwater reserves are defined as concentrations of groundwater which may be used economically either at the present time or sometime in the future. They are characterized by quantity and quality of water, conditions of extraction and treatment, and divided in the following types:

a. Renewable groundwater reserves, collected and maintained in one or several aquifers either from infiltration of precipitation or from natural recharge by surface water or karst;

b. Supplementary groundwater reserves, maintained from such measures as bank filtration and artificial recharge, mostly from surface water;

c. Groundwater deposit reserves, available in one or several aquifers at a fixed time. These are the result of large-scale withdrawals during open-pit lignite mining and are increasingly used for water supply.

The assessment of groundwater reserves complement and reflect the demands for water which lead to specific exploration projects. The procedures to be followed are chosen in consultation with actual and potential users and are managed jointly by the appropriate water authorities and hydrogeologic offices.

During the 1970s the development of efficient modelling and simulation procedures in groundwater management was a high priority in groundwater research in the German Democratic Republic. The theory and practice of groundwater hydraulics experienced a significant upswing. This development in the field of geohydrology reflects a general trend towards greater application of mathematics in all fields of hydrology.

Geohydraulic simulation methods are routinely applied in a wide variety of cases, permitting the simulation of complex subterranean flow processes.

Of special importance are the National Norms and Regulations (e.g., on the dimensioning of wells and assessment of pumping tests - which are based on the results of research and incorporate the latest states of the art. These norms are an important tool for the user, and promote the effective application of the results of research.

With the help of geohydraulic simulation models, it is possible to answer most questions relating to data on groundwater levels, flow direction, velocity, infiltration and tapping quantities.

The most pressing problems in managing groundwater in the German Democratic Republic are currently those of providing adequate drinking water and controlling the impact of open-pit mining.

4. Impact of open-pit mining

Open-pit lignite mining is an economic activity which entails severe environmental impact, especially related to water resources systems.

In the German Democratic Republic, the annual production of lignite amounts to more than 300 million tons per year (almost one third of the world output). For mine drainage, about 1.7 km^3 of water have to be pumped per year using about 10,000 dewatering wells. Of the whole stable runoff (subsurface runoff) of about 9 km^3 per year, the amount of mine drainage water exceeds 20 per cent.

In loose rock aquifers in particular, the mine drainage leads to the formation of large cone-shaped groundwater depressions. As a result, water supply wells often go dry, small rivers dry out or large ones lose part of their runoff due to the infiltration of water into the depression cone. The water balance of these regions is changed: the infiltration rate grows, subsurface runoff is reduced to zero. Flood peaks of rivers are lowered by increased infiltration and the water level of the rivers is raised by the continuous discharge of mine water into them. Agricultural crop production suffers from the lowering of the water table. To maintain stable crop production, supplementary irrigation becomes necessary which implies higher costs and higher agricultural water demand as compared to natural conditions.

After the completion of mining activities and the ending of the drainage operation there will be a rise in the water table. As a rule, this process is much slower than the initial lowering of the groundwater level. Thus, on cessation of mine drainage, groundwater levels rise, but only slowly, while the level of rivers drop drastically and abruptly.

During the post-mining period, however, as the water tables rises to its former level its depth under the soil surface might be less than in the pre-mining period. As a result, artificial drainage systems are sometimes necessary to protect municipalities and factories located in former mining areas. Agricultural land and forests often have to be drained.

These examples show that large-scale mining causes major impacts on the water resources systems and consequently on the regional environment as well as on other economic activities in the region such as agriculture, industry and municipal water supply.

Dewatering of complex aquifer systems requires the calculation of the spatial and temporal lowering process in advance, including the optimization of the necessary drainage works. In the course of the increased use of electronic data processing, models were applied on the basis of the finite difference method. Today, simulation methods for time- and space-discrete treatment of aquifer models with mining-hydrologic tasks are used for subjects of 3,000 to 3,500 space-discrete elements in routine running for the mining industry of the country.

Such models have been used for a long time in the prognosis of groundwater dynamics and of the water balance in these areas as well as for the design of open-pit mine drainage. Commonly, such models include the catchment area of a mine region with not more than three open-pit mines.

As a result of the opening of new and deeper open-pit mines the area influenced by the large-scale groundwater lowering becomes larger. Simultaneously the open-pit mines move closer together; from this follows a closer hydraulic interrelation in this territory. For the prognosis of the regional groundwater dynamics under mine pumping it is necessary to apply models of larger dimensions and of higher precision. The larger dimension of the models include the modelling of longer time periods (usually the whole period of mining activity in this region) and an increasing variability of geometric data and geohydraulic parameters. This new problem requires greater skills in applying computer technology, especially in using microcomputers.

Large-scale hydrogeologic models are being developed by the coal mining industry. The areas to be modelled cover about 5,000 km^2. In some cases it is necessary to include five water-bearing beds in the model. In order to be able to solve the tasks for this size of model it is necessary to compute systems with 15,000 to 17,000 elements. This presupposes that equation systems with the same number of unknowns can be solved economically by electronic data processing.

The development of large-scale models is prerequisite for computer-aided decision support systems for water-management policies in mining areas, which assume a great number of degrees of freedom of the hydrogeological process model and demand a great number of computations with high precision. Such a computerized decision support system can be an efficient tool for the analysis of various water use problems in mining regions, keeping in mind the vast amounts of information and the complexity of the processes involved. As an example, the decision support system called MINE has been developed at the International Institute for Applied Systems Analysis, Austria, in collaboration with institutes in the German Democratic Republic. The test region is located in the eastern Lusatian lignite district in the south-east of the German Democratic Republic.

5. Protection and monitoring

Groundwater resources must be protected against excessive use and ultimately exhaustion. They should be developed only to an extent to which they are recharged by natural processes. Where there is overexploitation, it should be compensated by artificial recharge (e.g., by irrigation systems or recharge basins, wells and trenches). They should be used economically and purposefully, with drinking water given the highest priority.

(a) Groundwater protection

To secure the protection of groundwater resources against degradation and contamination, it is necessary to maintain the self-purification capacity of the soils and groundwater zone. It should be recognized that subsurface systems react extraordinarily slowly. Some harmful effects become evident years or decades after the cause. Rehabilitation of contaminated soils or aquifers, if possible at all, frequently requires effort over several generations. Therefore, protection activities must not be limited to the presently exploited groundwater resources but should be extended to groundwater resources in totality.

For the German Democratic Republic the problem of protecting groundwater resources against excessive use and degradation is of specially great importance. The population is supplied with drinking water mainly through groundwater development (75 per cent) and groundwater will remain the main source for drinking water for the foreseeable future. There is no appropriate alternative in sight. The interface between fresh water and salt water is in most parts of the German Democratic Republic at a depth of 100-300 m and accordingly the use of groundwater is allowed only to the extent to which it can be naturally or artificially recharged. The use of groundwater has reached very high levels already and the demand for drinking water increases by 2 to 3 per cent a year. Further development of groundwater resources has become more complex and costly than ever. Groundwater pollution is becoming ever more noticeable, not only locally but also on a regional scale. Water supply authorities are confronted with many problems due to the increasing diffuse and local contamination sources accompanying the progress of industrialization and urbanization.

Essential to groundwater quality control is the definition of drinking water protection and reservation areas and bans on land use. Two basic tasks are involved. The area to be protected for a given groundwater resource must be clearly defined and the optimum use and management of that area must be adopted.

The computer-aided solution of the first of these tasks uses groundwater isochron maps for definition of the nearby protection zone (protection zone II) and for differentiation within the distant protection zone (protection zone III).

This dimensioning concept is incorporated into the revision of the National Norm "Drinking water protection" of the German Democratic Republic (TGL 43850) which will ensure its broad application in the water management of the country.

(b) <u>Monitoring</u>

Monitoring of groundwater levels and quality is essential for sound decision-making on technologic and economic measures as well as legislation for the rational management of groundwater resources.

Scientific approaches to observation are fundamental and should consider both natural and man-made impacts on groundwater. The monitoring of quality to protect against contamination is of special significance at this time.

The concept of monitoring extends beyond the scope of simple observation. A scientifically based programme for continuing observation of the systems state, i.e., the variations of significant dynamic process characteristics, must be included. Analysis and scientific explanations of the systems-state changes known to have taken place in the past (epignosis) will lead to reliable prognosis of these changes.

The value of knowing such information at the time of prognosis is, of course, expected to justify the expenses spent on monitoring. More specifically, the value of information will be greater than the expenses of monitoring. Therefore, monitoring and use of information obtained by it (e.g., for decision-making and control) are to be considered jointly in the planning process.

None the less, observation remains the mainstay of monitoring. The recent development of observation programmes based on modelling and simulation still depend on knowing:

a. Why the measurement is made,

b. What is being measured,

c. Where the measurement is made,

d. When (how often) the measurement is made,

e. How the measurement is made, and

f. Who is doing the measuring.

Only then can the appropriate staff and equipment for effective observation programmes be provided with maximum efficiency and economy.

Verifying the model conception and quantifying the parameters are essential to a scientific measurement programme as well as prognosis of the systems state.

Such prognosis is the basis for decision-making, control and optimization.

Groundwater observation stations are included in the baseline measuring network, the environmental monitoring network or in special networks designed for specific purposes.

The baseline net as national groundwater measuring network serves the long-term and regular observation of climate-induced changes in the subsurface part of the hydrological cycle and hence the groundwater resources.

Monitoring networks provide long-term and regular observation of the groundwater regime supplying the waterworks and in catchment areas with strong human impact. This category of measuring network is subdivided into regional monitoring networks and object-related ones.

The number of groundwater measuring points belonging to the national observation network is about 11,000. The average density of measuring points is one per 10 square kilometres.

6. Development

In the German Democratic Republic, groundwater is, as in most European countries, the main source of supply for drinking water. Groundwater plays a comparatively minor role in agriculture and industry. Public water supplies take about 20 per cent directly from surface water, 20 per cent is drawn from surface water indirectly through bank filtration and artificial recharge; the remaining 60 per cent is from natural groundwater.

Considerable differences exist in how groundwater is tapped between the northern part of the German Democratic Republic and the southern part. Whereas in the north tapping of groundwater from unconsolidated formations prevails (71 per cent), water supply in the hard-rock regions of the south is about evenly distributed between the various forms of water resources. Besides reservoir water (26 per cent) and water from unconsolidated aquifers (20 per cent), captations of streams and rivers (7 per cent), deep wells (25 per cent) and springs (22 per cent) are the water sources. The stronger dependency of the two latter sources on the precipitation pattern may result in local shortages in dry periods.

Future development may be characterized by a standstill of direct development of surface water and by limited extraction of groundwater by conventional means.

Thus, the increase of groundwater resources availability through unconventional means comes to the fore.

Development in the field of groundwater abstraction is marked by the predominant use of vertical filter wells with submersible motor pumps. Traditional forms of extraction continue, such as infiltration galleries, siphon wells and horizontal filter wells. Artificial recharge methods in wide use include quasi-natural infiltration on permanently or temporarily flooded areas, and infiltration in systems of basins (regenerable sand basins, grass-covered basins).

Acceleration of the infiltration rate is the subject of ongoing investigations aiming at improvement of these technologies as the basis for extended use of groundwater.

Groundwater abstraction in waterworks is controlled by means of measured water level or of calculated balances. Under complex conditions and in large-scale waterworks, computer-aided control systems are in operation.

C. Administration and management

1. Agencies involved

Responsibilities for questions on the use of groundwater are shared by several national and public institutions. The Ministry of Environmental Protection and Water Management through the many water authorities is responsible for monitoring, planning, and managing the drinking water supply. Its Institute of Water Management undertakes research in groundwater balancing, monitoring and management. The Ministry of Geology, through its Organization for geological research and exploration (GFE) and its Central Geological Institute, explore and develop reserves and carry out hydrogeologic research and mapping. The Ministry of Public Health monitors groundwater quality. The Ministry of Coal and Energy oversees the dewatering of open-pit lignite mines.

The Mining Academy of Freiberg gives training in theoretical and applied hydrogeology, groundwater protection and isotope applications. Its special field of research is the application of environmental isotopes in groundwater exploration and protection.

The Dresden University of Technology gives training in water management and hydrology. A post-graduate course in groundwater has been offered since 1976. Research in soil and groundwater science is directed at the fundamentals of geohydraulics and migration processes, including their application in the fields of groundwater monitoring and management.

2. International cooperation

The German Democratic Republic has continuously participated in a number of international projects in the field of groundwater resources assessment and management.

Important contributions were made to the International Hydrological Decade and to the different phases of the International Hydrological Programme (IHP) of

UNESCO. Joint contributions were made to regional projects of IHP as, for instance, maps for Eastern and Central Europe on the principal water balance components and on subsurface runoff.

In March 1987, at the International Symposium on Groundwater Monitoring and Management, held in Dresden, 109 papers were presented and published as Symposium Proceedings. Of these, 54 were selected for inclusion in a special volume of the International Association for Hydrological Sciences, published in 1988.

The International Association for Hydrological Sciences (IAHS) and the International Association of Hydrogeologists (IAH) in the International Union of Geological Sciences (IUGS) represent the international forum for questions concerning groundwater and dealing with special problems in developing countries, in arid and semi-arid zones, with geothermal waters and groundwater protection. The German Democratic Republic cooperates closely with these organizations.

Consulting services for groundwater problems have long been available. Among others, the Ingenieurbüro Thiem in Leipzig was engaged in exploring for groundwater on a scientific basis long before the Second World War.

Two consulting firms offer services on groundwater problems to clients abroad. Agro-Consult Dresden (ACD) is geared to consulting and engineering in the agricultural and food sectors, and Industrie Consult Berlin (ICB) does the same in the industrial sector. In fact, both give high priority to problems related to the effective use of water, soil and energy resources.

The permanent staff of these consultancy firms, each with a water management division, are supplemented by short-time assignments of specialists from national research institutes and water-management authorities. In this way, ACD and ICB set up problem-oriented teams of professionals for each consulting task, including field investigation, laboratory experiment and computer-aided modelling.

References

1. Individual authors

Busch, K.-F. and L. Luckner. Geohydraulik. VEB Deutscher Verlag für Grundstoffindustrie, Leipzig, 1972.

Freydank, E., H. Glander, E. Glos and G. Glugla. Karten der Wasserhaushaltsgrössen für das Gebiet der DDR. Ein Beitrag zum Internationalen Hydrologischen Programm der sozialistischen Länder Europas. Z. Meteor., 33-4, p. 244-257, 1983.

Jordan, H. and H.-J. Weder (Eds.). Hydrogeologie. VEB Deutscher Verlag für Grundstoffindustrie, Leipzig, 1988.

Lauterbach, D., K.-F. Busch, L. Luckner and K. Tiemer. Rational use of groundwater resources in the German Democratic Republic - problems and ways for solving them. United Nations Water Conf., Mar del Plata (E/CONF.70/TP 60), 1977.

Luckner, L. and W. Schestakow. Migrationsprozesse im Boden- und Grundwasserbereich. VEB Deutscher Verlag für Grundstoffindustrie, Leipzig, 1986.

Orlovsky, S., St. Kaden and P. van Walsum. Decision support systems for the analysis of regional water policies. Final Report. IIASA, WP-85-25, Laxenburg/Austria, 1986.

2. Collected works

Atlas Deutsche Demokratische Republik. AdW der DDR (Ed.), Haack, Gotha, 1981.

Grundriss der Geologie der Deutschen Demokratischen Republik. Akademieverlag, Berlin, 1968.

National Report of the German Democratic Republic to the UNESCO-WMO-Conference on Hydrological Problems in Europe. National Committee for Geodesy and Geophysics of the GDR, Berlin, 1973.

Proceedings of the International Symposium on Groundwater Monitoring and Management. Dresden, 23-28 March 1987. To be published by The International Association of Scientific Hydrology (IASH).

VI. THE FEDERAL REPUBLIC OF GERMANY

Area: 248,667 km^2 Population: 61.1 million

A. Physiography and climate

The Federal Republic of Germany comprises the States of Bavaria and Baden-Wurttemberg in the south, the Saarland, Rhineland-Palatinate, North Rhine-Westfalia and Hesse in the centre, Lower Saxony and Schleswig-Holstein in the north, and the city states of Bremen, Hamburg and Berlin. The country extends over approximately 900 km in a north-south direction and 300-450 km east-west.

A broad variety of geomorphologic regions is known in the Federal Republic of Germany ranging from high mountain areas to flat coastal lowlands. These are:

a. Part of the Alps mountain range in the extreme south, forming a deeply dissected area with high peaks (maximum 2,970 m);

b. Smoother mountain ranges and hilly uplands (Bavarian, Bohemian and Franconian Forests in the south-east, Black Forest in the south-west) and rugged, formerly peneplained, massifs (Rhenish Mountains, the Harz);

c. Extensive Mesozoic plateaus with cuestas and hilly ranges in southern and central Germany;

d. Flat graben zones (upper and lower Rhine grabens); and

e. The wide glacial lowland area in the north between the Mesozoic hills and the North and Baltic seas.

The Federal Republic of Germany is a densely populated and highly industrialized, but also agricultural country. The land use is distributed as presented in the following tabulation (Keller and others 1978-1979):

Use	Percentage
Agriculture	54
Formerly agriculture, now abandoned	1.2
Unusable	2.7
Uncultivated land area (marsh, bog)	0.6
Forest	28.9
Streams, lakes and reservoirs	1.8
Buildings, houses and yards	4.6
Roads, motorways and railways	4.6
Parks, cemeteries, sports, aviation and military	1.5

The major European drainage divide between the North Sea basin and the Mediterranean and Black Sea basin runs across southern Germany along the Swabian plateau. The Danube drains the south-east corner of the country, while the bulk of the territory is included in the catchment areas of the rivers Rhine, Ems, Weser and Elbe and numerous coastal rivers draining into the North and Baltic seas. The mean discharge data for the drainage basins of the Federal Republic of Germany and the inflow to, and outflow from, the country are presented in table 23.

Table 23. Federal Republic of Germany: Mean discharge of river drainage basins, 1931-1960

Basin	Area within country (Thousand km²)	Mean discharge (m^3/s) (1) To country	(2) From within country	(1 + 2) Outflow from country
Danube	56	653	777	1 430
Rhine	101	1 229	941	2 170
Maas	4.0	-	33	33
Ijssel	3.5	-	26	26
Ems	13	-	108	108
Weser	41	40	320	360
Elbe	17	680	150	830
Coast area	12	-	90	90
Total	247.5	2 602	2 445	5 047

The climate in the Federal Republic of Germany is changeable and is determined by Atlantic and continental air masses, with prevailing westerly influences throughout the year. Precipitation is fairly evenly distributed during the year. The percentage of snow, which is generally high in winter (November-March), increases with altitude above sealevel, reaching almost 50 per cent of all precipitation in the Alps. The mean annual amount of precipitation increases, on the whole, with altitude, ranging from less than 500 mm in the northern upper Rhine graben to more than 2,500 mm in the Alps. In the mountainous regions the values generally exceed 1,000 mm, while 600 to 800 mm are common in the lowland and hilly areas.

Mean annual evapotranspiration varies between 450 and 600 mm, the bulk of the area lying in the range of 450 to 550 mm. Mean air temperatures in the coldest season are around 0° to -3° C (January) and in summer between 14° and almost 20° C (July), with minima in the mountainous regions and maxima in the Rhine valley.

The water balance in the Federal Republic of Germany as projected by S. Clodius in 1986, for the year 2000 is illustrated in figure 27.

Figure 27. Federal Republic of Germany: Water balance for the year 2000

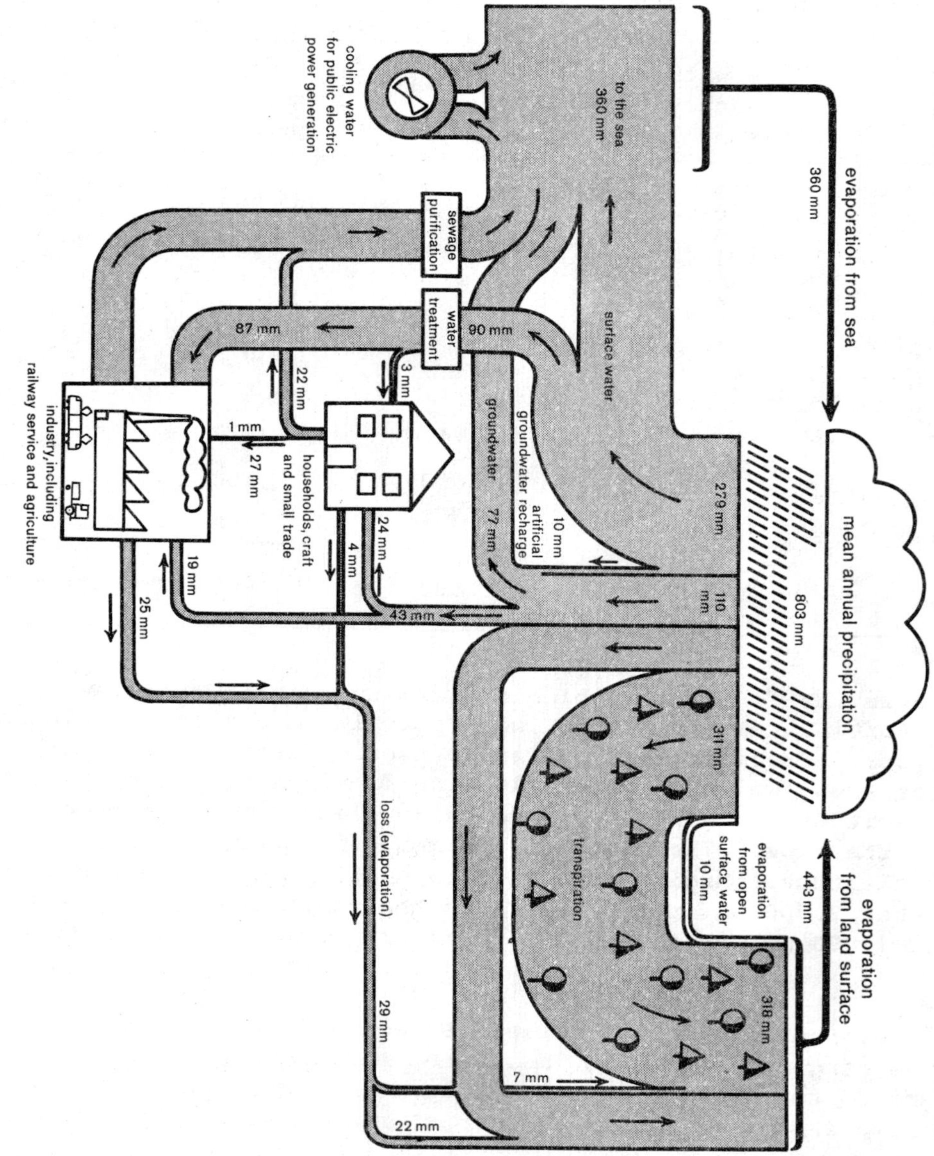

B. Geology

An outline of the geology of the Federal Republic of Germany is shown in figure 28. Large outcrops of Variscan and pre-Variscan basement complexes are known in the southern and central parts of the Federal Republic of Germany. These are the marginal areas of the Bohemian Massif (Bayrischer Wald, Oberpfalz, Frankenwald, Fichtelgebirge), the Black Forest (Schwarzwald), the Odenwald, the Spessart, the Rhenish Massif and the Harz. The structures of the basement generally trend south-west and present a zonation, perpendicular to the trend, that is characterized by a general decrease in orogenic stress, folding and metamorphism towards the north and north-west.

The basement outcrops in the south are built up predominantly of metamorphic rocks and granitic intrusions. In the Rhenish Massif and the Harz, however, the thick (several thousand metres) sequence of Devonian and Carboniferous clastic rocks is only slightly metamorphosed, but on the southern flanks is steeply folded and faulted. Locally, limestone, basic and acid extrusives and, in the Harz, granitic intrusions are encountered. On the northern slope of the Rhenish Massif the more gently folded beds of greywacke, slate and sandstone contain coal seams, giving rise to deep mining in the Ruhr area and the area of Aachen.

The post-Variscan cover, locally reaching a thickness of several thousand metres, can be split into:

a. A post-Variscan Paleozoic and Mesozoic cover, mainly composed of consolidated sedimentary rocks, and

b. A Cenozoic section comprising Tertiary and Quaternary unconsolidated sediments and volcanic deposits.

Following the Late Carboniferous (Variscan), folding and uplift of the Rhenish and Bohemian massifs, into extensive upland areas, greatly influenced the post-Variscan sedimentation. Thus, accumulations of thick series of detrital rocks of the Carboniferous and Permian (Rotliegende) associated with predominantly acid extrusives are found in troughs north and south of the Rhenish Massif.

The post-Variscan Paleozoic and Mesozoic sequence is represented by rocks of continental, marine and evaporite facies witnessing a rather uniform, continuous sedimentation. Six evaporite units are recognizable regionally in the Federal Republic of Germany, the largest being the Permian (Zechstein) evaporites forming salt diapirs in the northern part of the country and giving rise to "salt tectonic" deformations.

In southern Germany, the Triassic and Jurassic series are almost flat, forming a cuesta landscape. It is a complex alternation of strata, up to 1,500 m thick, of clastic rocks (sandstone, siltstone, claystone mainly of the Buntsandstein, Keuper, Lias and Dogger) and of calcareous rocks (limestone, marl, locally evaporites, of the Muschelkalk, Keuper and Malm). In the Saxonian hills of southern Lower Saxony and Hesse the rocks are mainly Triassic and are deformed in small synclines and anticlines and faulted in graben and horst blocks.

Cretaceous clastic and limestone rocks crop out mainly in the transition zone between the central uplands and Saxonian hills and the northern lowland, where they form anticlines and synclines, the largest of which is the Basin of Munster.

Figure 28. Federal Republic of Germany: General geology

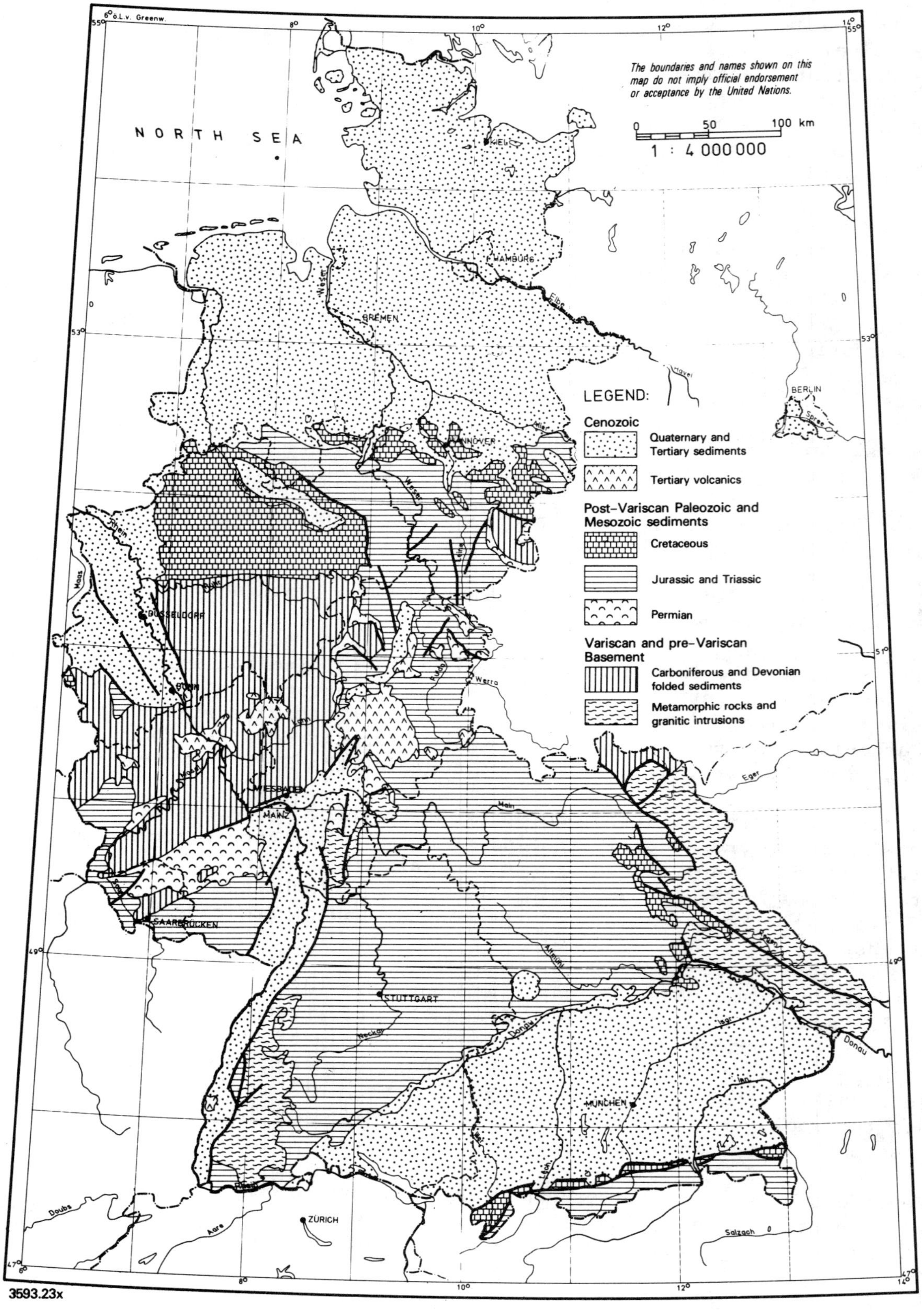

The southernmost part of the Federal Republic of Germany lies within the northern zone of the Alpine orogenic belt, where Cretaceous, Jurassic and Triassic rocks are steeply folded and overthrusted in nappes, faulted and dissected. The nappes moved far to the north during the Tertiary so that they even lie over the southern rim of the adjacent Tertiary molasse trough, which is filled by more than 4,000 m of debris from the Alps.

The Cenozoic era, which introduced a fundamental change in tectonic and sedimentation conditions, was characterized by block faulting, graben formation, volcanism, large-scale erosion and fairly discontinuous and heterogeneous sedimentation.

The general uplifting of central and southern Germany is associated with the formation of local basins and deep grabens (depression of Hesse, upper and lower Rhine grabens). The grabens are filled by a thick sequence of marine and terrestrial, mainly unconsolidated deposits (locally more than 3,000 m in the upper Rhine graben), in places with thick layers of lignite (e.g., in the southern lower Rhine basin west of Cologne). Tertiary basalts and associated volcanic products occur mainly in the Vogelsberg area and in the Rhenish Massif (the Westerwald).

Quaternary processes, mainly the Pleistocene glaciations, have greatly transformed the pre-Quaternary topography. Four Pleistocene glaciation periods are known in Germany in which huge continuous ice sheets originating from two major centres, in Scandinavia and the Alps, covered large areas of the country. In the north the glacial drift extends up to the Rhenish and Saxonian upland region. It contains Scandinavian and abraded and redeposited local material. Deep subglacial erosion channels, incising locally more than 300 m deep into the Tertiary sequence, are evidence of an elaborate Pleistocene relief, which was largely reduced to a plain during the Holocene but which today still shows such remnants of glaciation as terminal moraine hills, swampy depressions, large outwash plains and loess belts.

C. Hydrogeology

1. Background

Discoveries of Stone Age sites show that the first settlements with the earliest forms of water supply were along rivers and brooks and at lakes and springs. The funnel-shaped wells of the Iron Age, which were lined with wickerwork for stability, are early technical measures taken in order to obtain groundwater. Wells dating from the early Christian era in the North Sea coast area have a lining of mud and grass sods to prevent them collapsing. This method was still used by the Vikings in the eighth and ninth centuries.

During the second and third centuries, the Romans practised relatively efficient methods of well construction using timber lining in loose material and brick lining in solid rock (e.g., in Saalburg Castle north-west of Frankfurt). Springs were exploited and the water led by pipe to urban areas, e.g., the Roman pipeline leading from the Eifel to Cologne (length 95.4 km, difference in height about 370 m, discharge 175 to 230 l/s).

The Middle Ages were characterized by an increasing use of surface water. In areas where the use of surface water was not possible, deep wells were dug by conventional mining methods (Nürnberg Castle: 54 m deep, Wülzburg Castle at Weissenburg, Middle Franconia: 166 m deep). At first, buckets and hand winches were used to raise the water; later wooden pumps were installed. A well inspectorate was responsible for hygiene in the cities; in some of them well associations were also established.

By the end of the eighteenth century, drilling methods, which had so far been used only in mining, were being applied in the construction of wells. Around 1850, drilling was carried out with water circulation. In the second half of the nineteenth century, most cities established their own central water supplies, partly using surface water and partly using water from wells. The basic principles for this work were largely established by A. Thiem (1880). Owing to lack of knowledge and inadequate cleansing methods, there were several devastating epidemics (Hamburg 1892: cholera, 9,000 deaths; Gelsenkirchen 1901: typhoid fever, 350 deaths). The technical experience gained in this period led to intensified groundwater exploration and exploitation in the following decades.

In the meantime, most of the fundamentals of hydrogeology had been worked out and are dealt with by Keilhack (1896) in his Textbook of Practical Geology. In 1901, Herzberg, and nearly at the same time Ghyben, developed fundamental ideas on the behaviour of the interface between fresh water and salt water, knowledge of which is vital for groundwater development on islands and in coastal areas. The technical aspects of groundwater development were dealt with by Bieske (1958, 1960, 1965) in his Manual of Well Construction.

The 1:500,000 General Hydrogeological Map at scale 1:500,000 by Grahmann and others (1952-1958) covered the whole Federal Republic of Germany for the first time. Later, the Hydrological Atlas of the Federal Republic of Germany (Keller and others, 1978-1979) was published, as well as maps at 1:1,000,000 of the Groundwater Occurrences in the Federal Republic of Germany showing productivity, quality, and vulnerability to contamination (Aust, Vierhuff and Wagner, 1980).

In 1969, Bentz and Martini published their Textbook of Applied Geology, and in 1984, Bender's Applied Geosciences appeared; the latter includes an up-to-date account of the most important hydrogeologic methods.

2. Principal aquifers

The sequence of rock units ranging from the Precambrian to the Quaternary according to their aquifer characteristics, regional extent and role in groundwater production can be subdivided into three main complexes. They are:

a. The Cenozoic aquifer complex comprising the Tertiary and Quaternary rock units;

b. The Mesozoic aquifer complex consisting of the rock units from the Triassic (Upper Permian included) to the Cretaceous; and

c. The crystalline Paleozoic basement complex composed of rocks of the Variscan and pre-Variscan basement and some of Early Permian Age.

The complexes are described below from base to top, which is also from least productive to most productive. The lithology of the main aquifers is shown in figure 29. The availability of groundwater is presented in figure 30.

Figure 29. Federal Republic of Germany: Lithology of principal aquifers and mean annual precipitation

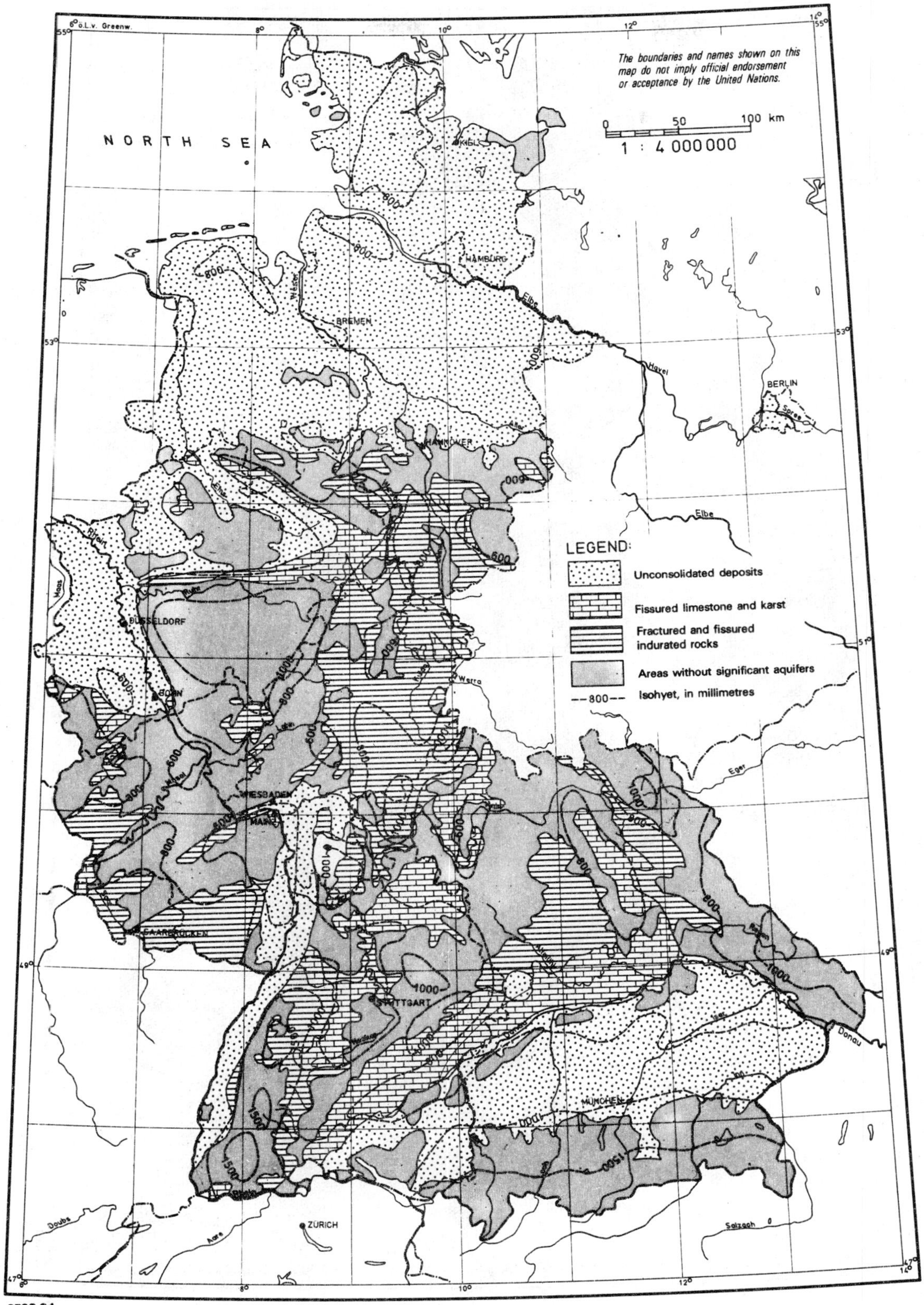

3593.24x

Figure 30. Federal Republic of Germany: Availability of groundwater

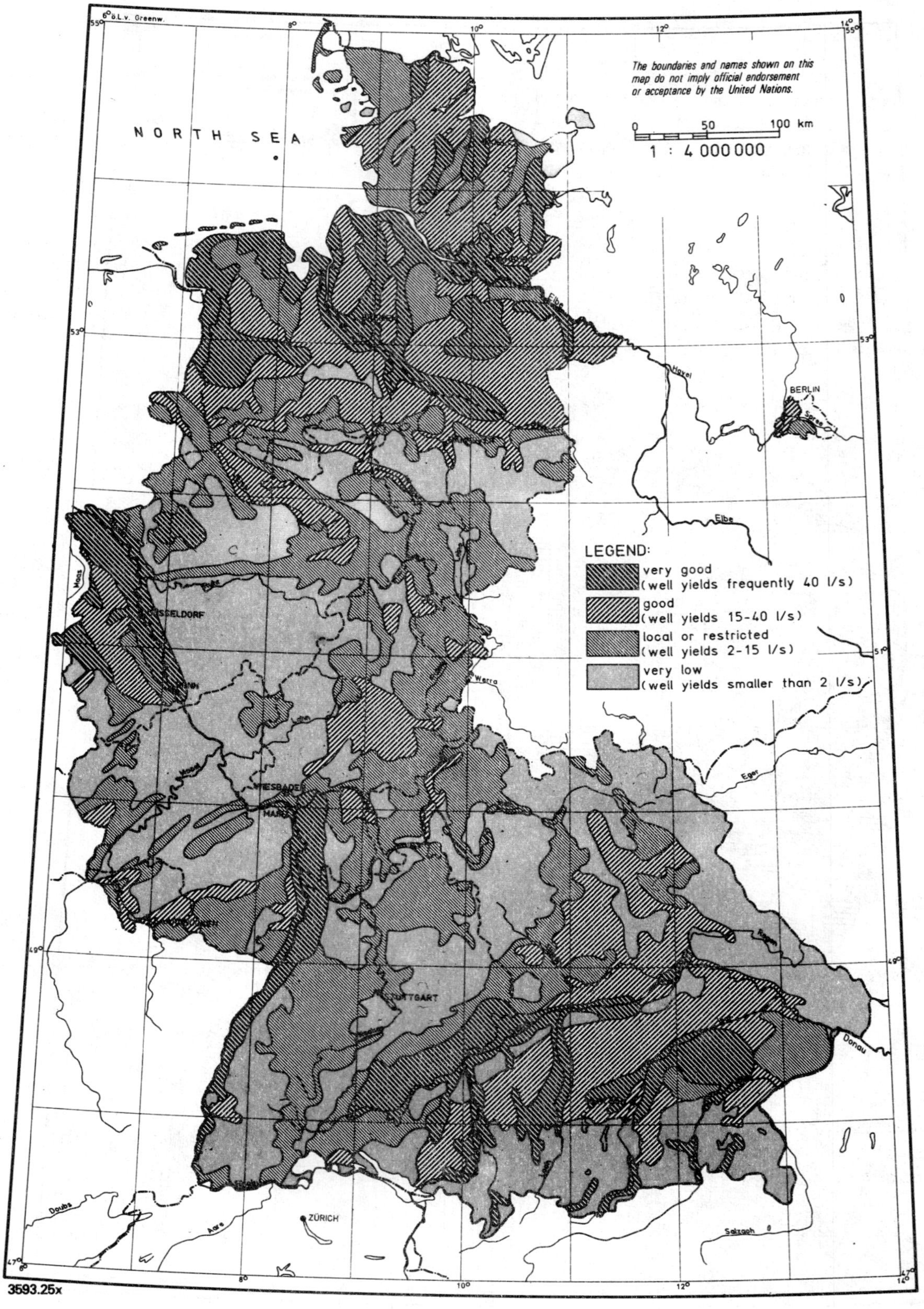

(a) Crystalline Paleozoic basement complex

The basement complex comprises various crystalline and metamorphic rocks of the Variscan and pre-Variscan basement, and post-orogenic sedimentary and volcanic rocks of the Lower Permian (Rotliegende). The permeabilities of these rocks are generally low.

In the highlands and massifs where the Variscan basement crops out, groundwater occurs in a thin zone of weathering and in joints and fractures near the surface, a circumstance that makes these highlands rather poor in groundwater. Well yields seldom exceed 2 l/s.

Deep-reaching faults situated at the margin of the crystalline Paleozoic massifs locally provide higher yields. Still more favourable conditions are known from quartz dikes and prominent fractures in basic extrusive rocks of the Rhenish Massif where well yields of more than 10 l/s are encountered. Middle Devonian reef limestones of the Rhenish Massif, which are strongly karstified, offer occasionally even higher well yields.

(b) Mesozoic aquifer complex

The aquifer complex comprising rock units of Upper Permian to Cretaceous consists of a sequence of flat-lying continental, marine and evaporite platform sediments. In the area of the outcrops in the central uplands of the Federal Republic of Germany it forms a broad landscape of cuestas.

As a sequence of aquifers, aquitards and aquicludes, the complex constitutes a multilayered aquifer system. Since nearly all the rocks are consolidated, the aquifers owe their permeability mostly to fractures and joints. Depending on the differing facies, thickness and fracturing, the aquifer properties vary in all ranges.

The best aquifer properties occur in the limestone members of the sequence. Permeability is commonly enhanced by karstification. In most of the sandstones the permeability depends on the degree of fracturing. The pelitic and evaporite members of the sequence generally act as aquicludes.

Separated aquifers can be interconnected hydraulically by change of facies, discordances or blockfaulting. The multilayered hydraulic systems formed in such a way may have great extent. Special structures such as folds and areas of subsidence are related either to the diapiric movements of the Zechstein rock salt or to the subrosion of the intraformational evaporites. The base of many of the hydrologic systems is formed by the relatively impermeable crystalline Paleozoic basement. The aquifer properties of the several formations are as follows:

i. Upper Permian (Zechstein)

The formation of the early stages of the platform sedimentation in the northern and central part of the country contains a series of limestone, which is highly karstified and represents at least locally an important aquifer. The few existing karstic springs have a high discharge (Rhume spring, 1 to 5 m^3/s).

ii. Lower Triassic (Buntsandstein)

The Buntsandstein formation is 500 to 1,000 m thick and consists almost throughout of terrigenous clastic deposits. Development of facies gives rise to a tripartition: the lower Buntsandstein beds consist mostly of claystone, the middle part of fine-grained to coarse sandstone, and the upper part is also clayey but contains occasional intercalations of evaporites (mostly sulphates).

Groundwater flow occurs predominantly in joints and fractures. In the Middle Buntsandstein, however, porosities can contribute to the flow. Depending on the degree of faulting, fracturing and porosity, well yields range from 10 to 150 l/s.

iii. Middle Triassic (Muschelkalk)

The Muschelkalk forms the first calcareous sequence of the Mesozoic cycle, has a thickness of 100 to 250 m and is, like the Lower Triassic, tripartite. Both lower and upper parts consist of well bedded limestone representing aquifer zones. The limestone, especially in the lower aquifer zone, has a relatively high permeability which is due, at least in part, to karstification.

Mudstones and evaporites prevail in the middle part of the Muschelkalk, the result of a regressional phase of the sedimentation. Groundwater, having been in contact with the middle Muschelkalk, can be mineralized. Where the evaporites have been leached completely, as in some southern parts of the country, wells can yield up to 50 l/s of fresh water.

iv. Upper Triassic to Middle Jurassic (Keuper, Lias, Dogger)

The Keuper through Dogger series is again predominantly clastic, being characterized by a facies differing in time as well as space. Permeabilities generally are low to very low. As a product of a regressional phase of the sedimentation, the Upper Triassic (Keuper) contains gypsum deposits beside coarse delta deposits (Schilfsandstein, Stubensandstein). The yields of wells in the sandstones range from 5 to 20 l/s.

v. Upper Jurassic (Malm)

The Malm formation consists of 200 to 600 m of limestone beds giving rise to an important regional aquifer. It has a wide continuous extent in southern Germany and constitutes a large hydrologic system.

The outcrop area is situated in a mountainous range north of the Danube. South of the river up to the northern margin of the Alps the aquifer is confined under a thick cover of Tertiary sediments (Molasse basin).

The limestone is strongly karstified, especially in the outcrop area. Creeks and rivers vanish into sinkholes. Part of the groundwater is discharged in large karstic springs (Aach spring, 8 m^3/s). The deep flow reaches the river Danube as the final base of exit of the hydrologic system.

The occurrences of the formation in northern Germany, because of their limited extent, are less important as aquifers.

vi. Cretaceous (Kreide)

The Cretaceous formation consists of a predominantly marine sequence with varied facies. It comprises limestone and marl beds but includes also consolidated and unconsolidated sandy sediments. The carbonate and sandy members of the formation behave as fair to good aquifers. Borehole yields range from 5 to more than 40 l/s. In the Paderborn area strongly karstified limestones form an important aquifer giving rise to karstic springs in the order to tens of cubic metres per second.

In the northern part of the Federal Republic of Germany, where they have a regionl extent, these aquifers play an important role in groundwater production.

(c) Cenozoic aquifer complex

The Cenozoic complex is composed of mostly unconsolidated clastic sediments of the Tertiary and the Quaternary which cover extensive areas in basins in both the north and south of the Federal Republic of Germany, and also in grabens and along river courses in the area in between. The deposits are mostly of terrestrial origin. Some are marine or show marine influence. Wide areas in the north are covered by glacial deposits.

The Cenozoic deposits are important aquifers. In many cases, the aquifers, although variable in extent, facies and permeability, are hydraulically interconnected forming extensive systems. At their base, interconnections exist also with Mesozoic or older aquifers. Drainage generally is provided by the rivers on a local scale. Locally volcanic rocks contribute to the multiplicity of the Cenozoic aquifer systems.

Tertiary sediments occur over vast areas, mainly in the south and north. In the south they are several thousand metres thick and form the Molasse basin between the Alps and the river Danube. Its lower part is marine and consists of claystone and sandstone. The upper molasse contains more permeable material such as limestone, sand and gravel of fresh-water origin. Wells yield up to 100 l/s.

In the north, the Tertiary deposits consist of sand and clay of shallow marine or continental origin. They are buried under thick Quaternary cover and therefore play no major role in groundwater production. Similar deposits occur in the middle of the Federal Republic of Germany in mostly narrow grabens.

Aquifers of volcanic origin appear in isolated occurrences in the central uplands. The largest of these is the extinct Vogelsberg volcano where wells in a basalt yield up to 30 l/s. In the basalt sheets of the Westerwald, waterworks with boreholes located in fracture zones deliver even 100 l/s. In the volcanic landscape of the Eifel where volcanic activity continued into the Quaternary, tuffaceous products also form important aquifers.

(d) Quaternary

The Quaternary deposits which make up the lowlands of northern Germany, the grabens of the upper and lower Rhine River and the Munich gravel plain are mostly excellent aquifers.

In the northern lowlands of Germany the sediments comprise fluvial and glacial deposits. Both highly permeable fluvial sand and gravel and poorly permeable moraine formations are widespread. Favourable conditions for large-scale groundwater production do exist in the subglacial erosion channels many of which are filled with coarser material. Waterworks in the northern lowlands produce large quantities of groundwater and contribute to the water supply of cities like Hamburg, Hannover and Bremen. Groundwater development, however, is often hampered by salinity in the lower part of the sequence which originates from subrosion of diapiric salt domes or from sea-water intrusion near the coast.

The lower Rhine graben contains fluvial sediments of the Rhine and the Maas rivers. Their permeabilities indicate that the aquifer has good potential (1×10^{-3} to even 5×10^{-3} m^2/s).

The Quaternary gravels and sands in the upper Rhine graben are up to 130 m thick. The permeabilities are similar to those in the lower Rhine graben.

The Quaternary deposits of the Munich gravel plain constitute one of the major groundwater sources in the country. The areal discharge is in the range of 20 to 25 m^3/s.

3. Water quality

Owing to the generally humid climate and surplus in the water balance, groundwater occurrences in the Federal Republic of Germany are mostly of good quality with regard to total salt content (see figure 31). The most important aquifers in the south and north of the country, as well as the small groundwater occurrences spread over large areas which can be used for local water supply only, contain groundwater with low salinity. However, some natural conditions of local or regional importance and man-made influences affect the quality of the water and restrict the use of the water resources in various ways.

Figure 31 shows the main features of the quality of groundwater occurrences in the Federal Republic of Germany with respect to its suitability for domestic purposes. The principal constraints are sodium chloride, sulphate, total hardness, iron, manganese and nitrate.

High content of sodium chloride in the water is found in most aquifers near the coast due to sea-water intrusion. Another source of salt, mainly in deep Tertiary and Quaternary aquifers in the north of the country and in the southern Rhine valley as well as in some Mesozoic aquifers, are evaporite deposits of various geologic ages.

The content of the ions which determine the hardness of the water depends mainly on the type of the aquifer. For instance, in the southern parts of the country built up by Triassic sediments containing carbonate and gypsum, hard or very hard groundwater is abundant but cannot be used for domestic or industrial purposes.

At some localities, aquifer and hydraulic conditions allow the mineralized groundwater to discharge naturally in springs many of which are developed and used for medical purposes. If the groundwater rises from great depths, thermal springs occur which may be fresh or mineralized. Famous thermal springs are located at the edge of the Rhine graben in southern Germany (e.g., Baden-Baden and Wiesbaden).

Figure 31. Federal Republic of Germany: Chemical quality of groundwater and suitability as supply

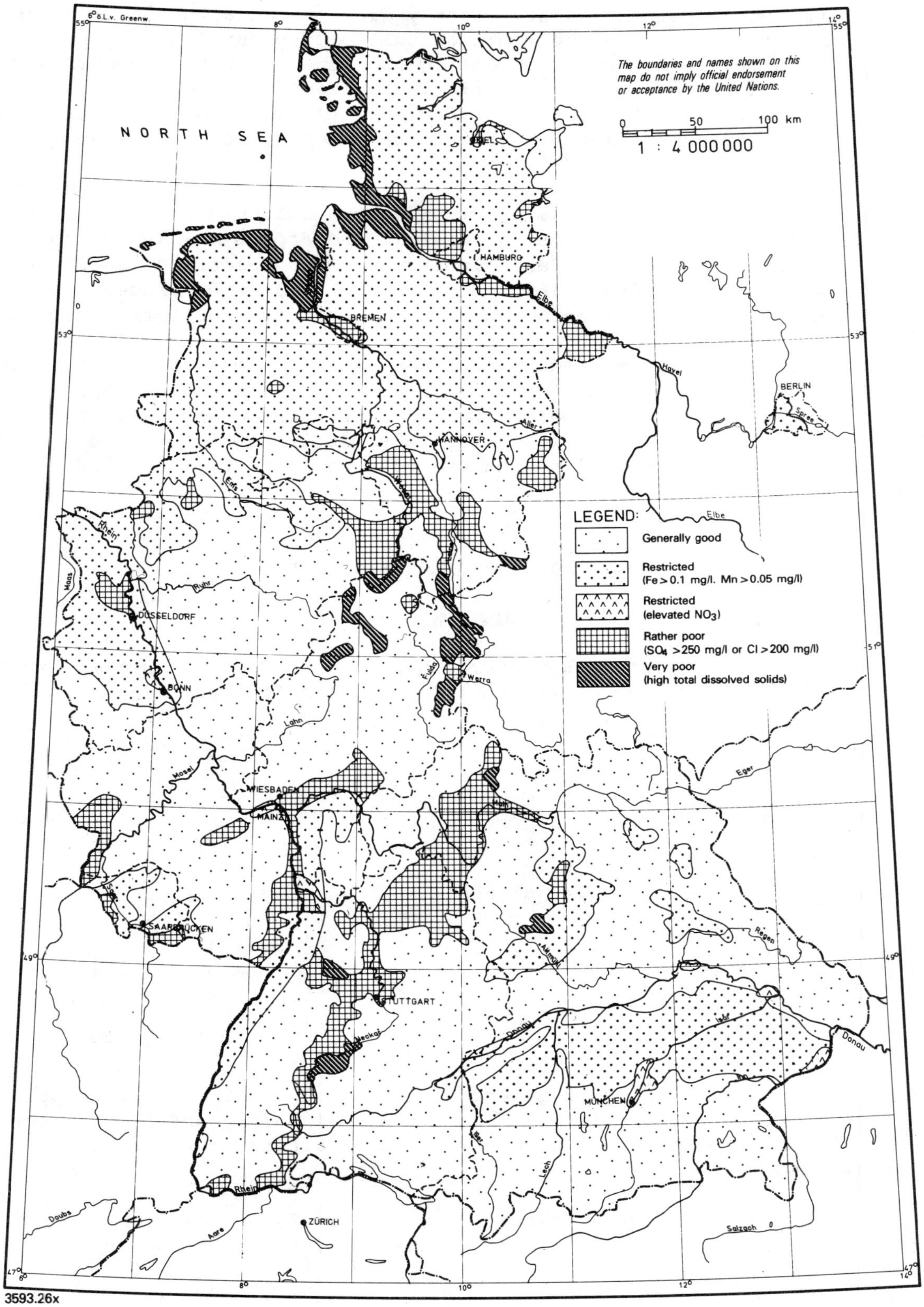

3593.26x

The large groundwater resources in both the north and south of the country are, in the majority, of low salinity and hardness, but contain high amounts of iron and manganese which need to be extracted before the water can be used.

The nitrate content of the water recently drew public attention when the European standard of 50 mg/l as the limit of suitability for drinking purposes was to be adopted. The nitrate content of many groundwater bodies is continuously growing. Sources may be mineral or organic fertilizers, air pollution or decaying natural organic material in the soil. Strong efforts are to be undertaken by waterworks to adopt treatment methods able to reduce the nitrate content of the water.

In many cases there are effects on groundwater quality which derive from the increasing amount of acid components in rainfall and depend mainly on the buffer potential of the soil and the aquifer. In areas of non-calcareous aquifers and soils with a naturally low alkalinity, "acid rain" has already caused a strong increase of aluminium and heavy metal contents in the water of springs and wells.

Owing to the widespread sewage systems and effective groundwater protection rules in the country, the hazard of bacteriologic pollution of developed groundwater resources is rather small. In areas of karstic aquifers, for instance, the use of vulnerable groundwater resources is strictly regulated, and in all recharge areas near producing wells, the use of organic fertilizers is severely restricted.

D. Groundwater development

1. Current status

Groundwater abstraction in the Federal Republic of Germany in the year 1981 amounted to 7,717 X 10^6 m^3 (about 7.7 km^3) for uses as follows:

Use	Amount (in 10^6 m^3)
Agricultural irrigation	139
Mining	1 487
Industry, including power plants	2 079
Private water supply	71
Public water supply	3 941
Total	7 717

The bulk of the production of groundwater is used for the public supply of drinking water. Large quantities are abstracted by industry and in the course of mining activities. The use of groundwater for agriculture, however, is comparatively small.

In community supply, groundwater amounts to over 70 per cent of the total resources utilized. The various sources of the water produced for the public water supply in the year 1983 are listed in table 24.

Table 24. Federal Republic of Germany: Sources for community water supply, 1983

	Natural ground-water	Springs	River bank infil-tration	Artificial recharge	Surface (rivers, dams)	Total
Million m^3	2 623	387	263	445	426	4 144
Percentage	63.3	9.3	6.4	10.7	10.3	100.0

In areas where unconsolidated aquifers prevail, groundwater production is above average.

2. Projected needs, interests and planning

The demand for community water supply is estimated to grow from 4 km^3 in 1981 to about 5 km^3 in the year 2000 and to 5.4 km^3 in 2010. These estimates are based on a growth factor of 3.3 per cent per year up to 1995 and of 1.4 per cent per year thereafter. By this it is assumed that the per capita consumption will grow from 135 l per day (1975) to 245 l per day (2010). In order to cover the requirement, groundwater production is expected to be increased substantially only if pollution can be controlled or reduced.

The demands of industry are expected to grow like that of the public water supply. However, more recycling of water may reduce this increase. A shifting from the high-quality groundwater used by industry to water of inferior quality may also reduce the strain on groundwater withdrawal.

In the agricultural sector no major change of groundwater utilization for irrigation is currently expected.

The abstraction of groundwater in the Federal Republic of Germany is not without effect on the environment. The flow of creeks can be diminished, marshes can dry up, or forestry and agricultural production can decrease. Further, utilization of groundwater can be endangered or impaired by activities like housing and industrial development, waste disposal, gravel digging, and open-pit or deep mining. The emerging conflicts of interests have to be regulated. In the Federal Republic of Germany this is done by water planning.

Water planning is carried out with the purpose of establishing an inventory of the water resources available, allocating the resources according to need, and providing for permanent availability of the resources. In that context hydrogeologic studies contribute to the water planning by assessing the groundwater resources by quantity, quality and availability; by defining and quantifying the interrelation of groundwater with other water resources (e.g., rainfall, runoff); and by defining the measures necessary for groundwater protection.

Groundwater resources assessment is generally done on the base of borehole data including the results of electrical logging and test pumping. Wherever a more detailed knowledge of the aquifer is needed, geophysical prospection methods such as geoelectrics, seismics and magnetics are applied.

In areas where hard-rock aquifers prevail, observation of the minimum surface runoff (base flow) can support the assessment of the groundwater quantities available. In karstic aquifers, tracers are used to define watersheds as well as flow velocities. In areas with unconsolidated aquifers the estimated aquifer yields are controlled by the determination of the areal rates of recharge as well as occasionally by aquifer modelling.

Responsibility for water planning is with the 11 states of the Federal Republic of Germany. The boundaries of planning units usually correspond to those of watersheds.

Although groundwater abstraction in the Federal Republic of Germany has attained high levels, not all natural potentialities are yet exhausted. This holds true mainly for the unconsolidated aquifers.

In less favoured areas where the groundwater reserves are limited, local conditions sometimes permit augmentation of the natural groundwater quantities. Such measures are river bank infiltration (induced recharge) and artificial recharge.

The technique of river bank infiltration, which increases the yield of aquifers by inducing recharge from rivers (and thereby enhancing the quality of the river water) is well established in the Federal Republic of Germany and practised at the rivers Rhine, Danube, Main and Neckar. Still other rivers offer potentialities. However, because of the contamination of the water by mineral or organic pollutants some of these rivers (e.g., Elbe, Weser) must be excluded from application of this technique.

Artificial groundwater recharge is usually done by making the river water pass through infiltration basins. Such installations exist in the valleys of the rivers Ruhr and Main, and in West Berlin. The installations are used for community water supply purposes as well as for industry. Favourable conditions exist in various other parts of the Federal Republic of Germany for like installations.

E. Administration and management

1. Agencies involved

In the Federal Republic of Germany the responsibility for water affairs is with the federal states. The central Government can impose only a framework legislation on water. Groundwater is included.

Groundwater investigations which are the concern of State Geological Surveys generally include the following:

a. Execution of hydrogeologic investigations related to the assessment and planning of groundwater resources;

b. Advising on groundwater use, management and protection;

c. Collection and documentation of hydrogeologic data; and

d. Evaluation and interpretation of groundwater data, publication of hydrogeologic maps and scientific papers.

Groundwater investigations carried out by the private sector (parastatal agencies, consulting engineers) are, comparatively, less extensive.

Groundwater development and management are in the hands of communities, water associations or enterprises liable to the public interest. Their activities are licensed and controlled by state water authorities. Groundwater protection increasingly becomes the concern of special environmental branches of the state administration.

Groundwater research in the Federal Republic of Germany is conducted by universities and such associations as the Deutsche Forschungsgemeinschaft (German Society for the Advancement of Scientific Research) as well as by Federal or State Government institutions, e.g., the Federal Institute for Hydrography (Bundesanstalt für Gewässerkunde), the Federal Institute for Geosciences and Natural Resources (Bundesanstalt für Geowissenschaften und Rohstoffe), the Federal Health Office (Bundesgesundheitsamt), and the Federal Environmental Agency (Umweltbundesamt). Moreover, several semi-public or private institutions carry out specific research programmes.

Today most research activity is linked to practical problems like contaminant migration, purification, protection against pollutants and quantitative assessment of groundwater resources. The findings of such research programmes are usually published.

Education in the field of hydrogeology, water resources planning or hydrogeologic engineering is provided in the main by the universities, technical colleges or institutions for higher education. Such specialized education normally lasts three to five years, depending on the level of qualification, e.g., technician, engineer, manager or scientist.

Training facilities in hydrogeology are offered in the Federal Republic of Germany by numerous institutions, universities and associations, above all the German Technical and Scientific Association for Gas and Water and the German Association for Water Resources and Land Improvement. Most of the training aims at improvement of the practical application of scientific work, e.g., optimization of hydrogeologic data collection programmes, monitoring networks and management schemes, the introduction of new technology for groundwater abstraction, treatment and supply, or training in methods of groundwater assessment and modelling.

On-the-job training in developing countries is an integral part of most of the technical cooperation projects of the Federal Republic of Germany, and numerous trainees from developing countries are sent to institutions in Germany, e.g., the Federal Institute for Geosciences and Natural Resources. Finally, a number of governmental agencies such as the Carl-Duisberg Gesellschaft or the German Academic Exchange Service (Deutscher Akademischer Austauschdienst) grant courses, scholarships and fellowships to individuals from developing countries.

2. International cooperation

In groundwater matters, governmental agencies of the Federal Republic of Germany cooperate in numerous projects with United Nations organizations, regional organizations and national authorities.

In cooperation with UNESCO, German agencies are engaged in the compilation and publication of the International Hydrogeological Map of Europe, scale 1:1,500,000, and contribute to projects of the International Hydrological Programme (IHP). In 1983, an International Symposium was held in Koblenz on Groundwater in Water Resources Planning, and a Workshop in Lahnstein on Groundwater in Rural Water Supply. Projects of other United Nations organizations (UNDP, WHO, UNICEF, IAEA) have, in many cases, benefited from the services of short- and long-term experts.

On the regional level, partners in cooperation were governmental and international professional organizations in Asia and the Pacific (ESCAP), Africa (ECA, CILSS, CIEH), Western Asia (ACSAD), and Europe (EEC). Worth mentioning is also the cooperation of European countries at biannual professional meetings on salt-water intrusion problems (SWIM).

Bilateral cooperation on the national level finally concentrates on projects in the frame of the technical cooperation with developing countries. A great number of projects have been assisted. Some deal entirely with groundwater, others include groundwater investigations as an essential part. The range of applications is wide, covering the fields of water supply (cities, rural areas, industry, agricultural irrigation) and groundwater investigations related to regional and water resources planning, water resources protection, environmental preservation, engineering projects and mining. The contributions were to take part in the investigations, to provide expertise, to assist in institution-building and to train personnel. Table 25 presents a list of the major bilateral groundwater projects in the field of technical cooperation with developing countries assisted by the Federal Institute of Geosciences and Natural Resources, Hannover, since 1958.

Table 25. Federal Republic of Germany: Groundwater projects assisted by Federal Institute of Geosciences and Natural Resources, 1958-1985

Country	Period	Activity	Region
Indonesia	1961, 1964	Groundwater investigations	Various islands
	1978-1985	Institution-building	Bandung
Syrian Arab Republic	1961-1962, 1966-1967	Community water supply	Western Syria
Cyprus	1963-1964	Assessment of groundwater resources	Kyrenia range
	1978-1985	Groundwater methodology and protection	Southern Cyprus
Afghanistan	1964-1970	Community water supply	Various provinces
	1971-1972	Assessment of groundwater resources for regional planning	Paktia Province
Brazil	1967-1970	Community water supply	North-east Brazil
	1971-1974	Hydrogeologic mapping	Bahia Province
	1974-1977	Assessment of groundwater resources for regional development	Piauí Province
Argentina	1969-1974	Assessment of groundwater resources for regional planning	Central Pampa region and Conlara valley
Malaysia	1974-1977	Institution-building, community water supply	West Malaysia
India	1974-1977	Institution-building, isotope hydrology	Hyderabad
Sudan	1976-1979	Assessment of groundwater resources for regional planning	Khartoum area, Red Sea coast
Djibouti	1979-1982	Assessment of water resources for development planning	Territory-wide

References

Aust, H., H. Vierhuff and W. Wagner. Grundwasservorkommen in der Bundesrepublik Deutschland. - Schriftenreihe "Raumordnung" des BMBau, 06.043: 64 S., 19 Abb., 3 Kt.; Bonn, 1980.

Bender, F. (Hrsg.). Angewandte Geowissenschaften. - III: 674 S.; Stuttgart (Enke), 1984.

Bieske, E. (Hrsg.) Handbuch des Brunnenbaus. - I, 452 p., 1958; II, 1,279 p., 1960; III, 315 p., 1965, Berlin (R. Schmidt), 1969.

Bentz, A. and H. J. Martini. Lehrbuch der Angewandten Geologie. - II, 2: 2251 S.; Stuttgart (Enke), 1969.

Clodius, S. Rahmenbericht zur Lage der Wasserwirtschaft in der Bundesrepublik Deutschland. - In Festschrift zu Kongress und Ausstellung Wasser Berlin 1968. S. 11 - 28, 4 Abb., München, 1968.

Grahmann and others. Hydrogeologische Ubersichtskarte 1:500,000. 14 Bl., m. Erl.; Remagen (Bundesministerium für Wirtschaft), 1952-1958.

Herzberg, B. Uber die Wasserversorgung einiger Nordseebäder. - Der Gesundheitsingenieur: S. 359, München, 1901.

Keilhack, K. Lehrbuch der praktischen Geologie. Stuttgart (Enke), 1896.

Keller and others. Hydrologischer Atlas der Bundesrepublik Deutschland. Atlas: 68 Kt., Erl.: 365 S., Boppard (Deutsche Forschungsgemeinschaft), 1978-1979.

Thiem, A. Der Versuchsbrunnen für die Wasserversorgung der Stadt München. Journ. f. Gasbel. u. Wasservers., 24; München, 1880.

VII. IRELAND

Area: 68,900 km^2 Population: 3.5 million

A. Physiography and climate

Topographically, Ireland consists of a wide central lowland, much of which lies below 100 metres elevation, surrounded by an irregular and discontinuous fringe of hills, with peaks in the range of 800 to 1,000 metres. Only 5 per cent of the country lies at elevations over 300 metres. Much of the central lowlands is underlain by Carboniferous limestone, covered with varied amounts of glacial drift and related Quaternary deposits. Bogs cover 17 per cent of the country. The surrounding hills are in part folded lower Paleozoic rocks, with older rocks in the north-west; there are also Caledonian and a little Tertiary granite; Armorican-folded Devonian and Lower Carboniferous rocks form the east-west trending hills and mountains of the south.

The Irish climate is temperate oceanic, and cool. Rainfall ranges from about 750 mm annually in the driest parts of the east and south-east, to over 3,000 mm in the limited mountainous areas of the west (figure 32). Rainfall is reliable and uniformly distributed throughout the year. Much comes from the south-west, that is, from the Atlantic, and is generally warm and conducive to vegetative growth. Temperatures are moderate, with a mean annual figure around 10° C. Mean daily temperatures at sealevel range from about 4° C, in January, to 16° C in July; it is rare for summer temperatures to exceed 23° C. The influence of the Gulf Stream, flowing northwards along the west coast of Ireland, is strong, and frost is almost unknown along a narrow coastal belt; inland frost can be severe. There are light falls of snow most winters, with heavy falls about once every 10 years. Evapotranspiration is high in summer, so that infiltration to groundwater occurs mainly in the winter.

Much of the central lowlands are drained by the river Shannon which has a low gradient; its entire catchment covers 14,100 km^2, or some 20 per cent of the country. Some 18,950 km^2 (28 per cent) are drained by the following rivers - Boyne (2,695 km^2) reaching the Irish Sea north of Dublin; Slaney (1,760 km^2) flowing over the Leinster Granite to reach the sea north of Wexford; the combined Barrow, Nore and Suir (9,300 km^2) breaking out where Armorican and Caledonian structures meet, and reaching the sea at Waterford; and the Blackwater, Lee and Bandon (5,190 km^2) draining due east between the Armorican fold mountains of the south. Many of the other streams are short, running from the coastal hills to the sea. A few are longer, as the Erne in the north (4,380 km^2) which partly drains from Northern Ireland; while in Donegal and Connemara, the complex geology gives rise to some peculiar drainage patterns.

For water resources development, use and management, Ireland has been divided into seven regions, with boundaries along watersheds, as on figure 32. This is most convenient for surface-water resources purposes; but it can also be used for groundwater resources studies. The Geological Survey of Ireland has prime responsibility for groundwater. An Foras Forbartha (National Institute for Physical Planning and Construction Research), the Office for Public Works, the departments of Fisheries and Forestry and of the Environment, as well as the Electricity Supply Board (ESB), all deal with different aspects of surface waters and deal to a lesser extent with groundwater.

Figure 32. Ireland: Annual precipitation, seven water regions

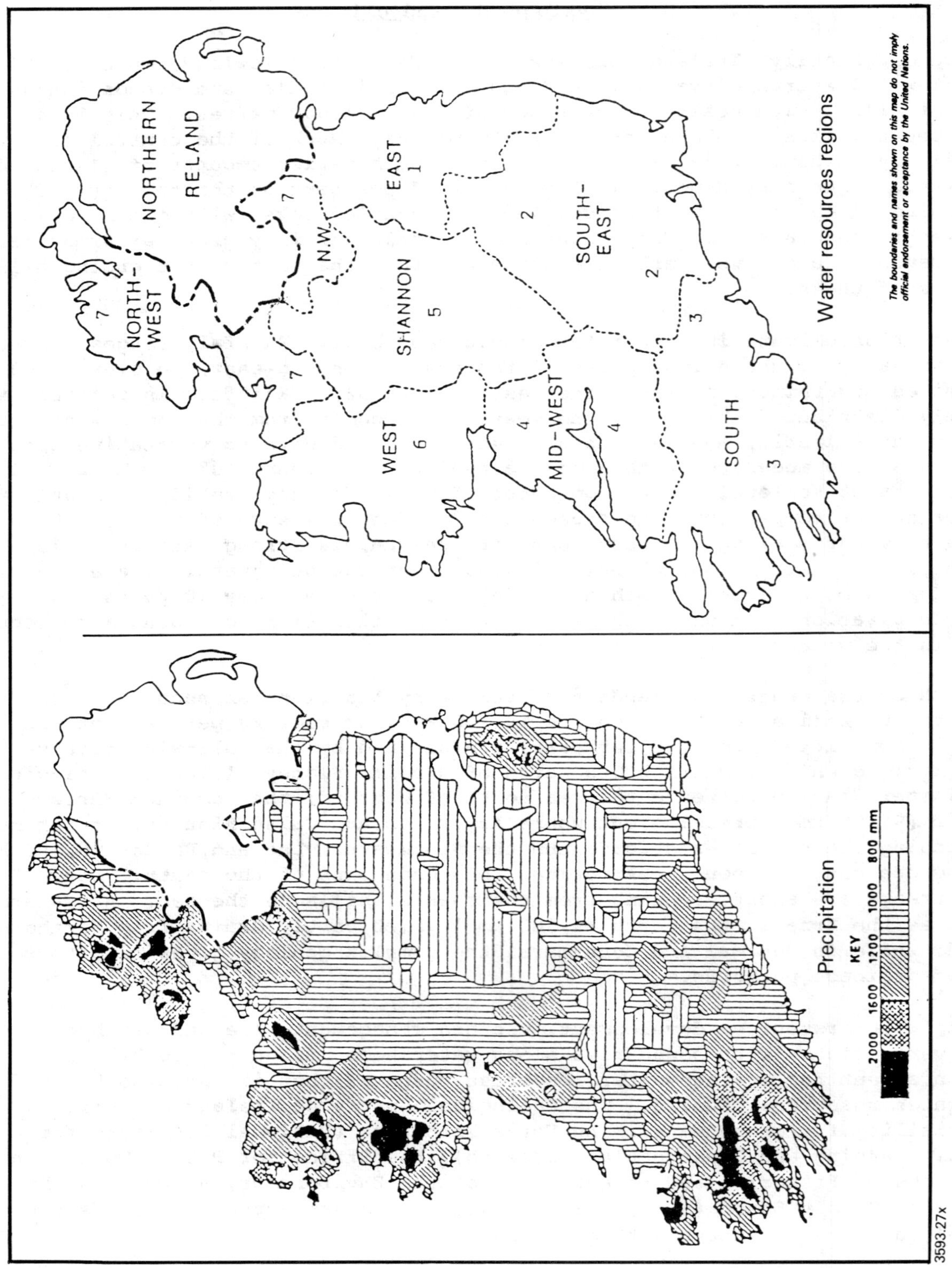

Groundwater development has been somewhat minimal largely because surface sources are abundant. Probably less than 2 per cent of the annual infiltration is extracted from boreholes and wells. However, if discharge of springs is considered to be groundwater (at the end of an underground reservoir) rather than as surface flow (at the beginning of a stream), then the amount of groundwater used for urban and domestic supply can be termed large. In this rather uncertain state regarding groundwater development and use, only rough estimates can be provided. Table 26 sets out relationships between population and availability of water.

The main components of the hydrologic cycle for Ireland are summarized in table 27. The volume of water falling on Ireland is nearly 78.7 km^3 per year. Of this, surface runoff has been calculated as about 51 per cent of the total precipitation. Groundwater infiltration has been tentatively calculated at nearly 14 per cent of the total precipitation. The balance is attributed to evapotranspiration, as 35 per cent of precipitation. However, as surface runoff contains an element of groundwater, the 51 per cent should be reduced by a considerable amount, and this amount added to the 14 per cent attributed to groundwater.

Climate, with emphasis on precipitation, is measured by the Meteorological Service, which maintains some 770 stations, about 90 equipped with automatic rain recorders. Figure 32 is the precipitation map of Ireland. The annual rainfall averages 1,100 mm, distributed rather uniformly over the 12 months. While monthly figures may vary sharply, the total annual variation almost never exceeds 30 per cent of the long-term average.

Only the driest part of Ireland (east and south-east) falls within the semi-humid region; most of Ireland lies within the humid region.

Figure 33 gives data on Irish precipitation, temperature and evaporation at Valencia Observatory, Cork University, and Phoenix Park, Dublin. Temperature and precipitation need no comment, although the fact that minimum monthly rainfall occurs in April is somewhat unexpected. The problem of evaporation and evapotranspiration is of more importance in studying the Irish water balance. The average evaporation from Class A pans at 11 stations scattered over Ireland was 641 mm; of this, 76.6 per cent occurred in the six summer months of April through September, and only 23.4 per cent in the winter months, October through March. This A Pan evaporation for Valencia is plotted on figure 33, and shows that for June, July and August, potential evaporation exceeded rainfall. Converting Class A Pan evaporation to evapotranspiration presents problems.

Turning to table 27, and allowing for the fact that Class A Pan evaporation is from a free supply of water all the year, the fact that the calculations by the Penman formula, find 404 mm as the annual evapotranspiration is quite acceptable. The Class A Pan evaporation of 641 mm per year should be reduced by 37 per cent to get the field figure for evapotranspiration of 404 mm. The fact that 76.6 per cent of evaporation occurs in the six summer months, indicates that evapotranspiration is also so distributed. This in turn implies that recharge to groundwater is small in summer, when evapotranspiration can easily exceed rainfall. In fact, in calculations of infiltration there is no groundwater infiltration in June, July and August.

Table 26. Ireland: Relation of population to availability of water

Aspect	Water resources region							All Ireland
	Eastern 1	South-eastern 2	Southern 3	Shannon 4	Mid-western 5	Western 6	North-western 7	
Area (km^2)	7 878	12 708	11 992	10 834	7 644	8 560	9 354	68 900
Population (1983)	1 373 000	482 000	520 000	294 000	247 000	261 000	263 000	3 440 000
Surface runoff (km^2)	3.595	6.269	7.427	5.629	3.034	5.872	7.307	40.033
Average runoff (m^3/s)	114.0	198.8	235.5	178.5	124.8	186.2	231.7	1 269.5
Average runoff per capita (litres per head per day)	7 174	35 635	39 130	52 457	43 637	61 639	76 117	31 885
Low flow runoff (m^3/s)	5.8	15.2	11.2	10.0	5.0	4.7	6.2	58.1
Low flow runoff per capita (litres per head per day)	365	2 725	1 911	2 939	1 749	1 556	2 037	1 459

Table 27. Ireland: Elements of the hydrologic cycle

Aspect	Water resources region							All Ireland
	Eastern 1	South-eastern 2	Southern 3	Shannon 4	Mid-western 5	Western 6	North-western 7	
Area (km^2)	7 878	12 708	11 922	10 834	7 644	8 560	9 354	68 900
Precipitation (annual) (mm)	954	1 009	1 383	1 003	1 141	1 205	1 281	1 142
Evapotranspiration (mm)	350	366	502	383	374	359	343	404
Surface runoff (mm)	457	492	623	519	515	686	782	581
Infiltration (mm)	147	151	158	101	252	160	156	157
Precipitation (km^3)	7.511	12.818	16.488	10.862	8.717	10.314	11.974	78.684
Evapotranspiration (km^3)	2.760	4.632	7.172	4.144	2.860	3.073	3.206	27.847
Surface runoff (km^3)	3.595	6.269	7.427	5.629	3.934	5.872	7.307	40.033
Infiltration (km^3)	1.156	1.917	1.889	1.089	1.923	1.369	1.461	10.804
Evapotranspiration (%)	36.7	36.1	43.5	38.2	32.8	29.8	26.8	35.39
Surface runoff (%)	47.9	49.0	45.0	51.8	45.1	56.9	61.1	50.88
Infiltration (%)	15.4	14.9	11.5	10.0	12.1	13.3	12.1	13.73

Figure 33. Ireland: Relation between precipitation, temperature and evapotranspiration at three locations

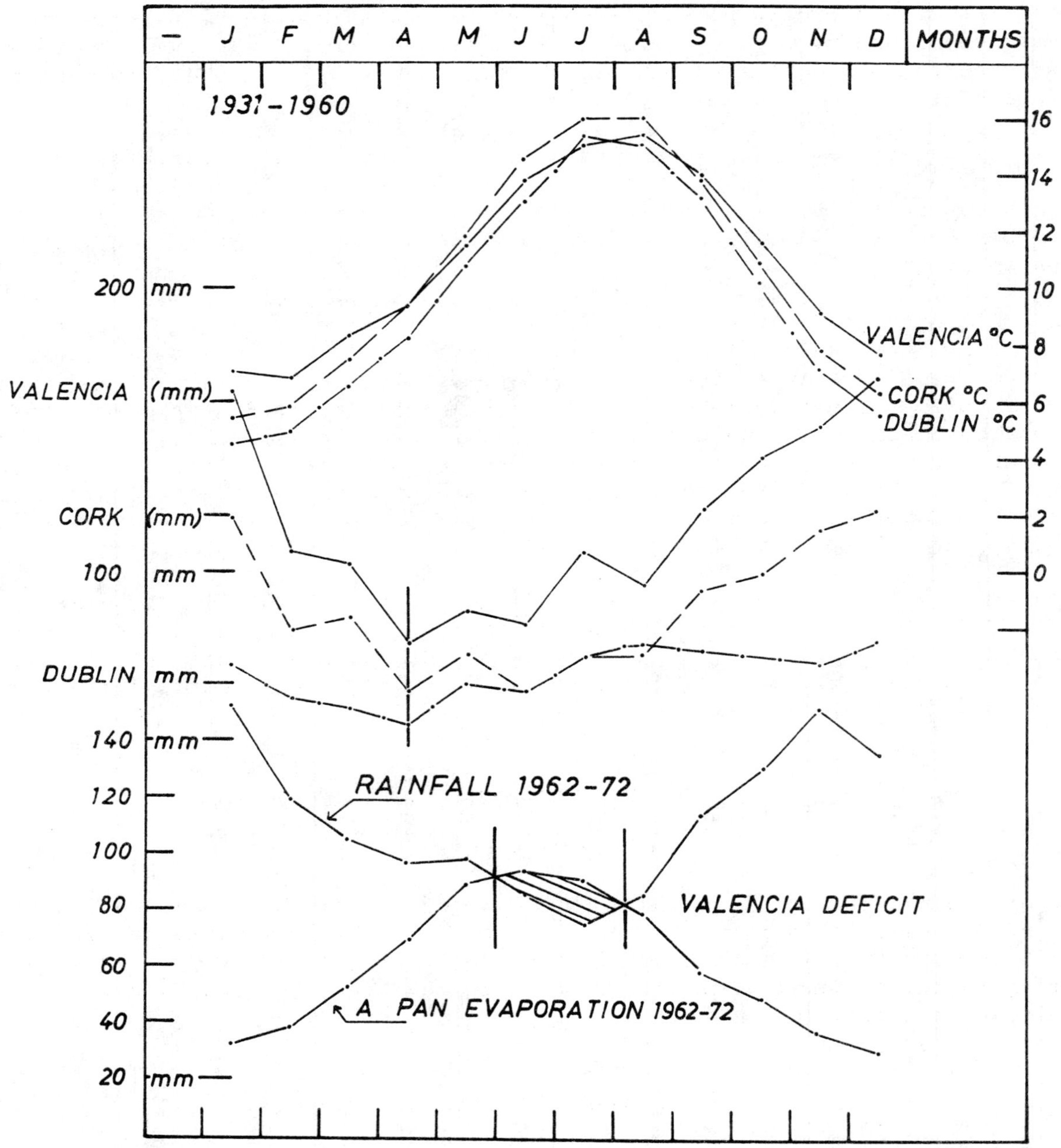

In Ireland, surface-water resources have been well studied by the Board of Works for a long period, with some emphasis on flood flow and arterial drainage. More recently, An Foras Forbartha has studied surface water in more detail, with emphasis on low flows, pollution and volume of water available for development. The ESB and local authorities also gauge flows as required for their own purposes. By May 1982, there were 536 autographic flow gauging stations, in both Ireland and Northern Ireland. The surface-water resources of Ireland are large, include many lakes, and are only slightly polluted; this abundance of surface water has delayed groundwater development, although many farms and rural areas rely on groundwater. An Foras Forbartha has published extensively, as for 1975-1978, and more data is being processed.

As already noted, Ireland is divided into seven Water Resources Regions, as shown on figure 32 and listed in table 27. The most recent figures for surface-water resources in relation to population and development potential are given in table 26, based on the June 1983 report by An Foras Forbartha, Review of Water Pollution in Ireland.

Table 27 lists the surface runoff for the seven Water Resources Regions and for the country as a whole. The percentage of total precipitation which forms surface runoff varies from 45 per cent in the South Region to 61.1 per cent in the North-West; while for the whole of the country it averages 50.88 per cent. These figures are based on flow gaugings and hydrograph analysis; they include all water flowing past the gauges of which some represent groundwater discharging into the lakes, rivers and streams from seepages, concealed springs and open springs. There may be some recharge to groundwater from the bed of streams, rivers and lakes; this is probably small.

When the low figures for groundwater are compared with the high figure for surface runoff, it is clear that some of the latter, estimated at 20 per cent at least, should be deducted from surface flow and added to groundwater infiltration. Making this transfer brings surface water down to 40.7 per cent and infiltration to groundwater up to 23.91 per cent.

B. Geology

1. Background

The Geological Survey of Ireland, established in April 1845, mapped all of Ireland and Northern Ireland in the 45 years to 1890. Field mapping was at the scale 1:10,560, and the resulting Memoires of the Geological Survey of Ireland still provides the definitive background data to all ongoing geologic and hydrogeologic research.

2. Structural geology

Pre-Caledonian tectonic structures are mainly deeply buried in Ireland, except in the Donegal-Connemara region and in a limited exposure around Carnsore Point in the south-east. The oldest identified structures are the gneisses of Inistrahull, dated 1,500 million years (m.y.) before the present (BP). In North Mayo, amphibolite facies gneiss striking north-west is dated 1,070 to 1,000 m.y. In the Ox mountains, there is reworking of older gneisses, so that hornblende from them has been dated 700 m.y.

The emplacement of the Carnsore Granite is dated about 550 m.y.

The Caledonian tectonic cycle (early, 700-460 m.y.; later, 590-400 m.y.) has had major influences on the geology of Ireland. During the Cambrian and Silurian, the north and south of Ireland were separated by the wide Iapetus (proto-Atlantic) Ocean. The plates converged during the Silurian, narrowing the Iapetus Ocean; and they met along what is now the deeply buried Iapetus Suture in the Silurian and Devonian rocks.

To the north of the Iapetus Suture, an early Caledonian orogenic cycle (700-460 m.y.) formed the Orthotectonic Caledonides, which crop out in the Dalradian and Murrisk zones, and a younger cycle (590-400 m.y.) which formed part of the Paratectonic Caledonides, lying to the south-east in the Murrisk Zone. To the south of the Iapetus Suture in Wicklow and Wexford there were extensive extrusions of submarine tholeiitic and calc-alkalic basalt and andesite, followed by rhyolite in the Ordovician. But the main expression of the Caledonian orogeny is the Leinster Granite (404 m.y.), covering some 2,000 km^2, and forming five large and a few smaller outcrops. The other main outcrop of Caledonian granites is to the west of Galway (some 600 km^2), comprising one main and four smaller outcrops.

The Hercynian or Armorican orogenic cycle affects the south of Ireland and varies from intensive east-west folding south of a line from Dungarvan to Dingle through decreasing folding northwards; it has little effect on the Midlands of Ireland, where the Devonian and Carboniferous beds are almost horizontal. Folding by Armorican compression tends to die out entirely to the north of the Iapetus Suture; but faulting, commonly block faulting, is not unusual and is commonly attributed to Armorican stress relief. In their strongest expression, the Caledonides strike north-east, while the Armorican folds strike east. However, in Central Ireland, the two strikes are nearly parallel.

Since the end of the Hercynian orogeny, tectonic activity has been slight, and mostly related to the opening of the North Atlantic. In the north-west, however, there were outpourings of the Antrim basalts.

3. Stratigraphy

(a) Pre-Quaternary

In Cambrian, Ordovician and Silurian rocks, marked differences exist between the sediments to the north and south of the Iapetus Suture. In the North-west Plate, Pacific and American graptolites, trilobites and shelly faunas have been found in Lower Ordovician rocks of the Murrisk Zone. In the South-east Plate, the Lower Ordovician fauna are of European (Acado-Baltic) provinciality. In the north, clastic sediments and submarine volcanics predominate, but there are also psammite and quartzite, pelite and thin limestone beds; there are great thicknesses of greywacke and shale. There was a basal tillite at the beginning of the Cambrian. Most of these lower Paleozoic sediments in the North-west Plate have been subjected to low-grade (greenschist and even amphibolite facies) metamorphism; this is not so in the South-east Plate. Here the rocks are greywacke, siltstone and quartzite, and extensive submarine volcanic activity affected the Ordovician.

The Devonian and Carboniferous of the upper Paleozoic (covering some

72 per cent of the country) differ considerably from the lower Paleozoic sediments, and the Caledonian granites. There was considerable erosion between the two eras, and the Leinster Granite was unroofed in places before the Old Red Sandstone of Devonian age was unconformably deposited. Much of the Old Red Sandstone was deposited under terrestrial, desert conditions. The Lower Carboniferous was deposited in a deepening sea which slowly expanded northwards throughout most of Carboniferous time; the rate of advance has been estimated at 30 km per million years.

The Devonian can be described under three regional areas. In the Munster basin, there are thick red-bed sequences, well exposed in the mountains of Cork and Kerry; near the top of the Devonian sequence there are some marine estuarine and deltaic beds in the extreme south of this Munster basin. In the south-central plain basin, the Old Red Sandstone outcrops as rims around the lower Paleozoic in the hills and low mountains. In the north of the country, there are scattered Old Red Sandstone outcrops; in the extreme north, the Old Red Sandstone facies extend into the Dinantian of the Carboniferous. The Iapetus Suture has been covered and is not reflected in the Devonian and Carboniferous sediments.

Lower Carboniferous formations occur widely in the central portion of Ireland, covering 53.3 per cent (36,850 km^2) of the country. Most of these rocks are Dinantian. Marine conditions were initiated in the south Munster basin in Late Devonian times with shallow water non-carbonate deposits followed by beds of calcareous mudstone to the west, and dark mudstone, chert and limestone to the east. Eventually dark muds were deposited in the Brigantian stage, with thick basinal sequences and thinner shelf deposits on the edges of the basin.

In the Central Region, vast sheets of carbonate-dominated sediments extend from the Cork-Kenmare line in the south to the lower Paleozoic massif in the north.

In the Northern Region, Carboniferous rocks are found in several basins between the Donegal-Connemara and the Longford-Down massifs. The North-west Basin is the main one, with subsidiary ones in North Mayo, Sligo syncline, Donegal syncline and Omagh syncline. Above the basal sandstone and siltstone, there are extensive beds of limestone, shale and argillaceous limestone.

The Namurian covers some 5,700 km^2. Westphalian rocks have very limited outcrops. In Namurian times, sand, silt and mud derived from northerly, western and south-western sources were deposited in subsiding troughs and basins separated by structural highs which received much less sedimentation. The most extensive Namurian outcrops are along the west coast from the Burren across the estuary of the Shannon to the Lakes of Killarney. The sediments deposited here were eroded from what are now the Appalachians, before the opening of the North Atlantic. The Namurian is based by the Clare Shale Formation, in an abrupt change from shelf limestone to black mudstone; this is overlain by the Ross Sandstone Formation and by the Gull Island Formation.

The Permian and Mesozoic deposits were either never present or have been eroded from the land areas of Ireland. Downfaulting at Kingscourt has preserved a Permian-Triassic remnant, the lower Kingscourt Gypsum Formation and the Upper Kingscourt Sandstone Formation, with a combined thickness of some 600 metres. Offshore oil investigations have revealed thick Mesozoic deposits in the Kisk Bank basin, the Celtic Sea grabens, the Fastnet basin, the Porcupine basin, the Donegal basin and the Rockall trough.

Towards the end of the Cretaceous, there was widespread regional uplift and local downwarping in north-east Europe. The ocean retreated during the Late Cretaceous and in the early Tertiary. The continental plate, comprising Rockall and western Europe, finally separated from the Greenland plate. There were extensive outpourings of plateau basalts; some 900 metres thickness of basalts were erupted, all in Northern Ireland. Dykes associated with these Tertiary basalts occur in the south, some quite recent. In the Tertiary also, complex granitic and basic intrusions occur.

(b) Quaternary

The Quaternary geology of Ireland is complex, dominated by the formation and subsequent melting of successive icecaps. Some 96 per cent of the country is covered by glacial deposits and bogs; only some 4 per cent, essentially the higher hills, is free of glacial drift. Their maximum thickness in few places exceeds 100 metres; thicknesses of 15 to 5 metres are usual. The topography was moulded by ice and subsequently reformed by water, both ice-melt and rainfall; this has resulted in a complex topography presenting many drainage problems, especially on the minor topographic scale. The Quaternary geology of Ireland has been well studied. The Irish bogs are also well studied, from the very early 1685 paper by King to that of Hammond (1981).

Only two glaciations are known with certainty in Ireland. The older, Munsterian, is now exposed only in Munster. Elsewhere, it is covered by the Midlandian glaciation, which covered much of Ireland excluding Munster; the last glacial maximum was about 15,000 BP. By then sealevel was some 130 metres lower than now. Over the subsequent 5,000 years the climate warmed, the ice melted, lakes and marshes formed as the liquid drainage asserted itself, flora and fauna migrated from the continent, and from about 10,000 BP, the Littletonian Stage came into being. There was isostatic uplift after removal of the ice load from the Baltic Shield region.

Several types of glacial deposits are, therefore, found in Ireland. The older boulder clay in Munster commonly has its carbonate rock dissolved away; it covers 13.5 per cent of the country. The newer boulder clay covers most of the Midlands, and in all some 27.5 per cent of the country. The Irish Sea ice, pushing material gouged from the sea floor onto the Wicklow-Wexford coast, is locally of importance. In many areas, especially in the north, the glacial drift, mainly boulder clay, has been moulded into drumlins. Glaciofluvial material, gravel, sand and silt, with some boulders and clay, was washed out by the melting ice; it gives rise to much complex micro-topography, and covers 19 per cent of the country, mainly in the Midlands. Moraines and eskers cover only small areas, but give rise to striking topographic features. In all, bog covers some 17 per cent of the country, but much of this is thin, known as blanket bog, and grows over the mountains in the areas of heaviest rainfall in the west. Thick bog, known as raised bog, is found mainly in the Midlands, where much of it grew from shallow post-glacial lakes. In many areas, these different types of drift are intermingled in a complex manner, difficult to map and untangle. The soils forming on them have been mapped in great detail by An Foras Taluntas (Agricultural Research Institute), as in the Atlas of Ireland (Royal Irish Academy, 1979).

C. Hydrogeology

1. Assessment of potential

Calculation of the overall annual infiltration to groundwater and of the capital reserves of groundwater held in Irish aquifers is still at a rudimentary stage. Three estimates of annual infiltration have been made; they are summarized in table 28. In 1979, Burdon and Aldwell estimated the average annual groundwater as just over 22 km^3; this was based on analyses of the regression curves on the main rivers. In 1982, EEC published the estimate by Wright and others of known and readily developable aquifers and groundwater resources; this totalled only about 3.5 km^3, and referred to less than 27 per cent of the land area of the country. In 1983, Burdon supplemented this by a hydrogeologic study of the rest of Ireland (primarily for hydrogeothermal purposes) and estimated an additional 7.35 km^3 of groundwater. Combining Wright (1982) and Burdon (1983) the groundwater resources of Ireland amount to some 10.8 km^3 per year. These discrepancies call for an explanation.

The 1979 estimate was based on analyses of river hydrographs by the classical method first set out by Maillet in 1905, as elaborated by Webber (1961) and others. An Foras Forbartha's (April 1978) two volumes, Hydrographs of Daily Mean Flows for 1975 and Tables of Daily Mean Flows for 1975 were, with the earlier (June 1977) publication (7 volumes) Summary of Hydrometric Records then the only publications dealing with the overall water resources of Ireland. The resulting figure of groundwater discharge, at 22 km^3 per year (or 31.8 per cent of overall precipitation on the mainland) is probably too high. In separating base groundwater flow from overland flow on the graphs, possibly too much was given to groundwater flow and too little to overland flow. In addition, this method of analysis contains an element of estimation of the groundwater resources behind the springs and seepages under analysis. Hence the more than 22 km^3 of estimated groundwater may have contained an element of groundwater resources (held underground) as well as the figure for actual discharge of groundwater. The extent of the error may be about 20 km^3.

The November 1979 report by Wright, Aldwell, Daly and Daly, published by EEC in 1982, dealt solely with groundwater which had been surveyed in varying detail and which could readily be developed and used. It is defined as surveyed groundwater. The total volume came to less than 3.5 km^3 per year. These volumes were obtained by multiplying the surveyed areas of the outcrops, both rock and unconsolidated Quaternary, by the estimated infiltration. This was calculated by estimating effective rainfall for each area, based on data from the Meteorological Service and An Foras Forbartha, measuring or estimating the permeability with infiltration factors of 80, 50, 20 and zero per cent, with special allowances in some areas, as karst conditions, sea-water intrusion potential and high runoff from rugged terrain. Based on these approaches, this surveyed groundwater occurred in a rock area of 15,098 km^2, had an infiltration of 181 mm (or 29.9 per cent of effective precipitation) and amounted to just over 2.7 km^3 per year. The surveyed groundwater in the Quaternary covered an area of 3,643 km^2, had an infiltration of 198 mm (or 35.6 per cent of effective precipitation) and amounted to 723 million m^3 per year. The total, as in table 28, came to nearly 3.4 km^3 per year. The total area covered by all these aquifers amounted to 18,741 km^2, or 27.2 per cent of the 68,900 km^2 of Ireland. Because in some cases, the Quaternary aquifer overlies a solid rock aquifer, the area of Ireland covered by these surveyed groundwater resources was about 25 per cent of the total area.

Table 28. Ireland: Three estimates of groundwater infiltration

Sources of calculation	Volume of infiltration to groundwater (Cubic kilometres per year)							
	Eastern	South-eastern	South-ern	Shannon	Mid-western	Western	North-western	All Ireland
Aldwell & Burdon (1979)	2.157	3.761	3.342	2.165	3.659	2.936	4.018	22.038
Wright <u>et al</u>., for EEC (1982)	0.209	0.764	0.628	0.502	0.488	0.650	0.213	3.454
Burdon (for Geothermal) (1983)	0.947	1.153	1.261	0.587	1.435	0.719	1.248	7.350
Combined Wright & Burdon	1.156	1.917	1.889	1.089	1.923	1.369	1.461	10.804

The December 1983 survey by Burdon was designed to estimate the groundwater resources of the rest (73 per cent) of Ireland, and to add these estimates to the accurately surveyed figure of nearly 3.4 km^3 per year. This groundwater is defined as estimated groundwater. The calculation was complex and the areas covered by the surveyed groundwater resources had to be excluded. The amount of infiltration to groundwater had to be estimated based on all available data. The amount of groundwater held in the Quaternary cover, and the amount infiltrating to rock aquifers where such underlaid the Quaternary, had to be calculated. The resultant figures for the seven Water Resources Regions are summarized in table 28, with a total for Ireland of 7.35 km^3 per year.

When the results of the survey by Wright and others and the Burdon 1983 estimation are combined, the total comes to about 10.8 km^3 per year, as in table 28. These figures are broken down by 13 rock formations in table 29, and by eight types of Quaternary cover in table 30. In turn, these combined figures have been transferred to table 27, and so are used in estimating the overall water balance in Ireland. It can be seen that the surveyed and estimated totals are low, summing only as 13.37 per cent of the annual precipitation. It is clear that too much caution has been used in making these surveys and estimates; a figure of 20 km^3 per year of infiltration to the groundwater (25 per cent of the annual precipitation) would probably be more nearly correct. However, there are two main considerations, and some minor ones, which indicate that the annual recharge to the groundwater is indeed less than might be expected.

There is no infiltration to groundwater in the summer months of June, July and August. This is because vegetative growth is then rapid and vegetative cover universal over Ireland. Temperatures are high and, though relative humidity is also high, transpiration and evaporation absorbs all the rainfall in these three months. This is particularly so for the three types of boulder clay and the drumlins, which cover more than 44,000 km^2, or 64 per cent of the country; they protect the underlying rock from summer infiltration. Comparisons between the effects of the 1975 and 1976 droughts in England and Ireland support the concept that summer infiltration is low (Burdon, 1978).

In winter, there is some rejected recharge, preventing further infiltration to aquifers which in Ireland are shallow. For the rock aquifers, karstification of the Carboniferous (infiltrating 67 per cent of the rock groundwater) seems not to exceed 100 metres. In only one area (the Curragh) are the glaciofluvial deposits (infiltrating 55 per cent of the Quaternary groundwater) thicker than 100 metres; elsewhere, thickness of less than 35 metres is usual. With strong infiltration in the months October to January, some aquifers are saturated and reject further infiltration; this leads to drainage problems as noted by Burdon (1984).

Although high evapotranspiration in summer and rejected recharge in winter appear to be the two main reasons why groundwater recharge is on the low side in Ireland, other influences also lead to low infiltration. Weathered rock is rare, stripped by the glaciers, except above the level of the ice sheet. The granite, metamorphics and also most of the Cambrian, Ordovician and Silurian sediments have low infiltration; they cover some 17,200 km^2 or 25 per cent of the country. Infiltration into the Devonian is also problematic, and it covers 12,800 km^2. Almost nothing is known of the hydrogeology of river beds and lake bottoms, which may be regions of either discharge or infiltration.

Table 29. Ireland: Annual infiltration to principal bedrock aquifers

Rock formation	Area of outcrop (km^2)	Volume of infiltrating groundwater (m^3/year X 10^6)		
		Surveyed	Estimated	Total
Granite	4 125	2	118	120
Basic intrusives	42	-	3	3
Volcanics (Ordovician)	846	115	1	116
Marble	43	-	3	3
Quartzite	1 081	-	16	16
Metamorphics	5 135	-	105	105
Cambrian	727	-	7	7
Ordovician	4 015	-	137	137
Silurian	3 218	-	112	112
Devonian	12 792	271	753	1 024
Carboniferous	31 150	2 183	1 999	4 182
Namurian, etc.	5 700	157	276	433
Permo-Triassic	26	2	-	2
Ireland total	68 900	2 730	3 530	6 260

Table 30. Ireland: Annual infiltration to Quaternary aquifers

Quaternary formation	Area of outcrop (km^2)	Volume of infiltrating groundwater (m^3/year X 10^6)		
		Surveyed	Estimated	Total
Boulder clay - New	18 976	-	222	222
Boulder clay - Old	9 292	-	325	325
Boulder clay - Irish Sea ice	4 260	-	222	222
Drumlins (boulder clay)	11 597	-	170	170
Glaciofluvial	13 144	636	1 916	2 552
River alluvium	3 152	83	585	668
Moraines, eskers	1 147	5	265	270
Raised bog	4 778	-	117	117
Uncovered rock	2 544	-	-	-
Ireland total	68 900	724	3 822	4 546

In Ireland, the average rates and distances of flow for groundwater are not well known. Wright, Daly and Daly (1983) dealing with vulnerability to pollution, give seven travel time categories ranging from more than 20 years to less than a week. Burdon (Dec. 1983) constructed a rough model of Irish groundwater, dealing with the water in the uppermost 100 metres of aquifer. Maximum flow distances were 4 km, and average 2 km. All of Ireland is covered with a network of watercourses fed by seepages and springs. Average flow rates were of the order of 2 to 3 metres per day, so that total flow time averaged two or three years. The 100 metres of aquifer consisted of 10 metres of Quaternary cover, with 4.2 per cent porosity and 90 metres of limestone (karstified and non-karstified) with 1.2 per cent porosity; overall porosity was 1.5 per cent. The depth of annual infiltration worked out at 10.5 metres, and the years of infiltration held in the aquifer totalled 9.5 years, or 102.5 km^3, with annual infiltration held at 10.8 km^3 per year. The flow rate was some 735 metres per year, and the time to clear out the aquifer about three years. The permeability was calculated to be 2.43 darcys.

Springs in Ireland generally respond rapidly to high infiltration following heavy rainfall; this does not mean that the infiltrated water moves through the aquifer and reaches the spring in this short time. Pressure is transmitted quickly, almost instantaneously, through confined and semi-confined aquifers; it is the increased pressure which causes the increased spring discharge, not the arrival of the newly infiltrated water. In regions of karstic flow, where there are true underground rivers, flows at the rate of several kilometres per day occur; the infiltrated water physically and quickly reaches its discharge spring many kilometres away.

Man-influenced intrusion of sea water into coastal aquifers does not occur significantly in Ireland. Tides as high as six metres occur, and both the sea water and the overlying fresh water are affected by this phenomenon. Some turloughs, or karst lakes, whose levels are a function of phreatic surfaces, are affected by tidal phenomena, even at a considerable distance inland.

2. Principal aquifers

(a) General

Figure 34 shows the major aquifers and aquicludes of Ireland. It is based on sheets B-3 and B-4 of the International Hydrogeological Map of Europe (scale 1:1,500,000) compiled between 1974 and 1979, when knowledge of the hydrogeology was somewhat rudimentary.

In this section, the solid-rock formations and the unconsolidated deposits are presented, with their aquifer properties. The description of formations is based on the 1963 Geological Map of Ireland, with the Quaternary geology based on Singe (1950, 1979 and 1980). To facilitate transfer of information from the more general list of aquifers, aquitards and aquicludes of the legend of figure 34, table 31 has been compiled showing in general terms the equivalent rock and unconsolidated equivalents, as well as noting formations and unconsolidated cover not identified on figure 34.

Figure 34. Ireland: Hydrogeology

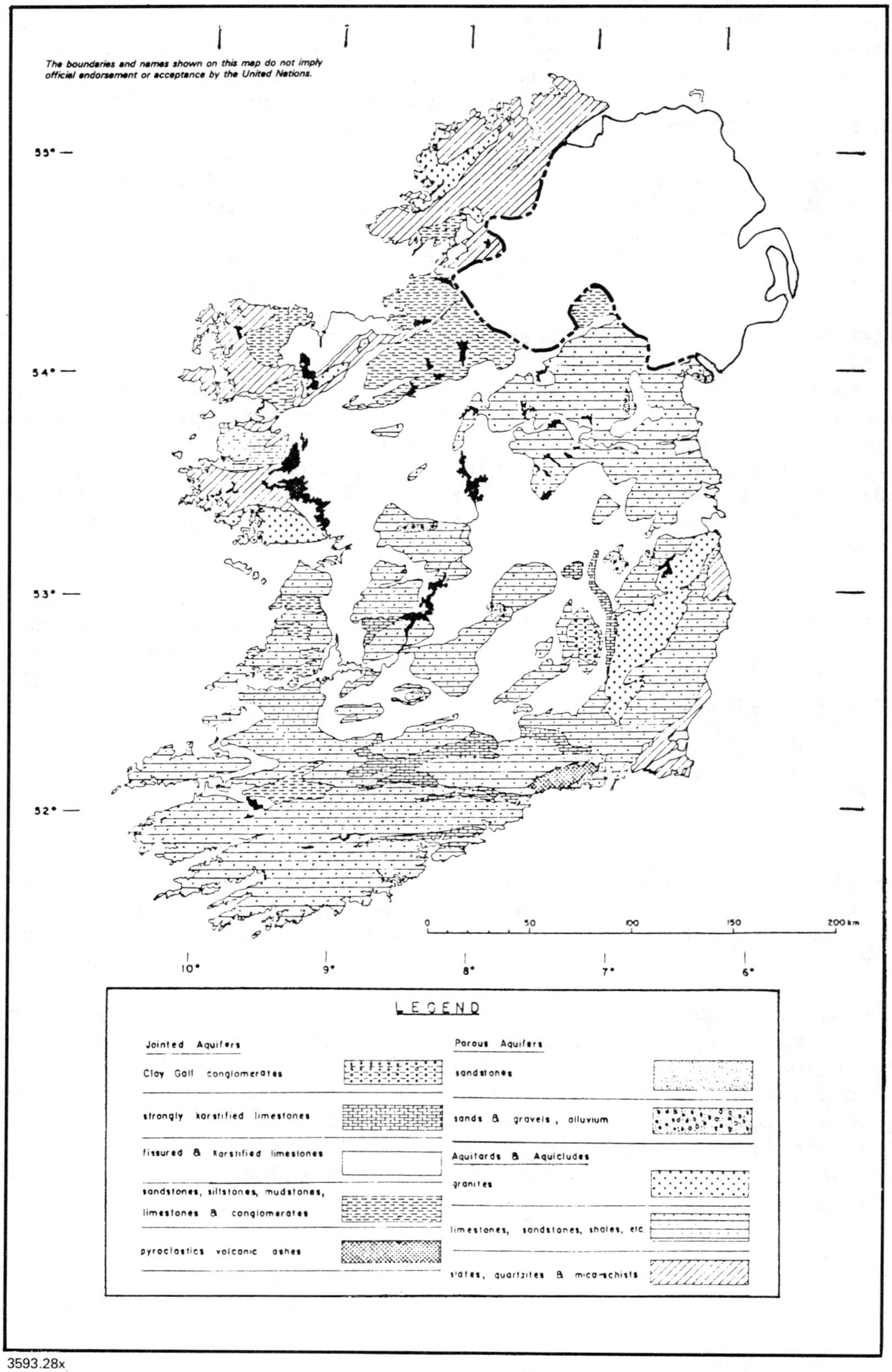

3593.28x

Table 31. Ireland: Lithology and distribution of principal hydrogeologic units

Hydrogeologic unit	Distribution
A. Jointed aquifers	
1. Clay Gall Conglomerate	Namurian/Westphalian, Castlecomer.
2. Strongly karstified limestone	Carboniferous of Munster synclines; Barrow valley.
3. Fissured and karstified limestone	Widespread; Central Plain, Killarney and Tralee, North Mayo, Donegal, Cavan-Louth, and minor areas.
4. Sandstone, siltstone, mudstone, limestone and conglomerate	Widespread; Tralee, Co. Clare in part, North Mayo, wide region from Co. Roscommon to L. Neagh.
5. Pyroclastic volcanic ash	Belt 110 km long in Wicklow, Wexford and Waterford.
B. Porous aquifers	
6. Sandstones	Small.
7. Sands and gravel; alluvium	Curragh of Kildare; some valleys; narrow belts bordering rivers.
C. Aquitards and aquicludes	
8. Limestone, sandstone and shale	Widespread; mainly Devonian in south, the Ordovician and Silurian in south-east, Midland hills and the Longford-Cavan-Monaghan horst, extending south into Meath.
9. Slate, quartzite and micaschist	Cambrian of Bray and Wexford; metamorphics of West Galway and West Mayo; much of Donegal.
10. Granite	Leinster, Galway, Donegal and Mourne Mts.

Tables 29 and 30 summarize our knowledge of the rock and unconsolidated aquifers and aquicludes of Ireland, and are the basis of the following descriptions. In so doing, most of the emphasis is on lithology and permeability; less attention is paid to structures, faults, fissures, dyke intrusions and other tectonic and structural factors which have less influence on behaviour of the aquifer, although important when siting production boreholes. Karstification is limited to the Carboniferous rocks, with minor aspects in the Namurian and the old boulder clay.

(b) Granite

Much of the granite areas (4,125 km^2) remained above the icecap, and were weathered during glacial times, so that deep weathered granite is not unusual. For the Leinster Granite, there is some hydrothermal and pneumatolytic alteration; the south-east extension of the granite, buried under terrestrial Devonian rocks, may be deeply weathered and contain much groundwater. There is considerable faulting and fracturing. At Turlogh Hill, calcite is a major component of the granite (either primary or secondary), and should undergo karstification and weathering to some depth. The other granites of Ireland have also their local peculiarities, and some useful amounts of groundwater may yet be found and developed in granite, using standard geophysical and hydrogeologic methods. Thus, groundwater has been developed successfully from felsite dykes and lenses in the Galway granite.

(c) Volcanic rocks

Ordovician volcanics (846 km^2) have some potential as aquifers, and have yielded fair amounts of high quality groundwater. They are listed under Pyroclastic volcanic ashes on figure 34 and in table 31. These metamorphosed Caradocian rocks occur in a belt 110 km long; they were submarine extrusives of which 70 per cent are rhyolitic. Much of their groundwater is high in sodium carbonate, with comparatively high potassium. Yields of 2-10 l/s have been obtained from several dozen boreholes in these volcanic rocks.

(d) Metamorphic rocks

Metamorphic rocks (5,135 km^2) occur in the West and North-West Water Resources Regions, and are quite varied. They are listed under Slate, quartzite and micaschist on figure 34 and in table 31. They include amphibolite gneiss on Inistrahull and the similar amphibolite gneiss of Grenvillian in North Mayo. High-pressure granulite facies occur in the Ox Mountains and around Rosses Point. These and other metamorphic rocks of Connemara and Donegal have had a complex history, and are much studied by petrographers. They occur in regions where precipitation often exceeds 2,000 mm per year, and form steep and rugged terrain, from which runoff is high and rapid. In many cases they are covered with mountain bog and glacial drift may be thin or absent. Although occasional interaction of petrographic, structural, weathering and topographic factors may favour limited occurrences of groundwater, in general, these metamorphic rocks contain none. The numerous small lakes, streams and springs are usually quite adequate for the small water demands of this region.

(e) Cambrian

Cambrian formations (727 km^2) are restricted to the East and South-East Water Resources Regions. They are included under Slate, quartzite and micaschist on figure 34 and in table 31. They are metamorphosed and generally impermeable.

(f) Ordovician and Silurian

Ordovician (4,015 km^2) and Silurian (3,218 km^2) formations may be classed together for hydrogeologic purposes. They are listed under Limestone, sandstone and shale on figure 34 and in table 31. They occur mainly in the south-west, as cores to many of the hills and mountains in the south and Midlands, and in the horst which occupies much of Longford, Cavan and Monaghan and into Down. These rocks are mainly impervious, yields range mostly from 1 to 10 litres per minute, but good siting in the Longford-Down massif can result in yields of 10-15 litres per minute.

(g) Devonian

Devonian, or Old Red Sandstone (12,792 km^2), is extensive in the south, outcropping under most of Cork and Kerry, and into Waterford. It surrounds or forms all the hills and mountains of the Central Plain. It is buried under the Carboniferous throughout the Central Plain. The Devonian is included under both aquitards, Limestone, sandstone and shale and under jointed aquifers, Sandstone, etc. on figure 34 and in table 31. As indicated in table 29, they contain a little surveyed groundwater (271,000 m^3 per year) and some estimated groundwater (753,000 m^3 per year). The rocks are mainly conglomerate, coarse and fine sandstone and silt; also some clay. They are essentially terrestrial deposits, laid down under arid to semi-arid conditions; shallow lakes occurred in which clay and silt were deposited. They were folded and slightly metamorphosed during the Armorican orogeny; in places the sandstone is almost true quartzite. In many ways, they resemble the sandstones of the Nubian Facies of the Sahara and North Africa which lead many to think that deep boreholes in the Midlands would tap abundant groundwater under high pressure in the Devonian. In its outcrop areas, it is possible to site good boreholes, but resistivity geophysical surveys, as well as careful hydrogeologic investigations, would be necessary.

(h) Carboniferous

Dinantian rocks (Tournaisean plus Visean) of the Lower Carboniferous (31,150 km^2) are extensively represented. They constitute also the best rock aquifer of Ireland and are listed under two headings on figure 34 and in table 31: Strongly karstified limestones and Fissured and karstified limestones. As in table 29, it is estimated to receive some 4.2 km^3 per year, or 67 per cent of the groundwater infiltrating into the whole of rock aquifers.

The highly karstified limestones (No. 2 of table 31) are located mainly in the Barrow basin and area (with Caledonian trends), and in the east-west Munster synclines (with Armorican trends), of the Blackwater, Lee and Bandon rivers. The limestone aquifers of the Suir have been less investigated. The basal Carboniferous beds include grey mudstone and sandstone (Kinsale Formation) overlain by calcareous mudstone and sandstone with carbonates (Courtmacsharry Formation); but they abruptly pass up into thick limestone, with some chert and shale, overlain by massive Waulsortian mudbank, or reef, limestone in the upper Courceyan and subsequent Chadian stages. These limestone beds were strongly folded in the Armorican orogeny, and subsequently have been so deeply eroded that only the lower portion of the synclines are now left. They occur in the river valleys of the south-east and south and, being structurally shattered and jointed, they have been karstified (caves are common) and they are excellent aquifers. Boreholes with yields up to 200 l/s are not unusual. The structures can even favour deep groundwater circulation, of which the warm spring at Mallow (and possibly those at Meelin and Gneevegulia) represent deep penetration of the groundwater along thrust zones. In the South-East Water Resources Region, recent detailed work by Daly (1982) has identified many specific limestone aquifers, as the Edenderry (upper Visean) Oolite, the Crosspatrick and Kilkenny-Carlow Dolomite and the Cullahill Limestone, with the limestone of the Suir valley and the Waterford coastal zone; the limestone graben in the extreme south-east of Wexford is a well-studied and developed limestone.

The fissured and karstified limestone aquifers (No. 3 of table 31) are located mainly under the Midlands of Ireland. On figure 34 they show up clearly. On table 29, they are included in the Carboniferous. Most known thicknesses exceed 1,000 m, except on the edges of basins. Following basal transition from Devonian to marine Carboniferous, beds of bioclastic limestone with subordinate shale pass upwards into solid clean limestone and are overlain by Waulsortian or reef limestone. While Waulsortian limestone continued in the south, in the north they are succeeded by bedded limestone (Chadian), with calcareous mudstone and cherty argillaceous limestone in the basins of the Shannon trough and the North Dublin basin. There was volcanic activity in Limerick during the Chadian. In the three succeeding stages, beds of dark, argillaceous cherty biomicrite and shale were deposited with distinctive facies towards the edges of the basins; there were some reef mudbanks in the Asbian. In the Brigantian, there are bedded dark mudstone and argillaceous biomicrite, and also evidence of tectonic activity and submarine slumping. Farther north, beyond the Longford-Down horst block, Carboniferous marine sedimentation did not commence till the Chadian. In the Asbian, there were shallow seas, so that evaporite sequences occur; much of the groundwater in Cavan and adjoining areas is derived from gypsum and other anhydrite lenses. There are also many sandstone sequences, which were deposited in the shallower coastal waters of the late Dinantian seas.

Most of these Carboniferous strata are not folded, and dips steeper than 5° are rare. But they are mainly well-jointed, and in places are faulted-fractured. Most are karstified, but to depths rarely exceeding 100 metres. They were protected from karstification by permafrost and solid ice cover during parts of the Quaternary; now they are in most cases protected from karstification by the little or no infiltration through the boulder clay and drumlins which cover much of them. Although there have been local investigations of karst processes, there has as yet been no overall study of karstification in Ireland. In the Midlands, there are few concentrations of groundwater flow; in the west, sinkholes and submarine streams occur. In most areas, boreholes will yield 5-20 l/s; but in some areas, where there are beds of black limestone and shale or other argillaceous limestone, yields of less than 5 l/s are usual. Much of the reef Waulsortian limestone is dolomitized; fissuration is localized, and although yields of about 20 l/s are usual, dry boreholes can occur.

The Namurian, including lower Westphalian (5,700 km^2) is generally a poor aquifer, with the exception of the Clay Gall Conglomerate (No. 1 of table 31). Its rocks tend to form plateaus, with normal hydrologic and hydrogeologic effects. The Namurian includes mainly shale, with sandstone in places, as in the Clare trough extending from the Burren to Killarney; although these formations are slightly calcareous, limestone sequences are not common and thicknesses there exceed 1,000 m. The Westphalian succeeds the Namurian with no stratigraphic unconformity; it accounts for some aquifers, as the Clay Gall Conglomerate (No. 1 of table 31) in the Castlecomber plateau, which has been extensively studied (Daly and others,

1980). Most of the Westphalian and all succeeding strata are missing, for erosion set in and lasted a long time.

In general, the Namurian does not yield good supplies of water, except where structural or related factors break up and fracture the shale.

(i) Permian-Triassic

Permian and Triassic rocks (26 km^2 per table 29) occurs at Kingscourt, and contains much gypsum-anhydrite. It occurs in a down-faulted graben, and the underlying limestone beds could yield much groundwater but they have not been explored.

(j) Quaternary

Quarternary formations cover most of the country. Although some 2,500 km^2 is uncovered, clay, sand and gravel were deposited by ice or rivers and lakes following the ice melt and bogs cover all the rest of Ireland, as in table 30. These are not shown on figure 34; they are listed in table 31 under Sands and gravel alluvium. Glaciofluvial outwash, river alluvium and limited moraines and eskers account for about 3.5 km^3 per year infiltration, or 77 per cent of all groundwater held in Quaternary aquifers.

Boulder clay is of three origins: there are old boulder clays (Munsterian Glaciation), new boulder clays (Midlandian Glaciation) and boulder clay brought in by the Irish Sea ice. With the drumlins, which are almost entirely of boulder clay, they cover some 44,125 km^2, or 64 per cent of the country. Boulder clay is generally not thick, except in the drumlins; elsewhere thicknesses in excess of 20 metres are rare. Because the old boulder clay has been decalcified, and almost nowhere contains limestone debris, it would seem that the boulder clay is not, or was not, entirely impermeable. Its behaviour under periglacial conditions is not known. It is considered as infiltrating limited amounts of precipitation through its soil cover, commonly low relief, and related factors. But the main effects of boulder clay are to reduce or stop infiltration to the underlying rock, and to act as an aquiclude confining groundwater in the underlying rock. It is usual to drill through 10 metres of boulder clay and locate water under 5 to 10 metres of head in the underlying rock aquifer.

Glaciofluvial deposits, outwash from the melting icecaps, are the major porous, inter-granular, aquifers of Ireland. They arise from the two glaciations, but those from the Midlandian, spreading over much of the Central Plain, are the most extensive. The farther from the melting ice fronts, the finer were the deposits. Towards the west, they thin out, and may be absent in parts of Mayo, Galway and Clare, where they are commonly replaced by blanket or, less commonly, raised bog. In many places, they give rise to small-scale broken topography, including kettle-and-kame, terminal moraines and eskers, all likely to hinder runoff, to form ponds, and to induce high infiltration. The glaciofluvial deposits are generally not thick, but in certain areas, as the Curragh of Kildare, they exceed 100 metres.

River alluvium (some 3,150 km^2) was deposited by post-glacial rivers. It contains much groundwater, in part recharged from the rivers and lakes. Eustatic and isostatic changes over the past 15,000 years have affected the gradients of Irish rivers in complex ways; along the east coast, past overdeepenings and changed courses have resulted in unexpectedly thick river alluvium filling old, and in places abandoned, river channels. The practice of using Ranney collectors and filtering river water for urban supplies is not followed in Ireland as it is in continental Europe, but wells and boreholes are common and useful yields are obtained.

Moraines and eskers tend to be linear features. The former were deposited where halted ice sheets melted, and lateral moraines bordering glaciers are less usual in Ireland. In the hills, however, local glaciation produced all types of moraine, as in the Leinster Mountains and around the Cork-Kerry local icecaps. Eskers were deposited by rivers flowing under the icecap; some of this was under pressure flow. Both forms have high infiltration capacity and give rise to many of the larger springs of Ireland. It is not known how deeply they are incised into the ground; so the volume of their groundwater content is somewhat uncertain.

Raised bogs (4,780 km^2) occur mainly in the Midlands. Many of them originated in shallow lakes formed at the end of the glaciation, filled first with some lacustrine marl, invaded by vegetation, filled to form, first, fens and then bogs as the organic debris raised the vegetation to where it was nurtured only by precipitation, as at Mulqueen and Burdon.

Mountain, or blanket, bog covers much wider areas, in particular on the west where precipitation is heavy and well-distributed. In all, bogs cover 17 per cent of the country but the blanket bog has only limited impact on the hydrogeology.

D. Water quality

The Meteorological Services provides comprehensive data on the hydrochemistry of precipitation in Ireland. Although the inland chloride content is small (4.1 mg/l at Birr, 1971-1980) it is high at Valencia on the coast, 21.3 mg/l for 1971-1980. In January 1976, the chloride content amounted to 51.3 mg/l. Acid rain is almost unknown, except to a limited extent in the extreme south-east, when wind and rain are from the east. Tritium has been determined at Valencia since 1952; it has been used to a limited extent in Irish hydrology (Williams and Bowen, 1973).

There has been no overall, comprehensive, study of the hydrochemistry of Irish groundwater. Likewise for pollution though, in general, Irish groundwater is mainly unpolluted. The 1983 report by Wright and others, Groundwater Vulnerability and Quality in the Republic of Ireland, prepared for the EEC, is most comprehensive, using some 700 analyses. But it deals only with the surveyed groundwater (3.454 km^3 per year) as in the previous report for the EEC (Wright and others, 1982), and not with the total groundwater (10.804 km^3 per year) presented in this report. From it, table 32 has been compiled, utilizing 564 samples to cover the Lower Carboniferous (Dinantian) rock aquifer and the combined Glaciofluvial and river alluvium aquifers.

In table 32 the number of samples indicate the extent to which these two major aquifers are investigated in the seven Water Resources Regions; the 133 samples for the limestone aquifer of the North-West Region include 100 samples for the Lower Limestone of Monaghan. The glaciofluvial and river alluvium is composed mainly of carbonate minerals; hence the two systems of aquifer have similar hydrochemistry. Variations in total dissolved solids (TDS) and total hardness are generally due to variations in the length of time the groundwater has been in contact with the aquifer; thus water from the unconsolidated aquifers has somewhat lower TDS and total hardness than water from the rock aquifer. In some samples, however, from the North-West Region, groundwater has been in contact with evaporite lenses in the limestone, and has dissolved much calcium sulphate from the gypsum and anhydrite.

Table 32. Ireland: Hardness and total dissolved solids of water in principal aquifers

Water resources region	Number of samples	Total Dissolved Solids (TDS; mg/l)			Total Hardness (as mg/l $CaCO_3$)		
		Average	Max.	Min.	Average	Max.	Min.
Lower Carboniferous (Dinantian) aquifers							
East	45	400	785	200	310	500	166
South-east	98	409	554	280	321	400	172
South	87	403	700	270	286	394	194
Mid-west	32	418	800	200	289	450	148
Shannon	37	>500	-	-	291	424	178
West	31	>500	-	-	326	470	252
North-west	133	495	2 799	46	335	1 085	24
Total	463	425	2 799	46	308	1 085	24
Glaciofluvial and river alluvium aquifers							
East	45	424	666	300	270	504	136
South-east	39	347	548	199	268	378	86
South	11	280	480	200	140	348	60
Mid-west	2	432	-	-	271	-	-
Shannon	4	>500	-	-	322	324	100
Total	101	371	666	199	254	504	60

A summary of an earlier study by D. Daly (March 1979) of the hydrochemistry of Irish groundwater is given in table 33. It is based on 120 samples, and gives the characteristics of nine aquifers. Water from the Old Red Sandstone (6 samples) is typical of quartz-dominated aquifers; the sandstone aquifer of Castlecomber is somewhat similar. Unfortunately the sodium and potassium were not determined for the Ordovician volcanic rocks; from other work they are known to be high. The other limestone aquifers call for no comment. The total hardness at 299 mg/l obtained by Daly agrees well with the 308 and 254 mg/l obtained by Wright (1983).

Table 33. Ireland: Ionic composition of water in nine aquifers

Aquifer	Ions, in mg per litre							Total hardness, as $CaCO_3$ (mg/l)
	Ca	Mg	Na	K	Cl	SO_4	TDS	
Old Red Sandstone (Cork)	9	6	..	..	28	..	92	46
Sandstone (Castlecomber)	34	14	38	2	15	1	252	212
Volcanics (Waterford)	24	18	..	..	32	29	240	130
Upper Visean limestone	101	16	8	2	21	4	388	315
Dolomitized reef	85	22	..	..	27	75	478	302
Limestone (Wexford)	85	22	29	4	40	14	420	303
Courceyan limestone	176	39	..	..	31	254	736	502
Limestones (Cork)	98	7	18	2	34	11	381	275
Sand/gravel; calcareous	86	14	12	3	26	--	393	309
Average	77	17	21	3	28	49	376	299

Some work has been done on trace elements. Iron and manganese in small concentrations are common and because they are unwanted in domestic water supplies, are often investigated; where present, remedies are proposed. For their medical interests, work has been done on fluorine and iodine. Selenium is also studied by An Foras Taluntais, for it affects horse breeding and some animal production. An Foras Taluntais has also made detailed work on as many as 20 trace elements held in soils; this has not been converted to hydrochemical terms.

The extent of pollution in groundwater is given in An Foras Forbartha's June 1983 publication Review of Water Pollution in Ireland and in other publications, including Lennox and Toner (1980) and Toner (1984) on which the following observations are based. Groundwater pollution is neither important nor widespread in Ireland although there are many localized problems stemming from the siting of wells and boreholes in urban areas or close to such waste sources as manure and silage pits or septic tanks. A general investigation of groundwater resources in the north-eastern part of the country indicate generally good quality in more than 200 wells and boreholes, although high nitrate content was recorded in a few, more specifically, concentrations in excess of 10 mg/l were recorded in 10 of the sources sampled, while another 30 sources had concentrations of nitrate ranging from 5 to 10 mg/l. In a survey of water quality in 45 groundwater sources in County Carlow, the most notable feature was the significant proportion (11 per cent) of samples in which the nitrate concentration exceeded the maximum permissible limit (11 mg/l) set by the EEC. Most publications emphasize the impact of agricultural activities, especially through the use of fertilizers, on the quality of groundwater.

Lately, the geothermal position in Ireland has been studied. With EEC support, geothermal and hydrogeothermal studies have been made, by various organizations, under the coordination in particular of the Geological Survey. The hydrogeothermal potential in Ireland could be significant only by using heat pumps to extract water in the range of 9.5° to 10.5° C, the average temperature of Irish groundwater. Only 55 springs with temperatures higher than 11° C have been located; they are all in the Carboniferous limestone aquifer. They lie in areas where structural conditions favour deeper than usual groundwater circulation. From investigations of two of the warmest springs, in Leinster (St. Gormans, maximum temperature, 22.0° and Kilbrook, 23.95° C), it was found that the flow of warm water was controlled by lunar and solar induced earth tides, compressing and expanding the deep aquifer; the cold water tended to follow the normal hydrologic cycle.

Most of this tepid groundwater has normal hydrochemical composition. That in the Celbridge syncline, however, penetrates deeply to the Furnes Member evaporite, from which sodium chloride and other salts are leached in limited amounts. A study has been made on the isotopes of a few of the tepid springs, including work on radon and radium, the five inert gases (He, Ne, Ar, Kr and Xe), as well as the stable isotopes of hydrogen and oxygen. It is hoped to continue these investigations; at present, work is confined to determining groundwater temperatures from observations on the springs of Ireland.

E. Training and research

The universities and regional technical colleges of Ireland give courses at several levels on different aspects of water to students in civil engineering, science, industrial engineering, as well as in geography (including hydrogeology and water resources) and fluvial geomorphology. At Sligo Technical College there is given a diploma in Civil Engineering (Water) and there are hydrogeologic lectures at Athlone.

The only postgraduate course of major importance in water is at University College, Galway. There, a one-year course is run which can lead to a longer course and award of the masters degree and even doctorates. The Master of Science course consists of Physical Processes in Hydrology; Groundwater Hydraulics and Geohydrology; Hydraulics; Fluid Mechanics; Hydrometry; Stochastic Hydrology; Hydrologic Systems; Numerical Methods and Computers; Water Treatment; Water Quality Modelling; Hydro-power Engineering. The course usually has about 15 students from developing countries, including Bangladesh, China, Ghana, India, the Islamic Republic of Iran, Jamaica, Lesotho, Liberia, Nigeria, the Sudan and the United Republic of Tanzania.

Many Irish professors and senior staff help in international courses in different fields of water. One spent a year in Canada, and has lectured at Kuala Lumpur, Roorkee and Nanking, and helped to establish water courses at the universities of Dar-es-Salaam, Tanzania and Lusaka. Another lectures each year as part of the International Courses at Delft and at Padua, and has given summer courses at Moscow University, at Jablonna near Warsaw, and at the East China College of Engineering Hydraulics at Nanking. Other contributions include lectures at the Institute for Water Research in Budapest and at a UNESCO Training Course in Nairobi, and acting as external examiner for hydrogeology at the University of Birmingham, at London and at Cairo. These activities are continuing (1982-1989) and becoming more diversified. Several Irish consulting firms have experience in developing countries on many aspects of water including hydrogeology and groundwater development, use and management.

References

Aldwell, C. R. and D. J. Burdon. Groundwater investigations in Ireland. Proc. Meeting on Hydrogeology in Ireland, Irish National Committee IHP, p. 9-70, 1979.

__________. Hydrogeothermal conditions in Ireland. Twenty-sixth Internat. Geol. Congress, Paris; Fossil Fuels, sec. 14.2, 1980.

Aldwell, C. R., D. J. Burdon and M. Sherwood. Impact of agriculture on groundwater in Ireland. Environmental Geology, vol. V, p. 39-48, 1983.

Aldwell, C. R., J. B. Day and W. Struckmeier. Explanatory notes for the International Hydrogeological Map of Europe, Sheet B-4, UNESCO, Paris; (Map, 1970-75), 1978.

Aldwell, C. R. and G. Wright. Groundwater in coastal areas of Ireland. UNESCO Malta Seminar on Selected Water Problems in Islands and Coastal Areas, 1978.

Archer, J. B. Lower Palaeozoic rocks of the northwestern part of the Devilsbit-Keeper Hills inlier, and their implications for the postulated course of the Iapetus Suture Zone in Ireland, Jour. Earth Sci., RDS, vol. 4, p. 21-38, 1981.

Bennett, J. R. P. and I. B. Harrison. Explanatory notes for Sheet B-3 of the International Hydrogeological Map of Europe. (Map, 1976-79), UNESCO, Paris, 1980.

Brock, A. Geothermal Energy in Ireland. Report to National Board of Science and Technology, 68 p., 1979.

Burdon, D. J. Contribution to discussion on "Effects of the 1975-76 Drought on Groundwater and Aquifers", in Scientific Aspects of the 1975-76 Drought, Royal Soc., London, p. 66-67, 1978.

__________. Final report of the Irish Geothermal Project, Phase I. Report to Geological Survey of Ireland, from Minerex Ltd., 285 p., appendices, April 1983.

__________. Irish groundwater resources in relation to geothermal energy investigations. Internal report, Geological Survey of Ireland, 276 p., December 1983.

__________. Hydrogeological aspects of agricultural drainage in Ireland. Report to Commission for Groundwater Protection, IAH Irish Meeting, June 1984.

Burdon, D. J. and K. T. Cullen. Hydrochemistry of Caradocian volcanics in south-east Ireland, and the effects of precipitation composition on the groundwater chemistry. Fourth Internat. Symposium on Groundwater, Catana, Sicily, 1980.

Burdon, D. J., D. J. Burns and S. Peel. Geothermal energy investigations in Ireland. EEC DG.XII and XVII, Munich, 10 p., December 1983.

Daly, D. Groundwater quality in Ireland. Internal report, Geological Survey of Ireland, March 1979.

Daly, D. Groundwater quality and pollution in Ireland, with particular reference to the effects of agricultural activities. Geological Survey of Ireland report to Irish Meeting of the Commission for Groundwater Protection, IAH, May 1984.

Daly, D., J. W. Lloyd, D. R. Misstear and E. P. Daly. Fault control on groundwater flow and hydrochemistry of the aquifer system of Castlecomber Plateau, Ireland. Q.J.E.G., vol. 13, p. 167-176, 1980.

Daly, E. P. Nitrate levels in the aquifers of the Barrow River valley. Internal Report, File 2.11.1, Geological Survey of Ireland, May 1981.

__________. Groundwater resources of the Southeast Industrial Development Region. Internal Report, Geological Survey of Ireland, 1982.

Dillon, N., G. Forde, A. Murphy and D. Duggan. Groundwater in County Wexford - A local authority view. Irish Group, IAH, May 1982.

Drew, D. Hydrogeology of north County Galway - south County Mayo, Lowland Karst area, western Ireland. Proc. Sixth Internat. Congress Speleology, 1973.

Foras Forbartha. Surface Year Book of Ireland, 1975. Vol. I, Hydrographs of Daily Mean Flow; vol. 2, Tables of Daily Mean Flows. Dublin, 1978.

__________. Review of water pollution in Ireland. Report by Water Pollution Advisory Council, 152 p., June 1983.

Geological Survey of Ireland. Geological Map of Ireland; 1:750,000 Scale. Third Edition, Dublin, 1962.

Holland, C. H., (Ed.). A Geology of Ireland. Scottish Academic Press, Edinburgh, 1980.

Irish National Committee for the International Hydrological Programme. Hydrology in Ireland. Contribution to IHD of UNESCO, 49 p., 1983.

King, W. Of the Bogs and Loughs in Ireland. Phil. Trans. Roy. Soc., vol. XV, No. 170, p. 142-145 (abbreviated version), London, 1685.

McCumiskey, L. M. Water resources of Ireland. Jour. Inst. Engineers of Ireland, Read 3 Dec. 1973.

Mitchell, G. F. The search for Tertiary Ireland. Jour. Earth Sci., RDS, vol. 3, p. 13-33, Dublin, 1980.

Mulqueen, J. Hydrology and drainage of peat. Report to Commission for Groundwater Protection, IAH, Irish Meeting, June 1984.

Naughton, M. Hydrogeological study of the Upper Kilmanagh River Basin, Republic of Ireland. Unpublished M.Sc. thesis, Univ. of Alabama, U.S.A., 1978.

__________. Groundwater resources in the North-East (Monaghan-Cavan-Louth) Region, 3 vols., Foras Forbartha and Geological Survey, internal report, 1981.

Naylor, D., W. E. A. Phillips, G. D. Sevastopulo and F. M. Synge. Introduction to the Geology of Ireland. Twenty-sixth Internat. Geol. Congress; guide to Excursion 058.C; Irish Nat. Committee for Geology, Royal Irish Academy, Dublin, n.d.

Ordnance Survey of Ireland. Rivers and their Catchment Basins, at 10 miles to one inch, Phoenix Park, Dublin, 1958.

Royal Irish Academy. Atlas of Ireland. Irish Nat. Comm. for Geography, 1979.

Williams, P. W. Limestone Morphology in Ireland. Irish Geographical Studies, p. 105-124; Queen's University, Belfast, 1970.

Williams, P. W. and R. Bowen. Geohydrogeological study of the Gort Lowlands and adjacent areas of western Ireland, using environmental tritium. Water Resources Research, vol. 9, p. 753-758, 1973.

Wright, G. R. Groundwater resources of southeast Cork. Internal report, Geological Survey of Ireland, 1978.

__________. Groundwater in the South Munster Synclines. Proc. 1979 IAH Internat. Meeting, Dublin; published by Irish Nat. Comm., IAH, 1982.

Wright, G. R., C. R. Aldwell, D. Daly and E. P. Daly (Editor, J. J. Fried). Groundwater Resources of the Republic of Ireland, published by Th. Schafer GmbH., Tivolistrabe 4, D-3000, Hannover, for the EEC DG for Environment, Consumer Protection and Nuclear Safety; with 12 maps of Series 1404, 1982.

Wright, G. R., D. Daly and E. P. Daly. Groundwater vulnerability and quality in the Republic of Ireland. Report for EEC under Contract U(82)178-(543) DG.XI. Geological Survey of Ireland, 1983.

VIII. ITALY

Area: 301,262 km^2 Population: 57.6 million

A. Physiography and climate

Most of the Italian Republic consists of a peninsula nearly 1,000 km long and 150 to 220 km wide that projects into the Mediterranean Sea. It is bounded on the north by the Alps and on its periphery includes two large islands, Sicily and Sardinia, and numerous small islands (totalling 50,136 km^2) clustered together in small archipelagos (Tuscan, Pontine, Campania, Eolie, Egadi, Pelagie and Tremiti groups) or isolated in the middle of the sea (Ustica and Pantalleria), and islets surrounding Sardinia (S. Antiaco, S. Pietro, Asinara, Maddalena, Caprera, Tavolara and Molara). A range of mountains and hills, the Apennines, stretches like a backbone across the peninsula and extends into Sicily. For administrative purposes Italy is divided into 20 regions, 95 provinces and 8,076 municipalities.

Italy may be divided into five main geomorphologic units: the Alpine zone, the Plain of Lombardy, the Apennines and the two large islands. Mountains cover 38.7 per cent of the national territory, hills 39.7 per cent and flatlands only 21.6 per cent.

Italy's principal river is the Po which flows across the Plain of Lombardy, taking in the waters of its tributaries which rise in the south slope of the Alps and north slope of the Apennines. All the other rivers rise in the crest of the Apennines. There are some small open plains along the seashores.

Given the configuration of the peninsula which, with its island appendages, extends from 36° to 47° north latitude, and given the continuous elevations, the local climatic changes are substantial and imprint particular characteristics on each environment. Just as the mean climate differs from arid in the southern regions to humid in the upper zones and those farther north, so local conditions are variable, fluctuating from year to year as much as 35 per cent above or below the mean. These fluctuations are influenced by the systems of the Alps and the Apennines, the changing state of the seas along Italy's extensive shores, the presence of currents in these seas, the constantly changing exposure of the slopes, and the presence, at least in north Italy, of large lakes.

Italy's annual mean precipitation ranges from 400 mm in certain southern coastal areas to more than 3,600 mm in some wide Alpine valleys. Perennial watercourses and watercourses that are almost always dry alternate in the various regions, just as zones with plentiful groundwater, scattered throughout the national territory, alternate with zones that are completely devoid of such water.

B. Geology

1. Lithology

The geologic formations of Italy can be classified as follows, according to their hydrological characteristics:

(a) Fractured rocks

The group of fractured rocks includes the basic or acid volcanic rocks of central Italy, Sardinia and the Iblei mountains in Sicily: ophiolite; phyllite; schist, mica schist and gneiss in the metamorphic districts of Calabria Peloritana, Sardinia and the Alps; and the mainly granitic intrusive rocks of Sardinia, Calabria, Liguria and the Alps.

All the carbonate rocks may be included in this group even if a distinction is made for rocks with more marked karst plateau features. The evaporites may also be included. All together, the land belonging to this group covers an area of 93,627 square kilometres and constitutes 31.08 per cent of the national territory.

(b) Porous rocks

Almost all sedimentary lands from the Miocene epoch to Recent belong to the group of porous rocks. River alluvium, alluvium fill in Quaternary depressions, the detritus, moraines and the transgressive and regressive sedimentary series found everywhere in the peninsula and in the larger islands constitute more or less extensive and highly important aquifers.

(c) Rocks of mixed permeability

Rocks of mixed permeability include: the basic volcanic rocks mixed with tuff or pyroclastic deposits present in central Italy and on Etna; some deeply weathered intrusive rocks, as in Sardinia or Calabria; and all the flysch series of the period from the Eocene to Miocene which are situated along the entire Apennine range and the larger lakes. Generally this group includes less important and commonly localized aquifers.

2. Geomorphic provinces

(a) Alps

In Italy, the Alps consist mainly of crystalline metamorphic rocks with ophiolitic intercalations. These are slightly permeable rocks in which the fracturing caused by strong tectonic forces generates a sizeable flow which is also abundantly replenished by the frequent heavy rains totalling more than 3,000 mm per year, with an average close to 2,000 mm, by the presence of glaciers and by the snow cover which replenishes the reservoirs and perennial watercourses of the zone shortly before summer.

On the rim of this arch lie huge and mainly calcareous sedimentary deposits, dating from the Triassic to the Cretaceous over large areas (Dolomites, Carso); they contain vast amounts of groundwater, especially where karst features are more prominent.

In addition, the detrital and alluvial deposits and the moraines, left over from the ancient glacial periods, form, on the lower edge of the range, extensive and substantial aquifers which are replenished by the perennial watercourses passing through them. The phenomenon of lines of springs which encircle the Plain of Lombardy in the north is characteristic of these watercourses.

(b) Lombardy Plain

The Plain of Lombardy is an immense trough which has been filled to the brim from the Pleistocene period to the present day. Its vast area (46,300 km^2) contains Italy's largest and most important aquifers. Its accumulated deposits (more than 1,000 metres thick) consist of clay and sand which alternate with shingle. Its permeability varies with granularity through the different strata, increasing near the outside where the coarsest deposits have settled. Salt water is found along the Adriatic seaboard and deep inside the ancient gulf, owing to the intrusion of the Adriatic or to residues from the ancient sea which formed the now-filled gulf.

(c) Apennines

All of central and southern Italy is affected by the Apennine mountain range which is, in a sense, the country's backbone. Three distinct sectors may be identified: a central sector with the highest elevations and two others consisting of marginal strips on the Tyrrhenian and Adriatic coasts.

Starting in the north, the Apennines run south-east and separate the Plain of Lombardy from the rest of the Italian peninsula in the regions of Liguria, Emilia and Tuscany where most of the land is of the flyschoid type, consisting of alternating patches of clay, marl and sand where only small local aquifers are to be found.

From Umbria and the Marches, the Apennines turn south-south-east, changing some of their own composition as is apparent from their outcrops, especially on the widely covered carbonate western slope with rocks dating from the epoch between the Triassic and Miocene periods. The range then continues in the same general direction, first reaching the central regions of Italy - Latium, Abruzzi, Molise and Campania - and then joining Basilicata in the south.

The presence of large calcareous outcrops, generally not of the karstic plateau type but deeply fissured and providing excellent and extensive aquifers, accounts for the abundance of groundwater in these regions where high-capacity springs yielding more than 1 m^3/s can be found.

Sedimentary Pliocene-Pleistocene flysch and limestone deposits in the less elevated zones and valley alluvium also constitute valuable aquifers.

Farther south, in Calabria, most rocks are of the intrusive or metamorphic type and therefore intrinsically impermeable, but severe fracturing and weathering may result in the formation of permeable superficial layers, many more than 50 metres thick, but these are heavily drained by the deep incisions that severe erosion has made into the flanks of the formations.

The land bordering the Apennines to the east may be subdivided into two subsectors: one runs along the coast from the Marches to Molise, then extends southwards into the Bradano trench between Puglia and Basilicata and ends in the Gulf of Taranto and the Puglia zone of the Gargano, the Tavoliere plain and the Salentina Peninsula. Most prominent are the Pliocene deposits of clay and mud with sandy lenses, many in contact with the alluvium of the valley floors, which form small local aquifers. In the Puglia subsector, the Gargano, the Murge and the Salentina Peninsula consist of karstic fissured calcareous rocks which are excellent aquifers but are meagerly replenished by the inadequate local rains and whose overexploitation has led to widespread sea-water intrusion. The Tavoliere plain connects with the strip of the first subsector and is then characterized by sandy deposits which form a sizeable aquifer.

Farther south along the east seaboard of Calabria there are only a few small alluvial deposits lying at the foot of hills and mountains, along the routes of the valley troughs and on the Sibari plain which also has the region's largest aquifer.

The coastal plain west of the Apennines, along the Tyrrhenian Sea, may be divided into two subsectors: one is the coastal area itself which connects with the hinterland plateaus (of Pisa, Grosseto, the Tiber, Pontina, the Vulture and the Sele), consisting mainly of Quaternary or alluvial deposits which form well-replenished large aquifers; the other subsector consists of volcanic elevations, running parallel to the Apennines from Tuscany to Campania and formed by permeable eruptive rocks which are excellent, amply replenished aquifers. Here and there surface deterioration of the volcanic rocks has resulted in the formation of impermeable terrane, as is apparent from the presence of numerous small lakes.

(d) Sicily

In Sicily the Apennines of Calabria extend over the entire north-east sector, the Province of Messina, with its flysch and metamorphic rocks which are only semi-permeable and therefore have few or no aquifers, into the alluvium of the valley floors and then continue along the north coast of the island across the Province of Palermo up to Trapani with a hydrologically important series of calcareous uplands (Madonie, Imerse, Panormidi, Trapanesi). The mainly calcareous upland plain of the Iblei rises in the south-east sector of the island; it has large, well-supplied groundwater reservoirs. In the south-east sector also there are calcareous outcrops such as the Sicani mountains which provide a good, although sparsely recharged, aquifer.

To the west, in the Marsala zone, there are sizeable sandy-calcaranitic Pliocene-Pleistocene deposits. To the east, the Plain of Catania has features that are, allowing for differences in size, similar to those of the Plain of Lombardy; actually it is a gulf filled in with sandy-clayey Quaternary alluvial deposits and is also the site of a small groundwater reservoir which in many places rests on salt water. Further to the east, between the Plain of Catania and the Nebrodi mountains, looms Etna, the highest volcano in Europe (3,340 m) which, with its vast outcrops spread over 1,300 square kilometres, holds Sicily's largest aquifers. Soils of effusive origin are also present, alternating with the carbonatite deposits of the Iblei, but they are severely weathered areas constituting a mixed aquifer of minor importance. Many of the islands forming a semicircle with Sicily (the Eolie, Pantalleria, Ustica) are volcanic, while the others (Egadi, Lampedusa) are of calcareous origin.

No noteworthy aquifers are to be found in the central part of the island where argillaceous soils of the evaporite series prevail.

(e) Sardinia

Most of Sardinia is crystalline terrane, consisting generally of thoroughly weathered metamorphic and volcanic rocks. Sardinia is of little hydrologic interest, the local aquifers being limited to zones with degraded surfaces and accumulated detritus. Aquifers of some importance are those composed of Nurra limestone, near Sassari, Iglesiente limestone and the Quaternary sand and clay of Cagliari's sandbank.

C. Principal aquifers

A schematic map of main aquifer units is presented in figure 39.

1. Alps

In the Alpine region the alluvial aquifers of glaciofluvial origin in the valley floors are noteworthy owing to the great thickness of the accumulated materials, their relatively high permeability and their steady substantial inflow from the snows on the summits and from abundant precipitation. Even though profusely recharged with water from neighbouring zones, these aquifers are not fully supplied. The main alluvial aquifers in the west are those situated in the Val di Susa, Val d'Aosta, Val Sesia, Val d'Ossola, Valtellina, Val Camonica, Valle Lugarina, Val Sugana, Piave and Tagliamento. They differ greatly in terms of thickness, as is characteristic of this type of aquifer, whether along or across the valleys and, at some points, may be 300 to 350 metres thick (Val Longarino-Adige), although in most places 50 to 150 metres thick.

Although the high Alpine crests are severely deformed by tectonic forces or fractured, their crystalline or metamorphic nature inhibits the formation of prominent aquifers except in zones where detrital material has accumulated. While the yield of the springs is moderate, seldom higher than 5 l/s, with some exceptions, they are useful in solving local water problems. The zones with limestone outcrops differ in importance. Generally they have been severely deformed by tectonic forces and have high peaks. As a result of severe fracturing and their karstic geologic features, these rocks constitute some of Italy's largest massifs. The most important Alpine carbonate rock are, from west to east:

a. The Maritime and Ligurian Alps, on the border with France, consisting of marlstone, dolomite and calcarenite, roughly broken into compartments; they cover an area of about 780 km^2. In the zones where limestone is prominent and karstic features are common and supply many small springs, 43 are known to yield more than 30 l/s, 14 have a yield of 100 to 300 l/s, 5 a yield of 300 to 500 l/s and one of 1,000 l/s, giving an aggregate yield of 6 m^3/s. The main outlet of the carbonate aquifer is in the vicinity of Cuneo, under an alluvial cover, with yield exceeding 2 m^3/s.

b. The Lombardy Alps (Orobie, Camoniche and Giudicarie) are deeply gashed by glacial erosion and have the largest lakes in northern Italy. The aquifers in that area are well drained by the watercourses that flow through them and are partly hidden by occasional impermeable heaps of detritus. The flow regime is moderate as compared with some of the 31 known larger springs which have an aggregate yield of 4 m^3/s; there are some boreholes where the water gushes out.

c. The Lessini and Brenta Alps (tableland of Asiago M. Grappa) connect with the Dolomites zone in the north-east. There are 35 springs with flow rates of over 30 l/s, of which 10 have yields of 100 to 300 l/s, one yields 300 to 1,000 l/s and one has a rate higher than 1,000 l/s. The aggregate yield is close to 3,300 l/s.

Figure 35. Italy: Hydrogeology

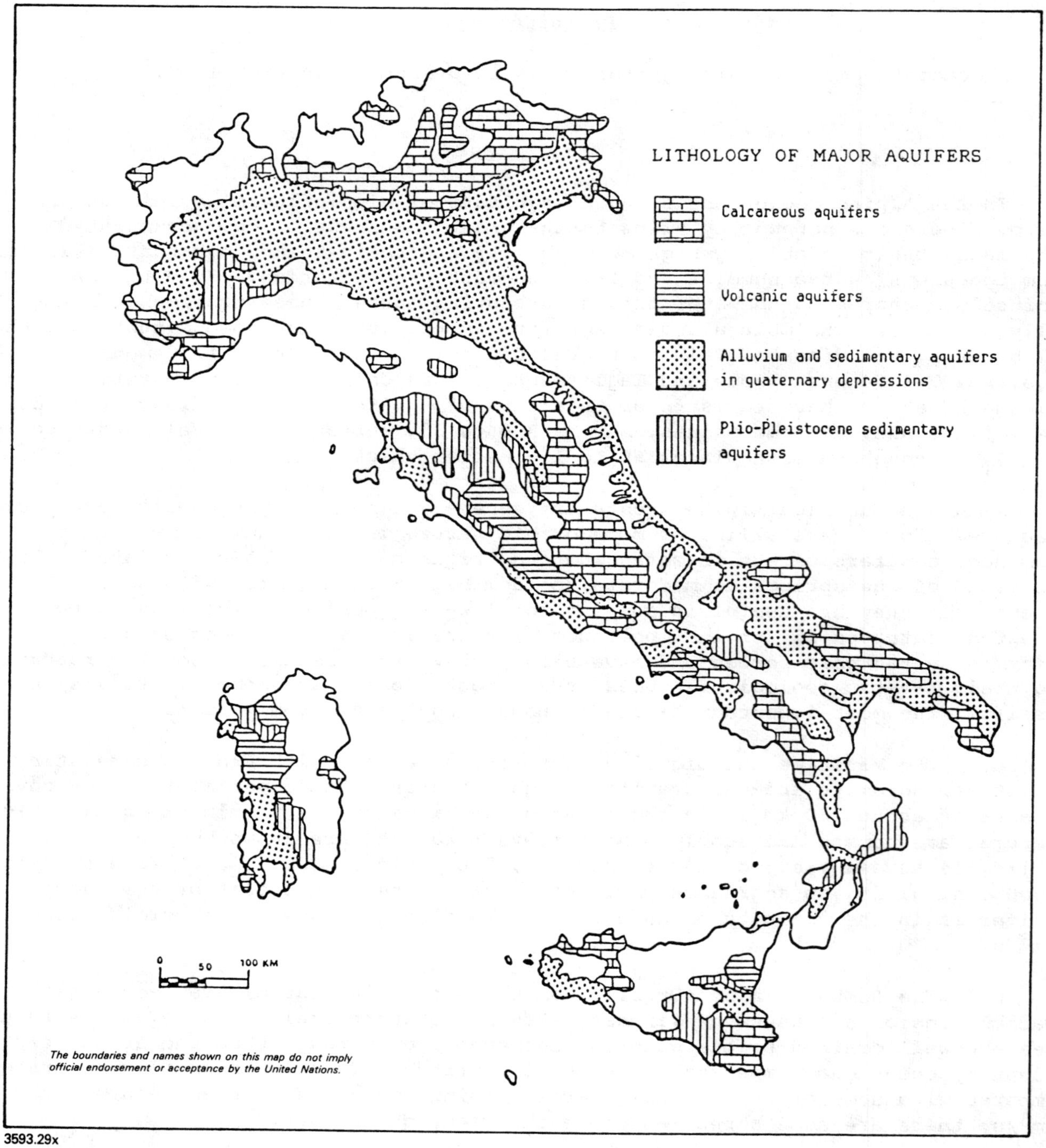

3593.29x

d. The Dolomites of Upper Venetia are famous for their rugged landscapes. They are rocky spikes of dolomite soaring aloft with narrow walls or with peaks surrounded by detrital deposits which conceal the abundant flow of water inside. There are 80 known springs with flows of more than 10 l/s in the provinces of Bolzano and Trento; their estimated total yield is close to 3.5 m^3/s.

e. The Carnia and Giulia Alps are also mainly of dolomite and marlstone. There are 48 known springs with a flow rate of over 30 l/s, of which 12 have yields of 100 to 300 l/s and 3 with yields of over 300 l/s (Porto Subiolo 3 m^3/s; Olivero 11 m^3/s). A good part of the water resources of the carbonate aquifer, however, goes towards recharging the sedimentary aquifer in the plain of Venetia. The aggregate yield of the springs is estimated to be 20 to 25 m^3/s.

f. The Carso plateau, in the far north-east of Italy, is famous because its name has been given to the phenomenon of karst. The waters have carved out of the surface and subsoil a typical topography, consisting of bare rock, deeply riddled with karstic pits (sinkholes, cavities, and grottos), which provides a vast amount of groundwater. This is where the underground stream of the Timavo River flows; farther east, in Yugoslavia, the famous grottos of Postumia are situated. Part of the aquifer flows into the sea through high-yield submarine springs.

Outcrops of effusive rocks exist in the east central sector of the Alps. They are very ancient (generally porphyry and porphyrite) and are situated in Val Sesia, near Lake Lugano, near Lake Maggiore and Lake Orta in Val d'Ega, Val Avisio and Val Sarentina, near Bolzano. They are virtually negligible aquifers of no more than local interest.

Hydrologically more interesting are the Tertiary and Quaternary volcanic rocks of the Euganean Hills and part of the Berici Mountains. There are smaller outcrops in the Lessini Mountains and on Mount Balbo. In the Euganean Hills 80 springs with an average flow rate of less than 3 l/s have been inventoried. In addition there are three known springs near Recoaro with an estimated average yield of 50 to 250 l/s.

2. Lombardy Plain

The Plain of Lombardy is the result of the filling of a tectonic depression caused by subsidence phenomena from the Tertiary period to the present day with more or less marked phases, especially during the Pleistocene period. With an area of 46,300 km^2, it covers 15.4 per cent of Italy, and constitutes 71.2 per cent of the flatlands. It is 420 kilometres long from the last Alpine outcrops in the west to the Adriatic Sea. Its width is from 90 to 270 kilometres. Folding and local isostatic upthrusts have moulded the bedrock and, in some places, even affected the Pleistocene terranes. Two ranges of hills, along which the pre-Pleistocene topography is visible, divide the Plain into four sectors.

In the west, where the ancient gulf was shut in, the Monferrato Hills separate the Plain of Cuneo from the Plain of Alessandria, and in the east the Berici Mountains and the Euganean Hills separate the Plain of Venetia from the rest of the Plain of Lombardy. There are smaller rises of the substratum, shown by drillings, between Pavia and the Adda River (S. Colombano), between Tortona and Valenza Po and between Ferrara-Mirandola and Carreggio. The Pleistocene deposits become progressively thicker from west to east and towards the centre of the ancient gulf, but they may differ depending on the extent of substrata settlement and on the locally more marked dips or rises. The Pleistocene deposits may be thicker locally, as much as 4,000 metres, but they are generally from 300 to 1,500 metres thick.

The older ground, which filled up the ancient gulf, belongs to the lower Pleistocene epoch: in continental (Villafranchian) facies on the perimeters, especially west and north of the Plain, and in pelagic (Calabrian) facies in the central and east zone. They consist of stratified sediments, with lenses of differing thickness, composed of shingle, coarse sand, mud and clay, commonly intercalated with peat or lignite in areas which earlier were the sites of lakes or marshes. The thicknesses range from about 1,000 metres in the Venetia zone to 100-300 metres in the west sector. These are highly permeable areas, well supplied peripherally by the watercourses flowing down from the Alps and Ligurian Apennines.

In the Villafranchian and Calabrian areas, and virtually throughout the Plain, sediments of the middle and upper Pleistocene epoch and some Recent alluvium have accumulated, differing considerably in thickness from place to place but being thicker towards the east and having increasingly finer particles.

At the foot of the uplands, on the borders of the Plain, the detritus and deposits accumulated during the glacial periods are prominent, whereas fluvial and deltaic deposits are more common in the centre of the Plain.

The watercourses flowing down from the mountains discharge, on the upper stretch, into the sedimentary aquifers bordering on the valley; farther downstream they cut into these aquifers, drain them and then return to feed the large central aquifer on the flatter stretch of the Plain. The gradual filling process of the ancient gulf has retained, in the lower part of the aquifer, salt or brackish waters which, as a consequence of the local isostatic movements or subsidence, are now found at various depths from zone to zone. Along the Adriatic coast the excessive extraction of water from the aquifer and of gas from the mineralized strata, in addition to promoting rapid subsidence in recent years (Venice and Ravenna), has led to the intrusion of sea water, and efforts are now being made to stop extraction of the water.

The aquifer complex consists of many strata in much of the Plain or of a single stratum in the more peripheral zones, with free or confined aquifer layers.

This aquifer which, roughly speaking, may be considered continuous, although it can be divided into distinct sectors, is certainly the most productive in Italy because of its volume and of its bountiful flow into the 40,000 wells dug there. In addition, the outflow of the Piedmont springs is abundant and, with the establishment of irrigation systems, it has enabled the Plain of Lombardy to become Italy's most productive agricultural zone.

In the upper zone of the Plain, in Piedmont, recent surveys have made it possible to ascertain the piezometric conditions which have shown average gradients of 0.4 per cent in the vicinity of Turin, 0.2 per cent near Novara and 0.5 per cent near Alessandria. The transmissivity of the aquifer, calculated on the basis of numerous capacity tests, is from 10^{-1} to 10^{-4} m^2/s and most frequently between 10^{-2} and 10^{-3} m^2/s.

In the Lombardy mid-zone of the plains, piezometric conditions are affected, locally, by heavy drafts for the supply of main urban centres (Milan). For the Province of Milan alone there are about 9,000 known wells from which an average of approximately 20 m^3/s are drawn. The piezometric slope dips from 0.2 per cent in the highest zones to 0.02 per cent in the more undulating area. Transmissivity is from 2 X 10^{-1} to 2 X 10^{-3} m^2/s.

In the Venetia part of the Plain the average piezometric slope drops from 0.5 to 1.1 per cent in the Piedmont zones to between 0.01 and 0.04 per cent in the coastal zone. The discharge from the watercourses on the upper part of the Plain is substantial and causes the piezometric surface to rise noticeably. The aggregate rate of flow of the watercourses into the alluvial aquifer is estimated at about 100 m^3/s. In the intermediate hilly zone this water reappears above ground and forms a continuous line of springs with an aggregate flow rate of over 150 m^3/s.

Both the volume and the outflow of the aquifer are substantial: there are more than 13,000 wells. In Venice area this oversupply has been found responsible for increasing subsidence and many wells have had to be closed.

In the lower Plain of Lombardy, in Emilia and Romagna, the piezometric surface tends towards the Po River, from 1 to 2 per cent in the hilly zones to 0.1 to 0.2 per cent in the lower zones.

The aquifer's recharge into the river system in the peripheral hilly zone is calculated to be about 10 m^3/s, whereas the estimated average drainage from the same rivers in the valley zone is 7 m^3/s.

The transmissivity in this part of the Plain varies from 5 X 10^{-2} to 10^{-3} m^2/s. The storage coefficient is from 10^{-2} for the surface strata to 10^{-4} for the deep zones where the stratum is under pressure.

On the edge of the Plain, in the Monferrato hills, there is a potentially productive aquifer, consisting of Pliocene or Astian sands stratified with silty clays, which is abundantly replenished. Its piezometric pattern shows that it is drained by the rivers that traverse it.

3. Apennines

Along the crest of the Apennine range, and between it and the volcanic uplands in the west, there are numerous depressions which were filled during the Quaternary period with more or less coarse sediments ranging in thickness from 100 to 300 metres in the central sector to as much as 3,000 metres in the southern sector. The most important of these depressions, now fertile plains, are, respectively, from north to south: those between Pistoia and Florence; between Terni and Rieti; the Tiber depression adjacent to Rome; the Sulmona depression on the east slope of the Apennines; the depression between the Lopini and Ernici mountains in Ciociaria; the depression of Upper Volturno in Campania; the Vello di Diano depression between Cilento and Lucania; and the Upper Agri depression in Lucania.

Along the Adriatic seaboard, on the Tavoliere plateau which is the biggest depression in this sector, the piezometric surface indicates a downward flow of the discharge to the sea with a gradient of 0.5 per cent with 7,300 wells under exploitation. Sea-water intrusion is increasing alarmingly.

Along the Tyrrhenian seaboard the depressions that affect the slightly inward-oriented sectors are: the plain between Pisa and Livorno (Leghorn); the marshy plains of Cecina, Piombino, Follonica, Grosseto and Orbetello; the plain surrounding Rome at the mouth of the Tiber; the Pontine marshes up to Cape Circeo; the Fondi zone; the mouth of the Garigliano; the Plain of Campania from the Volturno to Vesuvius; the plain of Sarno; and the plain of Pesto. In the Plain of Campania the estimated transmissivity is from 10^{-3} to 7×10^{-3} m^2/s.

The Sibari plain opens on the gulf of Taranto. It holds a multilayered aquifer with strata under pressure. The piezometric surface generally tends seaward, with a gradient of 2 per cent at the upper level and from 0.1 to 0.2 in the coastal sector. It indicates that the stratum is drained from the Crati River and fed from the adjacent watercourses. The aquifer's transmissivity is from 5×10^{-2} to 10^{-3} m^2/s.

The effective outflow is substantial and supports more than 5,000 wells and boreholes.

The sediments which filled these Quaternary depressions have a considerable range in thickness depending on their nature. Most common sediments occurring in the internal depressions are fluvio-lacustrian, sand-and-mud sediments with peaty stratifications, ranging in thickness from 50 to 150 metres and as much as 500 metres north of Florence; deposits of volcanic and travertine tuff are common, especially in the Latium depressions.

Fluvial detrital materials, coastal dune sands and, in the centre, volcanic tuff are the most common in the seaboard depressions; their thickness exceeds 400 metres in the Tuscan and Maremme plains, 200 to 300 metres in the Roman and Pontine plains of Latium, more than 1,500 metres in the zone of the Garigliano, and 3,000 metres in the Plain of Campania.

The valleys lying along the ridge of the Apennines branch off at right angles to the axis of the elevation. In the north, from Piedmont to the Adriatic, the watercourses flow into the Plain of Lombardy which they helped to form by feeding the Po River. They are valleys with substantial alluvium fill only in the last sector on the edges of the Plain. They are generally of moderate thickness and of rather low permeability and yield owing to the eminently flyschoid terrane. The most important aquifers in the west are those of Tanaro, Belbo, Bormida, Scrivia, Trebbia, Taro, Secchia Panaro and Reno. Along the south slope on this stretch of the Apennines, in Liguria, the few alluvial aquifers are of only strictly local interest, being no more than 20 metres thick and sparsely distributed.

There is one exception in this connection: on the border with Tuscany lies the Valla del Magra where the alluvium is 50-60 metres thick and subjected to sea-water intrusion along the shore due to extensive pumping.

Situated on the inland stretch of the Apennines, encompassing Tuscany, Umbria and Latium, are the tributaries of the Arno or its upper reaches (Val di Sieve, Valdarno, Val di Chiana, Val d'Elsa), of the Tiber (Val Tibarina, Val Nera, Tevere Medio), and of the Sacco-Liri which provide sizeable alluvial deposits. These deposits are normally not more than 50 metres thick but zones with higher accumulations have been found (Val Tiberina, over 150 metres; Val Nera, 60-100 metres). The discharge from these aquifers is steadily increasing. Around Lake Trasimeno alone, 2,300 wells have been inventoried.

Southwards in Campania, Basilicata and Calabria, the valleys along the crest are straight and short; they are almost devoid of alluvium capable of constituting aquifers of any, other than strictly local, utility.

Along the Adriatic seaboard there are more than 20 valleys with sizeable local alluvial deposits: they all run in parallel from south-west to north-east. Their thickness nowhere exceeds 60 metres and only their large areal extent makes these aquifers valuable for meeting the local water needs. Analysis of the alluvium indicates a steady increase in grain size with depth.

The prominent feature along the Tyrrhenian seaboard is the alluvial plains which have filled ancient depressions, whereas the watercourses discharging into them in the hinterland have no sizeable alluvium fill. In Calabria the alluvial plains of S. Eufemia, traversed by the Amato and Angitola rivers, and of Gioia, traversed by the Mesima and Marro rivers, are noteworthy; these two valleys are of great importance for the local economy thanks to the large volume of their aquifers which are 30 to 100 metres thick. These two aquifers feed more than 3,500 springs. Farther south, beyond the Strait of Messina, the "torrents" - very short and very steep watercourses which, in their last stretches, have accumulated substantial alluvial deposits even more than 100 metres deep along the coast - are of local importance; the erosion which generated these deposits still continues, and is so pervasive that man, in his efforts to avert alluviation of the plains adjacent to the watercourses, has made the beds slope down. These still moderate aquifers are being largely developed by more than 6,000 wells.

Ten or so valleys originating in the Apennines of Lucania flow into the Gulf of Taranto. Their alluvial deposits form a coastal plain (Metaponto) which has alluvial aquifers of considerable importance even though they are less than 100 metres thick and have poor permeability due to interstratified clay. Groundwater development there is quite limited.

Along the crest of the Apennines, most carbonate rocks occur as massifs or mountain ranges and only exceptionally as isolated outcrops. The first outcrops of some importance are those of the Umbro-Marches in the Metauro valley. These Mesozoic carbonate rocks, severely deformed by tectonic forces, are the site of an abundant karstic flow discharging into numerous springs or main watercourses of the zone. Contact with less permeable or with impermeable rocks creates local conditions which isolate the aquifers and generate artesian conditions downstream. Springs are very numerous. There are 27 known springs with yields of 100 to 1,000 l/s and 9 springs with yields higher than 1,000 l/s. Some of these springs are well known: those of Clitunno, 1.3 m^3/s; Sordo, 2.5 m^3/s; Narni, 11 m^3/s; Sta. Susanna, 5.5 m^3/s; Cotilla, 2.5 m^3/s; Capora, 4 m^3/s; Regina, 1.5 m^3/s; and Acquaoria, 1.5 m^3/s. All together the springs occurring in carbonate aquifers in Umbria, the Marches and Latium have a flow rate of about 45 m^3/s.

Many of the outcrops of the groundwater reservoir are concealed by lateral surficial deposits or by alluvium on the valley floor which are thus well supplied. The carbonate aquifer's discharge into the Nera River, which rises in the Sibillini Mountains, is estimated at 3 m^3/s.

Next come the outcrops of the Abruzzi and Latium Apennines where central Italy has its principal carbonate aquifers which, with their resources, secure the supply of the major cities of the two regions, including Rome.

Some of the calcareous outcrops, dating from the Triassic to the Miocene period, are bound by impermeable rocks which isolate them or generate perched aquifers. The connections between the various outcrops are not well known especially at depth, even though many high-pressure reservoirs have been discovered under the Tertiary and Quaternary covers.

The well-replenished aquifers and the widespread karstic systems give way to high-capacity springs such as those of Pescheria which supply Rome (18 m^3/s), those of Velino, those of Gran Sasso and, farther south, those of Cassino (approximately 20 m^3/s). Almost all the valley troughs drain part of these aquifers which, at their borders, overflow into the detrital-alluvial zones on their margins.

Calcareous outcrops extend all along the Latium Apennines with the Simbruini and Ernici mountains where the geomorphology is typically karstic. The drainage provided by the Tirino and Gizio rivers into the carbonate aquifers is estimated to be 17 and 35 m^3/s, respectively. Construction of the highway under Gran Sasso has resulted in drainage of the order of 2 m^3/s.

The flow lines in the Umbro-Latium carbonate rock massifs seem to be three in number: one in the direction of the bed of the Amiene River with a total flow of 46 m^3/s; one in the direction of the Cassino region with a total flow of 38 m^3/s; and a third to the Pontine marshes with a total yield of 25 m^3/s.

In the Apennines of Campania calcareous rocks occur more sparsely but, as a whole, they are the most important aquifers. The Matese mountains have a well-replenished aquifer with marked karstic features which discharges through large springs (Calore, Biferno). The calcareous uplands extend into the Picentini and Lattari ranges which form the Salentina Peninsula. The Picentini range has an aquifer of regional importance which discharges into the springs of the Sele (4 m^3/s), the Serino and the Apulia Aqueduct. Moreover, the aquifers in this part of the Apennines spill over into the permeable lands of the Campania hollow. The aggregate average yield of the springs fed by the Campania aquifers is close to 80 m^3/s.

Farther south, in the Apennines of Lucania, calcareous outcrops appear in a sequence of mountain complexes (Alburini, Cervati, Sirino) and end in the carbonate rocks of Pollino. These carbonate outcrops are surrounded and locked in by totally impermeable Tertiary terrane; they are the site of an abundant karstic flow which generates numerous springs, at least 10 or so of which have a yield of 200 to 700 l/s.

There are 260 logged springs with a total flow rate that varies in the course of one year from 8 to 12 m^3/s. The volume drained from the rivers (Noce, Lao, etc.) is estimated at 6 to 7 m^3/s.

Along the Adriatic seaboard calcareous outcrops appear in three distinct sectors: on the Gargano promontory, the Murge uplands and Salentina Peninsula.

The Gargano promontory consists of a prominent elevation containing an aquifer which gives way to a series of peripheral springs whose outflow spills into the sea. There are 27 known springs with a flow rate of more than 30 l/s, 14 with a rate of 100 to 300 l/s, and 5 with a rate of 300 to 1,000 l/s. The aggregate yield may be in the range of 4 m^3/s.

The Murge tableland, characterized by its bareness and obvious karstic geomorphology, holds a quite substantial reservoir of groundwater, amply supplied at various levels, but it also recharges many springs along the coast or spills directly into the sea. Both in this aquifer and in that at the end of the Salentina Peninsula, extensive sea-water intrusion occurs with a high rate of pollution.

The piezometric surface in the Murge tableland indicates a flow towards the Adriatic with a gradient of 0.1 to 0.3 per cent and a less marked gradient towards the Gulf of Taranto. Springs are few, but some have a sizeable capacity; the aggregate yield of the Murge springs is 7 m^3/s.

On the Salentina Peninsula, consisting of stratified and abundantly faulted calcareous rocks, there is a sizeable groundwater reservoir with an abundant karstic flow, the lower part of which suffers from salt-water intrusion. Intrusion of sea water occurs in some places where groundwater extraction is extensive.

The gradient of the piezometric surface (0.02 per cent) is minimal. There are 80 logged springs with an overall output of 10.4 m^3/s. The water in many springs is heavily salted.

Carbonate rocks along the Tyrrhenian seaboard are found either as isolated outcrops or in mountain ranges. They are everywhere a foot-of-the-mountain feature of the landscape and a factor in the local hydrogeology.

The first calcareous outcrops along the Tyrrhenian coastline, from the north, are the so-called Apuane Alps, famous for their marble caves, consisting of sharp and tectonized heights. Intensive fracturing has ensured a flow of water that is plentiful but of little local utility. Other isolated carbonate rock occurrences, alternately smooth and sharp, outcropping in the semi-permeable fields of flysch are found in the Colline Metallifere between the valleys of the Cecine and the Ombrone, in the marshy hills between Grosseto and Orbetello, in the heights of Mount Labbro south of Mount Amiata, and in the hills near Chianciano; all are rather small outcrops usually constituting heavily drained aquifers of little importance (only 24 springs with an output of more than 30 l/s are known). The deep waters circulating in the rocks are often in contact with hot masses which heat and mineralize them; such are the well-known mineral springs of Saturnia, with an output of some 600 l/s.

After the vast Latium depression that surrounds Rome, where soils of volcanic origin are prominent, calcareous outcrops reappear in the Lepini and Ausoni ranges, enhanced by a copious karstic flow replenishing the Pontine depression where large-scale land-reclamation and drainage works have been carried out. Spring outflow into the Pontine marshes is estimated at about 20 m^3/s. Farther south, the Ausoni range has a sizeable aquifer discharging from one side into the Liri River and spilling directly into the sea near Gaeta. After the Gulf of Naples, and between it and the Gulf of Salerno, the Latari mountains rise to form the Sorrenta Peninsula, a continuation of the Picentini mountains, and extend onto the Isle of Capri. The aquifer in these carbonate rocks feeds numerous springs (Castellammare di Sorrenta, Cava dei Tirreni) with a flow rate higher than 50 l/s, but part of it is discharged directly into the sea. The last calcareous outcrops to be found along the Tyrrhenian coast are those of Cilento near Cape Palinuro.

A series of "anti-Apennine" uplands of volcanic origin runs along the west coast of the Apennines and continues through Latium from Tuscany to Campania. Migration of the volcanic centres from north to south has confirmed that volcanic activity has been gradually phasing out; in recent times the only activity has been in the Naples district, the Phlegraean Plain and Vesuvius.

In many extinct volcanic craters, the progressive weathering of surface rocks has led to the formation of impervious soils and to crater lakes (Bolsena, Vico, Bracciano, Albano and Nemi).

The present spreads of outcropping lava have generally been found on pre-existing Cretaceous-Oligocene or Pleistocene and somewhat turbid, mainly argillaceous or flyschoid, impermeable soils, or on sand and tuff. The permeability of the volcanic rocks varies considerably and depends on their degree of weathering; it is greater in the upper strata, although it commonly varies in depth owing to the presence of interbedded layers, turned to clay by the rapid weathering that occurred during the quiescent phases of the effusive phenomenon, which has resulted in vadose water and springs at various altitudes. The peripheral watercourses are a heavy drain on volcanic elevations and their meanderings - everywhere on the periphery and in places where they are in contact with permeable surficial ground - amply replenish the aquifers contained therein.

From north to south the first volcanic elevation in Tuscany is Mount Amiata, all by itself in semi-permeable ground. On its periphery the aquifer replenishes some 20 springs, one of which (S. Flora) has an average flow rate of 700 to 800 l/s, 4 a rate of 125 to 250 l/s (Piancastagna, Fosso Caro, Peschiera, Olmicciolo), and 6 others a rate of 20 to 70 l/s.

Next comes the volcanic area of Mount Volsini, Cimini, Tolfitano-Ceriti and Sabatini, bounded on the east and south by the Tiber River. In this zone are many vadose reservoirs which replenish springs, even those with high yields (100 to 500 l/s); less productive springs are situated on the periphery of the volcanic outcrops. All together there are 41 springs with an output of more than 10 l/s.

The higher-capacity springs are situated on the periphery of Lake Bolsena. The most important springs are Vitozo (basin of the Lente River, 230 l/s), Vene di San Lorenzo Nuovo (Lake Bolsena, 270 l/s), Barano (Lake Bolsena, 105 l/s), Fabrice de Roma (Lake Vico, 200 l/s), Trianee (Lake Bracciano, 206 l/s), and Cavacoci (of the Treia River, 200 l/s). Many wells replenish the reservoir in the deeper aquifer.

There are outcrops in the volcanic terrane of the Albani and Latium Hills south of Rome. The hydrologic conditions in these smaller outcrops are similar to those of the preceding zone: the groundwater reservoirs feed numerous springs and are tapped by a large number of wells. The best known spring (Acque Vergini) has a flow rate of 1,100 l/s.

The last volcanic outcrops on the Italian peninsula are in the Neapolitan region: the Phlegraean Plain and the Plain of Vesuvius. The aquifers of the Phlegraean Plain are of minor importance because of the paucity of volcanic cover and of the relative impermeability of the land. Mount Vesuvius, on the other hand, is much more important thanks to the remarkable permeability of its rocks. The volcanic aquifer feeds most of the groundwater reservoirs situated in the underlying sedimentary ground which extends to its periphery and gives way to some springs. Numerous wells draw from the reservoir at lower altitudes.

Along the crest of the Apennines, where argillaceous and flyschoid terranes prevail, vast areas have no major aquifers. On the other hand, there are small aquifers, limited to marly or sandy intercalations of flysch, or to the more or less thick lenses of sand, conglomerate, sandstone and evaporites dating from the Miocene and Pleistocene periods. They are spread over small areas and give way to a few springs of low capacity that generally yield less than 100 l/s but are valuable because they feed the aqueducts of many small urban centres or villages in the hilly or mountainous zones of central Italy.

In Calabria crystalline and metamorphic rock aquifers of local importance give way to large numbers of small springs whose yield varies considerably in time. There are 2,662 inventoried springs with an average aggregate flow rate of about 10 m^3/s.

4. Sicily

In Sicily, depressions filled with Pliocene and Quaternary deposits include: the Plain of Catania, where predominantly argillaceous terrane alternates with sandy or conglomerate lenses and, along the southern border, with volcanic effusions; the plain of Lentini; the coastal plain of Lisimelie swamp; that of Priolo-Augusta; that of the Lisimelie plain near Syracuse; and that of Trapani-Marsala-Cape Granitola. Northwards a similar situation recurs in the Milazzo zone. Fills range in thickness from a few dozen to more than 300 metres.

In the plains of Priolo-Augusta, Lentini and Syracuse, the Quaternary aquifer has largely drained into the watercourses that run through it, but the deep carbonate aquifer is important in the same places. In the Plain of Catania varied local permeability conditions detract from the aquifer's importance, even where the Quaternary ground is in contact with Pliocene and Pleistocene volcanic rocks to the extent that these derive from submarine effusions and appear to be intensively weathered and argillaceous. The estimated transmissivity is from 2×10^{-3} to 10^{-4} m^2/s. The Quaternary aquifers of Milazzo and Marsala-Cape Granitola are more important and largely exploited.

Alluvial aquifers in Sicily's river valleys are of importance only in the north and north-east sectors. In the rest of the island they mostly include mud and clay, for the most part fed by brackish water from areas of the evaporite series. In the north and north-east sectors the watercourses have characteristics similar to those of the "Calabrian torrents": the first stretch is short and steep and followed by heavily alluviated stretches of valley with aquifers 10 to 80 or 90 metres thick developed by many wells but also by drainage galleries some of which were dug more than 100 years ago. The output of these aquifers is important for the local economy, but is always small.

Much more important in Sicily are carbonate aquifers, some of which are autochthonous (Sicani Iblei) and some allochthonous, and therefore intensively tectonically altered and deformed, calcareous outcrops which emerge in isolated masses surrounded by Tertiary and Pleistocene mainly flyschoid or strictly argillaceous rocks. Along the northern ridge lie the outcrops of the summits of the Madonie mountains whose aquifer replenishes the springs of Cefalu or Scillato (aqueduct of Palermo), and part of which spills into the sea. Farther west lie the outcrops of the calcareous Imerese mountains whose aquifer replenishes a few springs and most of which flow into the sea. Next come the rugged and heavily foliated uplands of Palermo. The aquifer in these uplands discharges into few springs and, more abundantly, into the Quaternary sandy-calcarenitic soils of the flatlands surrounding them. Part of the aquifer drains directly into the sea. At the extreme north-west part of the island lie the calcareous uplands of Trapani. Much of the water they contain flows into the sea; the rest feeds small springs or discharges into permeable Quaternary deposits on their periphery. All together the calcareous uplands replenish some 625 springs at a total flow rate of 6.8 m^3/s.

In the southern central part of the island looms the ridge of the Sicani mountains, completely surrounded by impermeable ground. The overflow of the aquifer feeds peripheral springs which constitute the only major underground resources in this sector of the island where 252 springs with an aggregate outflow of 2.5 m^3/s have been logged.

To the south-east lies the calcareous Iblei tableland, consisting of a nearly horizontal carbonate series, part of which is limestone and marl marked by severe faults and graben-and-horst features. The larger springs are on the south and east slopes of the tableland and, to the west and north, the waters of the aquifer discharge into Pleistocene formations which cover the peripheral limestone. The principal springs are those of the Scicli zone to the south and those of Syracuse to the east (Ciane, 600 l/s; Fons Aretusa, 250 l/s; S. Calogero, 200 l/s). All together 163 springs have been inventoried on the south slope, with a total yield of 3.3 m^3/s, and 168 on the east slope, with a total of 1.7 m^3/s. Groundwater development in the lower-level zones has led to the disappearance of many springs and has greatly diminished the outflow of others. The aquifer has lost its artesian characteristics almost everywhere and shows evidence of considerable sea-water intrusion.

In the space of 20 years, lowerings of as much as 100 metres and more have occurred in the piezometric surfaces. The situation that arose in the Priolo-Augusta zone, where the present piezometric surface is between 50 and 100 metres below sealevel, is well known in this connection; artificial recharge experiments have been made as to stem sea-water intrusion. During the various tectonic phases which have affected the carbonate sedimentation of the Iblei basin, submarine or sub-aerial volcanic effusions have been confirmed which have tended to interpose or cover the sediments at the depositing stage. The eruptive phenomena have subsequently drifted from south to north, and today the most ancient Cretaceous lavas are to be found in the south, near Pachino, and the most recent Pleistocene ones on the southern edge of the tableland, on the boundary with the plain of Catania. The soils range from impermeable to highly permeable and, being in close contact with the carbonate rocks, they join with those rocks to form a single aquifer.

Much more important are the volcanic rocks that constitute the Etna massif, the great Quaternary and still active volcano, Europe's biggest.

Subsequent volcanic eruptions have formed a lofty cone 3,340 metres above sealevel. Lava flows and pyroclastic accumulations spread over an area of more than 1,300 km^2, on top of sedimentary, mainly argillaceous, land already moulded by erosion and whose ancient troughs have now become the main routes for the water flowing onto the volcanic aquifer.

On three sides - north, west and south - the volcanic rocks are surrounded by clay and, where they come into contact with it, many springs of utmost importance for the local economy occur.

The volcanic rocks on the east side are in direct contact with the sea into which spills part of the aquifer water. The discharge of the waters of Etna started thousands of years ago and accounts for the development of an extremely bountiful agriculture. The formation of vadose reservoirs is attributable to the presence of argillaceous strata. Most of the discharge is via drainage galleries, some of them longer than 2 km, and only a minimum flows through wells, even though this pattern is spreading. Most of the wells are wide in diameter (2 to 3 metres) and 250 to 400 metres deep; boreholes are fewer. Actually they produce at the rate of 15 m^3/s, but the estimated total capacity of the aquifer is about 20 to 25 m^3/s.

Around Etna, there are 289 springs, and 38 drainage galleries of which 18 have average yields of 110 to 1,000 l/s. There are more than 1,200 wide-diameter wells and 189 boreholes.

Many of the small islands around Sicily (Eolie, Ustica, Pantalleria, Linosa) are also of volcanic origin but the fresh-water lenses they hold on top of the salt water is quite thin and the resource minimal or nil.

A feature of local, but considerable, interest is the presence of Quaternary sand and small calcareous rocks spread over various zones of the islands on the Pleistocene argillaceous covers. They are found in the Marsala-Mazzara del Vallo zone where the piezometric gradient is 1 to 1.5 per cent, in the zone of Piazza Armerina, on the Vittoria plateau, where 1,286 logged wells are situated with an estimated transmissivity of 3×10^{-3} to 10^{-3} m^2/s.

The aquifers in the evaporites of the Sicilian gypsum-sulphur series are even smaller. They are scattered strips, generally indicating severely deformed synclines, in which there are about 350 known springs with an aggregate average yield of 500 l/s. The waters of these springs are in some cases heavily mineralized. In the crystalline and metamorphic formations of the Peloritani mountains there are 321 known wells with an aggregate average yield of about 2 m^3/s.

5. Sardinia

The main depression of Sardinia, filled by sediments of the early Tertiary to Quaternary periods, is the Campidano di Cagliari. The Pleistocene and Recent areas consist largely of shingle, sand and mud up to 200 metres thick. The resultant aquifer is therefore of considerable importance for local agriculture and is expensively tapped by more than 1,600 wells.

The piezometric surface has gradients of the order of 0.2 per cent; one half of the ground-water reservoir flows towards the Gulf of Oristano and the other towards the Gulf of Cagliari.

Sardinian river valleys have small alluvial deposits but there are local accumulations in the Tirso, Fiumendosa, Isalle, Posado, Caghinas and Mannu valleys, up to 50 metres thick, and with aquifers of local importance.

Sardinia's calcareous rocks constitute sizeable aquifers. In Turrintano near Sassari, Miocene calcareous rocks contain a major aquifer, partly drained by the watercourses which, in Anglone, flow into the sea. There are more than 500 small springs with an aggregate average yield of 0.45 m^3/s, and a large number of wells situated on the periphery of the highlands.

In the extreme north-west part of the island the Mesozoic calcareous rocks of the Nurra are extensively tapped by numerous wells, which has resulted in the disappearance of many springs; part of the aquifer discharges into the sea. There are 43 springs with an aggregate average yield of 200 l/s.

In the east central sector of the island are various outcrops of calcareous spikes, the largest of which faces the Gulf of Orosei (Massiccio di Dorgali). These aquifers hold the only groundwater resource in the zone. Some springs have flow rates of 20 to 120 l/s. There are 155 springs with an aggregate average yield of 1,100 l/s. The quantity of water lost to the sea from these aquifers is from 400 to 700 l/s.

In the south-west, in Iglesiente, there are about 100 logged springs with a total average discharge of about 500 l/s. The calcareous outcrops are in contact with heavily mineralized zones and, owing to mining activities, it was necessary to clear the aquifers by means of draining and lifting operations involving the extraction of an estimated water volume of 1,600 l/s.

The entire west central part of the island was affected by volcanic activity. There are two sectors: one in which the acid volcanic rocks of the early and middle Tertiary period are prominent, heavily fractured and weathered; and the other in which the basic volcanic rocks of the late Tertiary period are prevalent. There are small aquifers, on the periphery of which about 2,000 springs are situated, but they all have small or very small yields, less than one litre per second and, in exceptional cases, from one to 10 l/s. The estimated total yield is 2 m^3/s, of which 0.5 m^3/s is for the acid and 1.5 m^3/s for the basic volcanic rocks.

Crystalline and metamorphic and the Paleozoic outcrops of east Sardinia are also of great importance. In the granite and gneiss there are 3,611 logged springs with an aggregate yield close to 1 m^3/s that changes considerably depending on the season. In the schistose and sandy outcrops are 3,060 springs with an aggregate average yield of 1.85 m^3/s.

D. Groundwater balance

In general terms, the water balance of Italy can be drawn tentatively as follows keeping in mind that rainfall regimes vary from one region to the other and from one year to the other. The estimated total average for precipitation is about 300 km^3 per year. Estimated losses through genuine evapotranspiration as a whole amount to some 140 km^3 per year.

The remaining balance of 160 km^3 should go towards recharging the watercourses and groundwater reservoirs. The relationship between surface waters and underground waters is extremely complex, as they flow in one direction and then in another, even repetitively in one and the same catchment basin. It is therefore estimated that direct infiltration and the river and reservoir exchanges feed into the Italian aquifers an annual average aggregate volume of 43 km^3. A substantial part of this volume is returned to the watercourses, another part spills directly into the sea, and the rest (about 13 km^3) is put to use.

A recent analysis of current water uses has made it possible to compile the following estimates of the quantities of water extracted from the various aquifers:

Aquifer type	Amount (in km^3)
Carbonate rocks	2.9
Volcanic rocks	0.9
Quaternary depressions	6.3
Alluvium	1.5
Pliocene-Pleistocene sediments	0.6
Crystalline and metamorphic rocks, flysch	0.8
Water use	
Community water supply	6.5
Industrial	1.7
Agricultural	4.6
Miscellaneous	0.2
Source of supply	
Springs	3.5
Wells and galleries	9.5

In Italy, groundwater resources are good mostly in the northern regions (Piedmont, Lombardy, Venice). Conditions differ considerably in the other regions in many of which groundwater consumption has reached or exceeded natural recharge.

Although groundwater extraction, whether based on catchments, on draining galleries, on dug or drilled wells, is steadily increasing without control or rational planning the fears rise for the future of the resource in quality, quantity and availability and for damage to the environment. Many studies and proposals have been made with a view to formulating a rational plan for the utilization and management of water resources, but no decision to that end has been adopted except in a few cases at the local level.

There are about 600 hydrogeologists in Italy, and the country has developed considerable expertise in the hydrology of karst, volcanic rocks (Etna, Apennine), metamorphic rocks (Sardinia) and coastal aquifers.

IX. LUXEMBOURG

Area: 2,586 km^2 Population: 369,000

A. Physiography and stratigraphy

The Grand Duchy of Luxembourg is a landlocked country situated at the juncture of the Federal Republic of Germany, Belgium and France. Towards the north and north-west it is linked to the Ardennes and towards the south and south-west it is linked to the gently rolling countryside of Lorraine. Two large geomorphologic units may be distinguished:

To the north, Oesling takes the form of a plateau (the former peneplain of the Ardennes) with an average altitude of about 450 metres. Several rivers flowing from north to south, such as the Our, or from west to east, such as the Sûre, have cut deeply into the plateau to form narrow enclosed valleys of some 100 metres relief. Oesling is made up of folded Lower Devonian sediments among which schist predominates.

To the south is Gutland with an average altitude of 250 metres. Its landscape is made up of cuestas in which the sedimentary layers dip towards the south-west, i.e., towards the centre of the Paris basin. It is made up of Mesozoic layers partly of Triassic and partly of Jurassic rocks.

The four cuestas which run from north-east to south-west are: the cuesta of Muschelkalk dolomites, the cuesta of Luxembourg sandstone, the cuesta of middle Liassic and the cuesta of the Dogger whose basement rock contains the iron ores known as "minette". Intercalated among the hard layers which have formed the high ground are outcrops of argillaceous-marly sediments.

Luxembourg has a sub-Atlantic climate with humid and cool tendencies. Although Luxembourg is a small country, fairly marked climatic differences exist between Oesling and Gutland with a striking difference between the city of Luxembourg (altitude 300 m), situated in the centre of Gutland, and the valley of the Moselle (144 m) at the eastern end of Gutland. Those variations are shown in table 34.

Nearly the whole of Luxembourg belongs to the Moselle River basin and therefore of the Rhine; only 51 km^2 are drained by the Chiers, tributary of the Meuse. The coefficient of runoff for the whole territory is 43 per cent. The wooded area of Luxembourg was 88,500 hectares in 1984. The principal geologic units are described in table 35.

Table 34. Luxembourg: Climatic data

Aspect	Oesling: Abbey of Clervaux	Gutland: Luxembourg City	Moselle: Grevenmacher
Altitude	454 m	330 m	144 m
Annual average temperature (1908-1975)	7.8° C	9.1° C	9.8° C
Average temperature, May-July (1942-1975)	18.1° C	19.3° C	20.8° C
Maximum absolute temperature, 1942-1975	35.5° C	35.5° C	37.0° C
Absolute minimum temperature, 1942-1975	-23.5° C	-23.1° C	-20.0° C
Average number of days of frost per year (1949-1982)	94	73	65
Number of days when the ground is covered with snow (1949-1975)	47	35	10
Average annual precipitation (1908-1975)	818 mm	780 mm	694 mm
Average annual precipitation, May-June (1906-1975)	219 mm	199 mm	186 mm
Average number of days per year with precipitation above 0.1 mm (1949-1982)	194	164	149
Average number of hours of sunshine per year	1 500	1 700	1 900
Direction of prevailing wind (1961-1975)	North	South-west	South

Sources: Technical agricultural services, Meteorological and Hydrological Service, Luxembourg, 1983 and Basic Programme of the State Planning Office, Ministry of Finance, Luxembourg, 1978.

Table 35. Luxembourg: Stratigraphy

Era	Period	Epoch	Lithology	
Cenozoic	Quaternary	Holocene	Clay, sand, gravel	
		Pleistocene	Clay, sand, gravel	
Mesozoic	Jurassic		Lorraine facies	Detrital facies
		Bajocian	Marly-limestone	Minette
		Aalenian	Marl	-
		Toarcian	Marl	-
		Pliensbachian	Marl	-
		Sinemurian	Marl	Sandstone
		Hettangian	Marly-limestone	Sandstone
	Triassic		Basin facies	Border facies
		Keuper	Marl, sandstone	Argillite
		Upper Muschelkalk	Dolomite	Sandstone
		Middle Muschelkalk	Marl	Conglomerate
		Bunter	Sandstone	-
Paleozoic	Lower Devonian	Emsian	Schist (Greywacke, quartzite)	
		Siegenian	Schist (Greywacke, quartzite)	

B. Hydrogeology

1. Regional aspects

Oesling is made up essentially of argillaceous-siliceous rocks. They are compact, of extremely low porosity and lacking in deep open joints. They do not allow the circulation or accumulation of water of any importance and thus cannot constitute aquifers.

Gutland, however, with its great diversity of rocks and alternating permeable and impermeable layers, with its less dramatic tectonics and the arrangement of layers in anticlines and synclines with a large radius of curvature, present favourable conditions for the occurrence of important aquifers.

Table 36 shows the aquifers of the Grand Duchy of Luxembourg, their importance for water supply, their stratigraphic position, their lithologic nature, the type of permeability, and their thickness and depth. The position of these aquifers is shown in the hydrogeologic map (figure 35).

Table 36. Luxembourg: Principal aquifers

Aquifer and relative importance	Stratigraphic position	Lithology	Permeability type	Thickness (m)	Depth (m)
Alluvium, minor	Holocene	Mud, sand, gravel	Pore	4-12	0
Corolline limestone	Bajocian	Limestone)			
Oolitic iron ore	Aalenien, Toarcian	Ferruginous sandstone)	Mixed	135-210	0-210
Upper Liassic sandstone, moderate	Toarcian	Marly sandstone)			
Middle Lias or Macigno sandstone, moderate	Upper Domerian	Marly sandstone	Mixed	0-100	0-200
Luxembourg sandstone, most	Upper Hettangian, Lower Sinemurian	Calcareous sandstone	Mixed	20-100	0-400
Reedy sandstone, minor	Middle Keuper	Micaceous sandstone	Mixed	0-50	0-100
Muschelkalk, major	Muschelkalk	Calcareous dolomite	Fractured	55	0-100
Shelly sandstone, minor	Muschelkalk	Sandstone, dolomite	Mixed	15-20	0-50
Buntsandstein, major	Upper and middle Bunter	Sandstone and conglomerate	Pore	0-250	0-800
Triassic border facies, major	Upper and Lower Keuper	Sandstone, conglomerate, dolomite)			
	Lower Keuper	Sandstone)	Mixed	50-250	0-800
	Muschelkalk	Dolomite)			

Figure 36. Luxembourg: Hydrogeology

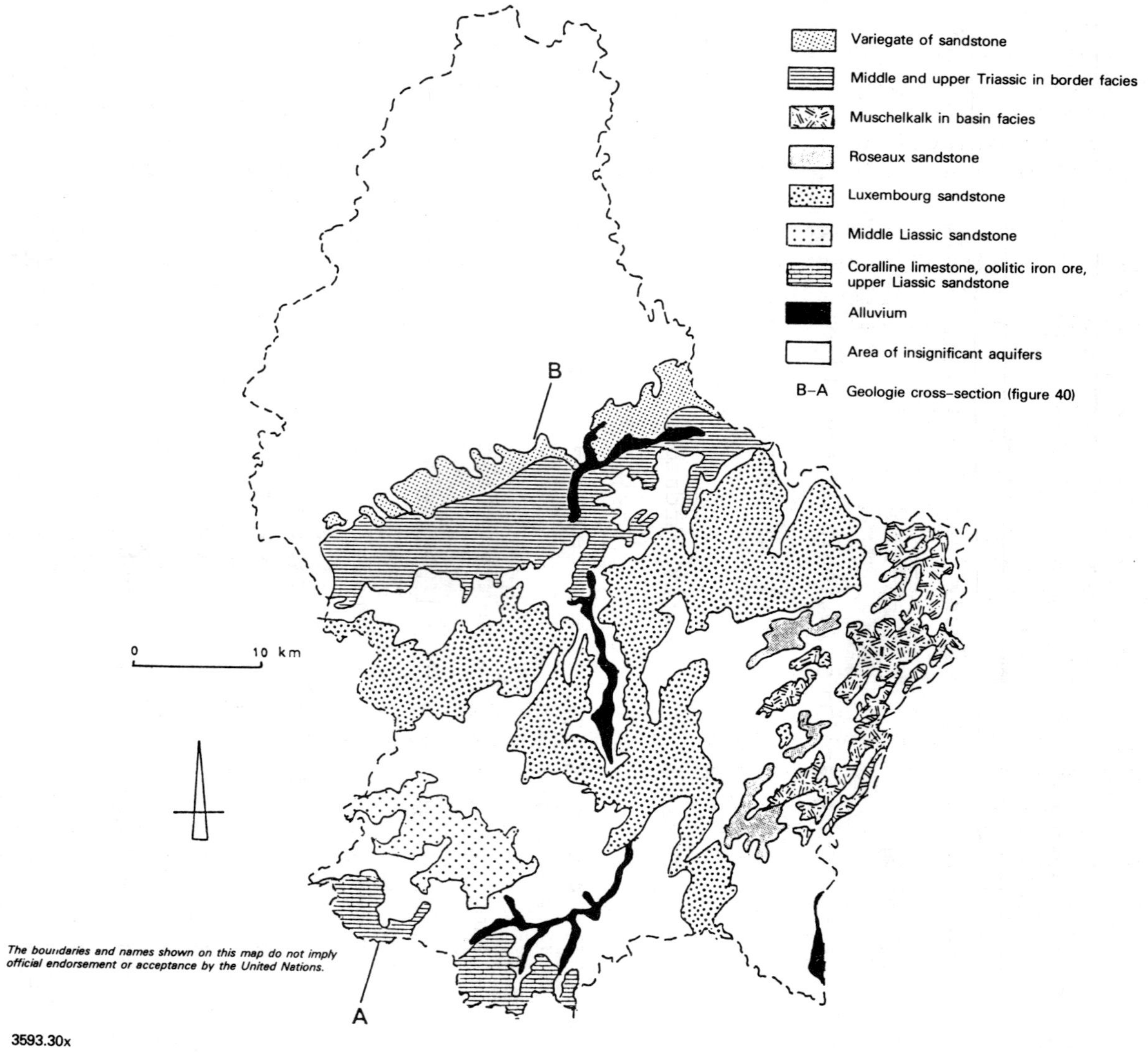

2. Principal aquifers

(a) Buntsandstein

The Buntsandstein (Bunter, Triassic) is made up of argillaceous-sandstone rocks in mid-basin facies and sandstone-conglomeratic rocks in border facies. Permeability is mostly of the pore-type.

From the point of view of chemical composition, three layers have been recognized: a fresh-water layer (total hardness: 20 French degrees), a layer of very hard water (total hardness: 60 French degrees), and a layer of mineralized water (total mineralization higher than 15,000 mg/l). In the first two layers, calcium carbonate hardness varies between 18.5 and 39.5 French degrees.

The layer of fresh water is in the border region north of Gutland where the Bunter is developed essentially in a border facies of sandstone and conglomerates. From west to east and north to south the hardness of the water gradually increases and finally is highly mineralized. The transmissivity determined for a well amounted to 9.3×10^{-4} m^2/s.

(b) Shelly sandstone

In its upper and lower parts, the shelly sandstone is made up of dolomitic sandstone. In the middle part, marl predominates. In view of this lithologic composition and modest thickness (15-20 m) it contains an aquifer of only minor importance and plays but a local role in the water supply. The total average hardness is 45 French degrees. The transmissivity is not known.

(c) Upper Muschelkalk

In the basin facies, the upper Muschelkalk composed of dolomitic limestone has a thickness of the order of 55 m. This aquifer contains the most important layer in the region of the Moselle and the lower Sûre but its utilization for supplies of drinking water is reduced because of its low filtering power. The total hardness of the water in this aquifer is between 45 and 50 degrees. The carbonate hardness varies between 21 and 30 French degrees. The transmissivity determined for a well amounted to 3.9×10^{-3} m^2/s.

(d) Roseaux sandstone

The roseaux sandstone takes the form of lenses of micaceous sandstone with clay cement. It has lateral extensions of several kilometres and fairly sharp variations in thickness. This aquifer is exploited locally. The total hardness of the water in this aquifer varies considerably, from 30 to 53 French degrees. The carbonate hardness varies between 22.6 and 36.5 French degrees. The transmissivity determined for a well was 6×10^{-4} m^2/s.

(e) Luxembourg sandstone

The Luxembourg sandstone is by far the most important aquifer in Luxembourg. It is a sandstone with calcareous cement, well-bedded, fractured and, because of the commonly uneven distribution of the cement, is of uneven consistency. Over the whole of the Luxembourg territory it covers the stage of the upper Hettangian and a part of the lower Sinemurian. Because of the general dip of the layers towards the south-west, it becomes submerged in a line from Steinfort to Luxembourg and then Mondorf, under younger formations, most of them impermeable. In this region the aquifer is confined and constitutes an important source of supply which has been little exploited.

The thickness of this aquifer which outcrops over 350 km^2 ranges from 20 to 100 metres. Of the drinking water derived from underground sources, 90 per cent comes from the Luxembourg sandstone aquifer. The chemical composition of the aquifer varies, according to the regions, from 10 to 35 French degrees for total hardness and from 7.3 to 27.5 French degrees for carbonate hardness. The average transmissivity is 10^{-3} m^2/s.

(f) Middle Liassic sandstone

In the western part of Gutland, the upper Domerian is developed in a sandstone-marly facies, the middle Liassic sandstone. This aquifer, which in places is 100 metres thick, has hardly been exploited so far and, therefore, constitutes a useful water reserve. The total hardness of the water from this aquifer varies between 10 and 40 French degrees depending on the region. The transmissivity is not known.

(g) Coralline limestone, oolitic iron ores, upper Liassic sandstone

In the past, there were three quite distinct aquifers separated by formations with low permeability. Following iron-ore mining operations both under and above ground, the three aquifers became interlinked and today only the deepest aquifer (upper Lias sandstone) is developed in two locations only. The total hardness of the water still being used is 15 French degrees and has a carbonate hardness of 11.6 degrees. The transmissivity is not known.

(h) Middle and Upper Triassic in border facies

In the region of the northern edge of Gutland, the Triassic is developed in a border facies of sandstone and conglomerates overlaying in discordance the impermeable Devonian basement. This is an aquifer with diverse permeability, with many layers, which contains partly Keuper and Muschelkalk in its totality. It is about 246 metres thick to the east, but only about 80 metres thick to the west. The total hardness of the water in this aquifer is from 10 to 40 French degrees. The transmissivity determined for a well is 5.75×10^{-3} m^2/s.

So far, this aquifer which constitutes an important reserve has been utilized only on the local and regional level. It was to be developed for use on the national level during 1989.

(i) Alluvium

Since the thickness of the alluvium does not exceed 8 metres, such aquifers are important only for supplying isolated farms with water.

C. Groundwater development

1. Use and quality

During a hydrogeologic study carried out in 1980, it was estimated that the aquifers above could yield about 70,000 m^3 per day. This estimate was based on two studies made in 1971, one on the Luxembourg sandstone aquifer and the other on that of upper Muschelkalk. Since then exploratory drillings have been made in the Buntsandstein aquifer and the Middle and Upper Triassic in their border facies and in the Luxembourg sandstone aquifer. The results show that the water supply is much higher than that estimated in 1980.

Luxembourg has recently established a reserve water supply based on the aquifers of Buntsandstein and Triassic in border facies and the Luxembourg sandstone aquifer. This supply involves 20 boreholes of an average depth of 150 metres and may supply between 50,000 and 60,000 m^3 per day over a limited amount of time.

All the localities of the Grand Duchy of Luxembourg are equipped with public water supply systems to which more than 99 per cent of the population are linked; the rest includes isolated houses supplied by private springs or wells which are not subject to compulsory hygienic control.

The average consumption of the water distributed by the public system is today 125,000 m^3 per day, which corresponds to about 350 litres per inhabitant per day, including industrial consumption. However, according to statistical data, the water consumption for domestic purposes only is about or is in the range of 170 litres per day.

Out of the 125,000 m^3 of drinking water distributed by public systems, two thirds is of underground origin, mainly (90 per cent) from the Luxembourg sandstone aquifer, whereas a third comes from surface water treated at the waterworks of the Syndicat des Eaux du Barrage d'Esch-sur-Sûre (SEBES).

Of all the groundwater utilized in Luxembourg, only 18 per cent comes from wells and springs. The water distributed conforms in its totality to the national legislation of 11 April 1985 concerning the quality of water for human consumption. This legislation defines the concentrations which should not be exceeded for physical-chemical and microbiological parameters and with respect to the minimum frequency of analytic tests.

Luxembourg is also facing the problem of the increasing nitrate content of drinking water caused by over-intensive agricultural activities in the areas of the aquifers. Today some 15 small rural localities, representing 0.7 per cent of the population of Luxembourg, are supplied totally or partially by a water whose nitrate content is higher than the prescribed level of 50 mg/l.

2. Agencies involved

Three ministries are concerned with different aspects of groundwater (with address):

a. <u>Ministry of the Environment</u> (with two offices)

   i. Commissariat général à protection du sol et des eaux
      43, Avenue Grande-Duchesse Charlotte, 1331, Luxembourg

   ii. Administration de l'Environment
       1a, Rue Auguste Lumière, 1950, Luxembourg

b. <u>Ministry of Public Works</u>

   Administration des Ponts et Chaussées
   Service Géologique, Division des Eaux,
   43, Avenue Grande-Duchesse Charlotte, 1331, Luxembourg

c. <u>Ministry of the Interior</u>

   19, Rue Beaumont, 1219, Luxembourg

References

Aluseau. L'eau potable au Grand-Duché de Luxembourg. Association luxembourgeoise des services d'eau (Aluseau), 1987.

Barthel, J., J. Bintz, G. Hansen and others. L'approvisionnement en eau potable du Grand-Duché de Luxembourg, Rapport interne, 1976.

Bintz, J. Les ressources en eau souterraines, Syndicat des Eaux du Barrage d'Esch-sur-Sûre, p. 146-149, 1971.

Bintz, J., A. Hary and Ad. Muller. Guides géologiques régionaux, Ardennes-Luxembourg, p. 135-188, Masson et Cie., 1973.

Comission des communautes Européennes. Bilan des ressources en eau souterraine du Grand-Duché de Luxembourg. Th. Schäfer GmbH D-3000 HANOVER 1.

Von Hoyer, M. Hydrogeologische und hydrochemische Untersuchungen im luxemburger Sandstein. Publication du Service géologique du Luxembourg, vol. 21, 1971.

Lahr, E. Temps et climats du Grand-Duché de Luxembourg. Ministère de l'Agriculture, 1964.

Lucius, M. Le Bassin Hydrogéologique du Gutland. Extrait Revue Technique Lux. No. 6, 1939.

__________. Les nappes d'eau souterraines du Luxembourg et leur utilisation rationnelle. Extrait Revue Technique Lux. No. 4, 1949.

Marque, P., G. Als, F. Hoffman and others. Luxembourg. Christine Bonnetone, Le Puy, 1984.

Neumann, C. Hydrogeologische und hydrochemische Untersuchungen im oberen Muschelkalk und Keuper Luxemburgs. Publications du Service géologique du Luxembourg, vol. 22, 1971.

X. MALTA

Area: 316 km^2 Population: 358,000

1. Physiography and climate

The Republic of Malta in the central part of the Mediterranean Sea consists of the islands of Malta, Gozo, Comino and two uninhabited islands. The distance between Malta and the nearest point in southern Italy is 93 km and its distance from the nearest point in the North African mainland is 290 km. Malta, the largest island, occupies 245.7 square kilometres and the second largest, Gozo, 67. At its longest, Malta is 27 km and is 14.5 km at its widest.

The climate is Mediterranean with a mild, wet winter and a long, dry summer. The seasonal rhythm is clearly marked, a wet winter invariably followed by a long dry summer.

Temperature variations range from 6° C in winter to 32° C in summer. The average daily sunshine hours range from about 5 hours in December and January to nearly 13 hours in July.

Because a large proportion of the potable water of the islands is derived from precipitation, the rainfall pattern is very significant and records have been kept systematically for over a hundred years. The annual total averages 470 millimetres. The wet season proper usually starts in October when rainfall is likely to be heavy. Most of the rain falls during November, December and January.

There are no perennial surface streams in Malta. Water flows along the bed of major valleys for only a few days after heavy downpours. About 3 per cent of the precipitation, measured over many years by runoff recorders located at exit points of major drainage areas, finds its way into the sea. The major surface drainage lines cross the island from sources near the western coast to enter the sea on the east. Surface water thus has ample time and opportunity to seep into the underground aquifers through fissures and permeable rock formations.

Many dams have been constructed across these drainage lines, primarily to retard storm discharge and so serve the dual purpose of increasing percolation to the aquifers and slowing down the rate of soil erosion.

Field walls are also kept in good condition whenever possible to retard soil erosion. Since time immemorial, the farmers of Malta have excavated underground tanks to store rainwater collected from the roofs of farmhouses, lanes and galleries. Tens of thousands of such wells are scattered over the island. In addition, roof catchments in towns and villages are mandated by law. Every house built in town or village has to include a sealed cistern with a capacity of 600 litres per square metre of roof space of the house.

B. Hydrogeology

1. Stratigraphy

The Maltese Islands lie on the eastern edge of the extreme North African oceanic shelf which extends from the coast of Tunisia on the west to the Ionian Sea on the east and from the Libyan coast on the south to Sicily on the north.

Here the shelf is predominantly calcareous, of fossil origin, at least from the Upper Triassic to the upper Tertiary. The Maltese Islands themselves are made of a sequence of calcareous rocks of Miocene age with the lower parts possibly of Paleogene age. These formations are locally covered by thin Quaternary deposits (figure 37a).

The stratigraphic sequence is summarized in table 37.

Table 37. Malta: Stratigraphy

Geologic period or epoch	Formation	Thickness (m)	Lithology
Quaternary	..	2-10	Fluvial silt, clay, conglomerate
Pliocene			
Tortonian	Upper Coralline Limestone	80-100	Calcareous, white or yellow, porous, compact
Miocene			
Helvetian	Greensand	1-12	Sandstone of various colours
Lower Helvetian	Blue Clay	0-65	Blue, yellow clay and marl
Burdigalian	Globigerina	30-230	Soft, yellow, argillaceous
Paleocene			
Aquitanian	Lower Coralline Limestone	150-225	Semi-crystalline

Figure 37a. Malta: Geology

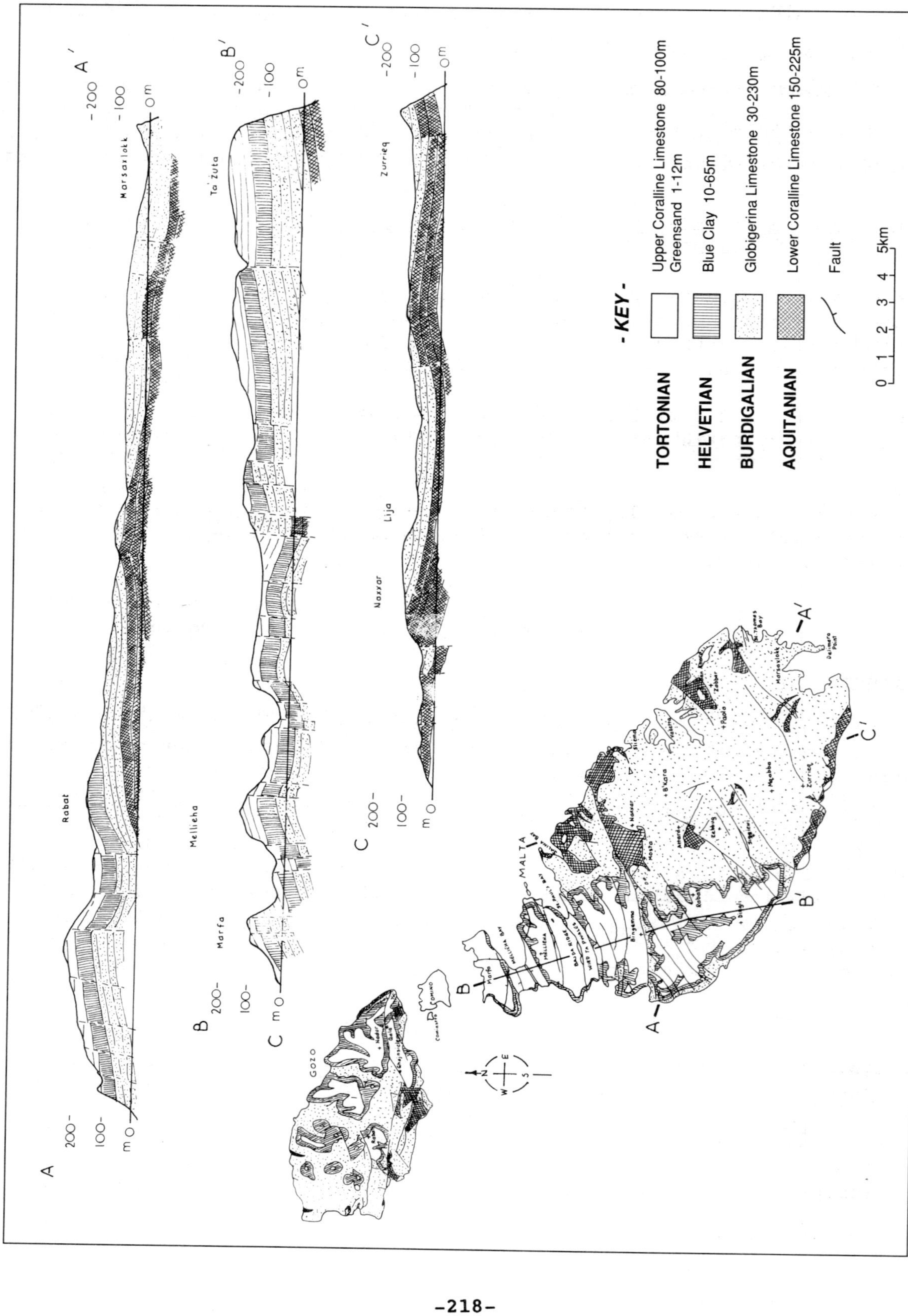

Oil exploration wells have revealed that underneath these outcropping formations successive layers are as follows:

Age	Depth (metres)
Oligocene-Eocene	530
Eocene-Cretaceous	800
Cretaceous-Jurassic	1 700
Jurassic-Triassic	2 400

This sequence is made up of limestone, dolomitic limestone and dolomite, with dolomitization increasing with age.

The strata are generally fissured and nearly horizontal with a slight regional dip to the north-east. A major escarpment trending east-north-east divides the island into a lower northern part and an elevated southern part. This escarpment is the surface expression of the Victoria Lines fault.

A series of small faults run more or less parallel to the Victoria Lines fault. A few exceptions, notably the Maghlaq fault at Ghar Lapsi trends east-south-east.

These faults are basically tensional in origin, controlled largely by Pliocene to Recent upheaval and movement of the Sicilian-Maltese micro-plate with respect to the main African plate.

2. Aquifer systems

(a) Perched aquifer

Two main aquifers are found on the Maltese Islands. A perched aquifer is situated in the porous upper Coralline Limestone which lies directly above the impermeable Blue Clay formation (figure 37b). A thin layer of permeable greensand may also be found at the base of the upper Coralline Limestone which outcrops mainly on the western parts of the island. No salt-water intrusion in the aquifer south of the Victoria Line is possible because it is everywhere located well above sealevel. To the north of this line, however, the aquifer is in many places in contact with sea water and intrusion is a problem. Private extraction (by farmers) exploits this aquifer almost to the full, leaving only a small fraction for the public supply which come largely from gravity springs and underground galleries (4-2 km of galleries collect 10,000 m^3 per day).

(b) Mean sea level aquifer

An aquifer in which the fresh-water table is slightly above sealevel is the most important and accounts for about 75 per cent of the total public supply. The aquifer occurs in the pores and cracks of the Globigerina and lower Coralline Limestone. This fresh-water body is in a saturated zone but owes its existence to the fact that every winter rainfall adds more fresh water to the underground store than is dissipated by direct discharge around the coast. There is no sharply

Figure 37b. Malta: Hydrogeologic cross-section

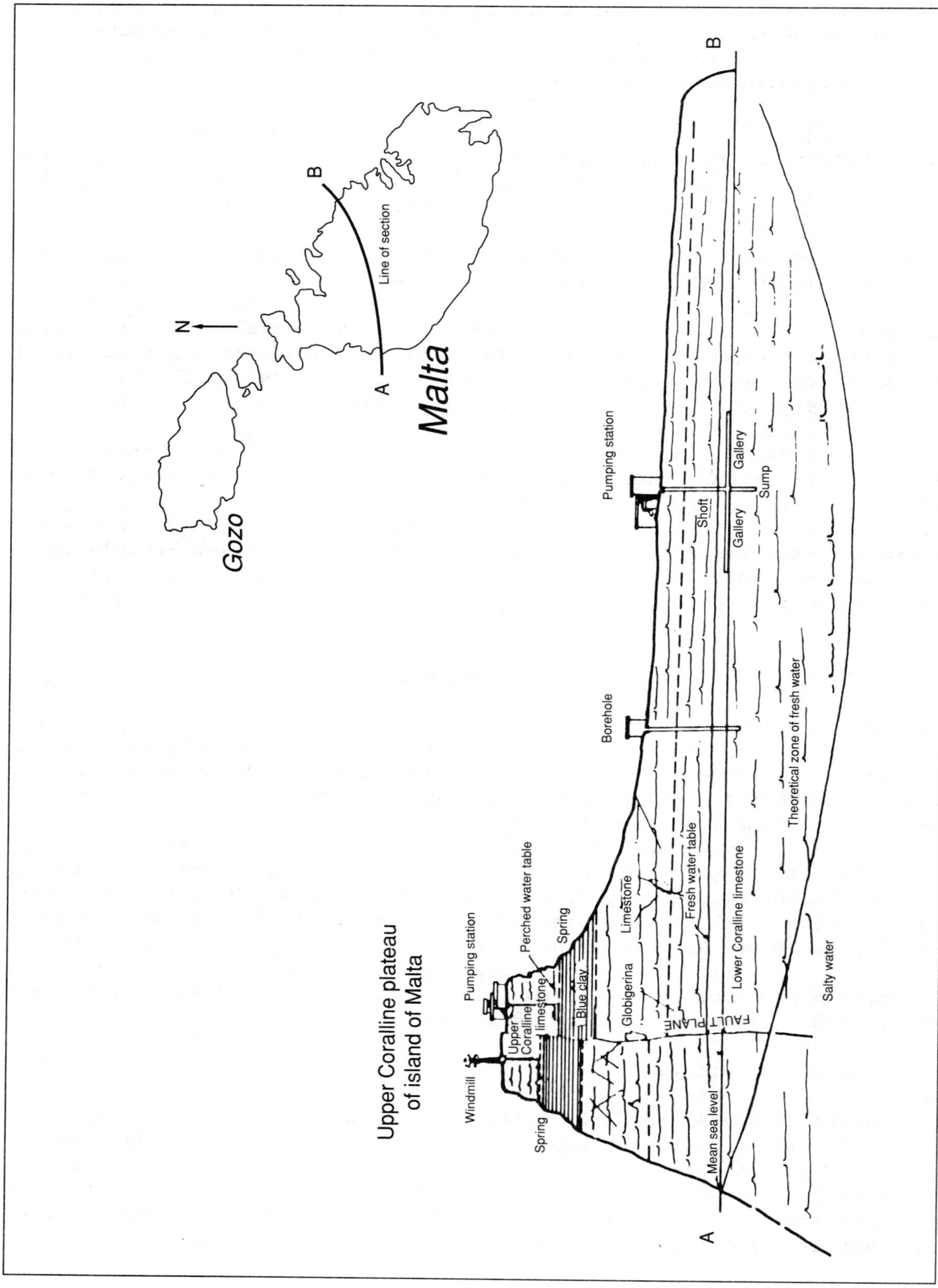

defined plane of separation between the superficial fresh water and the saline water underneath. The equilibrium of this lens is in state of flux as it depends on the fluctuations in rainfall, tides and other factors. The model which best defines this aquifer is the Ghyben-Herzberg lens in which for each metre of fresh-water head above sealevel, the salt water is displaced 40 metres below sealevel. Thus, in large areas in the central part of the island, the fresh-water body extends from 2-3.5 metres above sealevel to 80 or 140 metres below.

Recharge to the aquifer is predominantly through fissures in the overlying Globigerina Limestone. Porosity and permeability are mostly fissure-dependent. Extraction from the aquifer is by a system of underground galleries situated near the central axis of the island and also by boreholes farther away from the centre.

Sea-water intrusion into the mean sealevel aquifer present a permanent problem since rock permeability is mainly through fissures. A series of gauging boreholes scattered over the aquifer are used to monitor the lens and adjust pumping rates to keep intrusion to a minimum. Recent investigations have shown that the coastal fresh-water interface is sharp and goes for some distance beyond the coast. It has also been observed that because the recharge infiltration is slow, the size and shape of the lens do not change appreciably with fluctuations in yearly rainfall but is heavily dependent on extraction rates.

In recent years the major subsurface discharge points of this aquifer yielding brackish water began to be exploited for reverse osmosis treatment. One major discharge point provides about 3.8 million litres per day of treated fresh water and another discharge point is being investigated for additional extraction.

C. Groundwater development

1. Background

Water supply in the Maltese Islands has always been a difficult undertaking. An early document dated 1536, drawn up by a Commission of the Knights of St. John of Jerusalem prior to their settling in Malta, reports on the water supply situation which they immediately recognized as a problem.

> "The water is salty and sedimentary; and the sweeter springs in the Island I think come mostly from wintry showers. Their source is not deep, for in summer they frequently dry up and always diminish. Drinking water comes from the rain (when there is any) which is preserved in cisterns and more frequently in ditches."

The local population of that time, numbering about 2,500, were established around these naturally occurring springs in the Rabat area but the Knights, whose activities were mostly maritime, found the areas around the good natural harbours on the south-east side of the island more suitable, and settled around them. By the time the city of Valletta was built, the need for water had risen and, besides, it was necessary to transport water from the Rabat area to this new city, a distance of some 15 kilometres. Several springs and galleries driven into the perched aquifer were united by subterranean conduits and thence diverted into a single channel to feed Valletta by gravity. This aqueduct, considered to be a fine example of its type of civil engineering, was started in 1610 and completed in five years. Stretches of this aqueduct still stand today.

By 1870, with the demand for fresh water increasing, the mean sealevel aquifer was exploited by the drilling of shafts and driving of galleries, so as to reach the fresh-water body lying over the salt water. This development was accelerated after 1945.

Reservoir construction was an important part of the water supply network especially before the advent of round-the-clock desalination of sea water. Records show that in 1886 the first sizeable roofed reservoir of 42,000 m^3 capacity was built. Before this, public water storage was in small scattered tanks in towns and villages. Every private house had its own cistern to collect water - this was a requirement by law, showing the importance to the country of water storage for use in the long and dry summer. Since the 1880s, covered storage has increased and has now reached 500,000 m^3.

The scarcity of naturally occurring water supplies and rapid increases in demand at certain stages of its development made it necessary to resort to desalination, with Malta being associated with various technologies since their early stages.

The first distillation plant was operating in Malta in 1881, having been installed by the Government to provide the local population and the British garrison with water, then in short supply. This small sea-water desalination plant was installed at Sliema.

More recently in 1966 a 3,800 m^3 per day multistage flash distillation plant was installed as part of a dual-purpose power-water plant; it was followed within two years by three similar units. These plants were among the earliest units of their size. The reverse osmosis sea-water plant with a capacity of 20,000 m^3 per day built at Lapsi in 1982 is still the largest in the world.

Transport of water to Malta from nearby Sicily was also resorted to in 1944 at an average rate of 20,000 m^3 per month. This exercise was carried out to combat a water shortage created by low rainfall coupled with the presence of large numbers of military personnel in Malta.

It is not surprising that waste control has been a recurring concern. Metered household services were made compulsory as far back as 1887. Leakage detection was organized on a scientific basis in 1970 and periodically modernized.

Health aspects have always been given due importance. The first recorded chemical analysis of the water supply in Malta is dated 1865. In 1909, following an outbreak of typhoid fever, Sir T. Zammit, Government Analyst, introduced the purification of public water supply by means of chlorine. This was within five years of the introduction of the method in the United Kingdom. Routine bacteriological and chemical analysis have been carefully undertaken since 1912.

The present sources of first class potable water in Malta are listed in table 38. Their average daily output for the period October 1984 to September 1985 shows the relative importance of each source in the water supply structure.

Table 38. Malta: Sources of potable water, 1985

Source	Production (thousand m^3/day)	Percentage
Galleries, pumping stations	37.5	43.3
Boreholes	19.0	22.0
Springs	1.7	2.0
Treatment plants	0.2	0.2
Brackish water reverse osmosis	3.4	3.9
Sea water reverse osmosis	19.6	22.7
Sea water, distillation	3.4	3.9
Transfer from Gozo	1.7	2.0
Total	86.5	100.0

2. Sources of fresh water

Extraction of groundwater from the aquifers is the major source of potable water and should be so for some time to come. Groundwater accounts for some 45,000 m^3/day, that is, two thirds of total production, and is derived from 36 km of galleries in the mean sealevel aquifer, 4.2 km of galleries in the perched aquifer and from more than 100 boreholes scattered throughout Malta. Data on production from 1966 to 1986 are given in tables 39 and 40.

Table 39. Malta: Water production by source, 1966-1986

Year	Galleries	Boreholes	Springs	Distil- late	Reverse osmosis plants	Other	Total
	(in million cubic metres)						
1966	13.40	1.51	0.49	-	-	..	15.40
1969	11.74	2.03	0.83	2.63	-	..	17.26
1970	12.88	2.08	0.57	3.09	-	..	18.63
1972	11.14	1.81	0.81	4.67	-	..	18.42
1976	13.07	5.08	0.73	3.05	-	..	21.93
1986	11.45	9.07	0.56	1.33	8.97	0.17	31.56

Table 40. Malta: Percentage produced by each source, 1966-1986

Year	Galleries	Bore-holes	Springs	Distil-late	Reverse osmosis plants	Other
1966	87.0	9.8	3.2	-	-	..
1969	68.2	11.8	4.8	15.2	-	..
1970	69.2	11.1	3.1	16.6	-	..
1972	60.5	9.8	4.4	25.3	-	..
1976	59.6	23.2	3.3	13.9	-	..
1986	36.3	28.8	1.8	4.2	28.4	0.53

The amount that can be extracted from underground is directly related to the amount of rainfall, which varies from year to year, and the recharge that percolates to the aquifer. There are other factors which are important in practice, mainly the quality of the water drawn up and the possibility of blending with other supplies which would render the mix acceptable.

(a) Galleries

The galleries in the mean sealevel aquifer are the most important and themselves account for about two thirds of total groundwater production. These gallery systems, of which there are 12 functioning, are mostly constructed within the lower Coralline Limestone which is the most porous formation at the mean sea water level. Groups of galleries constructed in a radial pattern converge to central sumps from where water is pumped to reservoirs. Because of this radial pattern and the fractured nature of the rocks, these galleries are susceptible to sea-water intrusion. Water from most of the galleries tends to rising salinity levels and whenever supply or demand permits, a reduction of extraction or pumping rate is practised to minimize intrusion. This reduction will, it is hoped, become a permanent feature as soon as other sources being developed become available.

(b) Boreholes

Since the 1970s, rather than further exploitation of the aquifer by gallery systems, an extensive programme of borehole drilling has been resorted to. This has produced a steady supply of water at relatively low cost. This method of exploitation permits more efficient management of the aquifer since extraction from a great number of points avoids local overexploitation and increasing salinity. There are at present 106 functional boreholes in Malta.

The most productive boreholes are located within the porous lower Coralline Limestone and have an average production rate of about 205 cubic metres per day. An analysis of borehole salinity figures has shown that 85 per cent of the water being extracted is below or around 500 mg/l. As soon as replacement water is available to meet requirements, extraction of the more saline water will be avoided.

Water is pumped from these boreholes by means of submersible pumps. It is fed directly to reservoirs whence it is distributed for public consumption. Chlorination at source and at reservoir outlet is practised.

In conclusion, groundwater extraction rates from the aquifers as now developed are high and a significant lowering of the static water level of the mean sealevel aquifer has been observed. Recently, some recovery has been achieved by reducing extraction from certain areas and more is planned for certain areas.

(c) Springs

There are no permanent springs in Malta. The only ones are gravity discharges from the perched aquifer and they flow only during the wet season. Most of these few springs are exploited by farmers between intermittent rains and flow from the rest is collected in open reservoirs.

This water, which accounts for 2 per cent of the water production in the island, is treated by the Water Works Department before being certified suitable for public use. It generally has high nitrate content and is carefully blended before use.

3. Use of brackish water

The occurrence of a fresh-water lens in the island mass is now well established. Rain water infiltrating slowly through the rock formations reaches the mean sealevel aquifer where it forms a lens floating over sea water. The lens is thicker in its central part because on reaching the aquifer the water starts a slow but steady lateral migration towards the coast. On its approach to the coast this fresh water mixes with sea water becoming progressively more brackish until it flows to the sea.

This coastal discharge is more significant in certain drainage areas. Such an area is the Marsa basin. A shipbuilding yard was being built at Marsa and brackish water was met with during the construction of the dock. This was at the time where tenders for the construction of a large 15,000-m^3-per-day sea-water reverse osmosis plant were being processed. The opportunity was taken to form a package deal covering two reverse osmosis plants, one sea water and one brackish water. The one that concerns this section is the brackish water desalination plant at Marsa.

Feed water at salinity 13,000 mg/l is obtained by a system of strategically placed wells. The feed water is pumped into a breaktank, treated and boosted to the inlet of the high pressure turbo-pump which in turn feeds the Dupont B-10 permeators. The turbo-pump is equipped with energy-recovery facility which reduces the process energy consumption by some 15 per cent. Product water flows out of the permeators and into the sea.

The plant has been operating since July 1983. The main problem has been in the feed water quality - expressed as its Silt Density Index (SDI) - which has been cause of frequent shut down of the plant during winter.

Feed water for reverse osmosis treatment must be practically free of suspended silt to preserve membrane life. If open feed water intakes are used, relatively expensive pre-treatment equipment is used to remove silt and certain other matter. This stage is avoided at the Malta Reverse Osmosis Plant by making use of the natural filtering properties of the limestone rock formations from which feed water is extracted through wells. In the case of brackish water derived from the mean sealevel aquifer, however, it has been observed that even slight but sudden fluctuation of the water table by tidal forces or variation in pumping adversely affects the SDI. The Marsa Reverse Osmosis Plant on occasions (mainly periods of heavy rains in winter) had to be shut down because of unacceptable SDI levels of the feed water.

4. Monitoring and control

The monitoring of aquifers, in particular the mean sealevel aquifer, forms an important part of water management in Malta. A number of small-diameter gauging boreholes scattered over the island are used to measure fluctuations in the lens and to provide warnings of overpumping and sea-water intrusion. These measurements are made manually but it is hoped to introduce some form of automation. Records of rainfall, temperature, runoff and humidity are also kept to enable adequate update of the yearly water balance of the Island.

The aquifers are protected by stringent regulations regarding sources of contamination such as farmhouses, sewers, cesspits, sea-water swimming pools, quarries, dumps and landfills. Periodic inspection of farmhouses, are, for instance, made to ensure hygienic disposal of solid and liquid waste. It has long been the policy to restrict the erection of such potential sources of contamination in areas near the seashore. Despite such measures, some sources of groundwater, particularly boreholes cannot be used or can be used only for irrigation because of chemical or bacteriologic contamination.

Full chemical and bacteriologic analysis of all groundwater sources are periodically made by the Water Department and the Health Department to ensure public safety. The commonest contaminants are chlorides and nitrates. Whereas the former can be readily controlled by adjustments in pumping rates, the latter present serious difficulties since the excessive use of high-nitrogen fertilizer is not easily controlled. Blending with other waters with low nitrate content is resorted to in order that otherwise unusable water is not wasted.

Most experts, both local and foreign, agree that Malta's main groundwater production points are overexploited.

There could, however, be areas where higher production could be achieved and a concurrent reduction made from other points. The need thus arises of having a mathematical model of the aquifer so that maximizing this comparatively cheap resource is achieved without harming the aquifer.

Steps are being taken in this direction and a project on that matter was prepared (1987).

XI. THE NETHERLANDS

Area: 41,160 km^2 Population: 14.7 million

A. General geography

1. Background

Only about 80 per cent of the Kingdom of the Netherlands is land area. The remainder consists of territorial waters, estuaries, tidal flats and inland waters wider than 6 m. From the early Middle Ages onward the inhabitants drained and reclaimed the wetlands and tidal flats covering the low parts of the country. Few parts of the world have been more thoroughly reshaped by human activity than the polders in the coastal zone. Large-scale reclamation of the higher grounds started with the introduction of fertilizer some 100 years ago. Former heath and inland sand dunes were turned into arable land and woodland. By 1980 cultivated land occupied 24,870 km^2; woodland and natural areas, 4,710 km^2; and towns and roads, 4,334 km^2.

The country is densely populated with an average of 412 inhabitants per square kilometre. The west Netherlands have the highest concentration, about 900, and the north and south-west (Zealand) the lowest, about 200. The Netherlands is administratively organized into 12 provinces.

The groundwater situation in the Netherlands is generally that encountered in low-lying deltaic areas. Land reclamation has effected changes in the hydrogeological conditions and there is direct relation between the groundwater situation and many aspects of economy, like the consequences of withdrawals on crop production and on land subsidence. A dense population and intensive agriculture in an economically well-developed country has affected the shallow groundwater quality.

2. Scope of water problems

The Netherlands belongs largely to the lower fluvial plains of the river Rhine and some smaller rivers like the Ems, Meuse and Scheldt (figure 38). The landscape and the subsoil also have the characteristics of deltaic areas; generally the land is flat; thick unconsolidated sediments form good aquifers and the groundwater levels are shallow. A distinct feature of groundwater quality is the occurrence of brackish water in large parts of the coastal zones. This situation poses particular problems to major sectors of society like agriculture, water supply and engineering practices. Some aspects are:

(a) Drainage of agricultural land

More than 1,000 years ago, the peasants of the coastal areas tackled the problem of high groundwater levels by cutting ditches discharging towards natural watercourses (Bakker, 1983). These first artificial drainage systems developed gradually into an intricate pattern of ditches, collector canals and separate polders, where sometimes the natural elements are hardly recognized. The present drainage system largely influences the groundwater situation.

Figure 38. Netherlands: Hypsometry and drainage

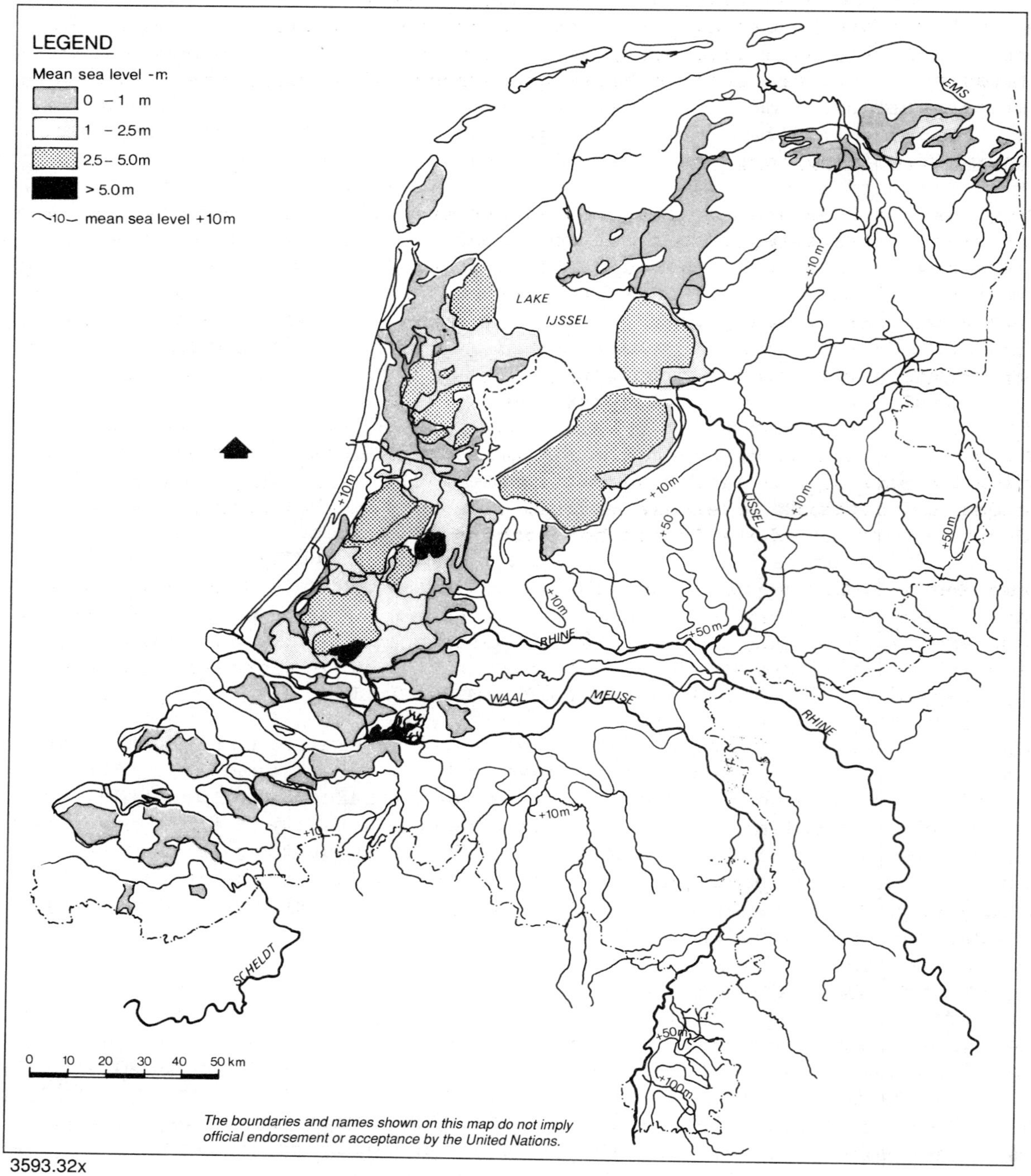

3593.32x

(b) Drinking water supply

Private water supply and public water supply, installed during the last 100 years, were predominantly based on groundwater use. Public water supply in the beginning encountered only problems in the coastal areas where fresh groundwater is absent over large areas. Fresh-water lenses underneath the dune ridge were heavily exploited until the dunes developed into important sites of artificial recharge with river water some 40 years ago. Currently, the effects of a lowering of the shallow water table on agricultural production, natural vegetation and land subsidence are playing a role everywhere; extractions of natural groundwater are restricted by a delicate balance between the economy of water use and effects on other interested parties.

(c) Engineering practices

The need to excavate for foundations of large buildings, sluices, docks, tunnels, etc., always requires the lowering of water tables and the removal of excess water. At an even larger scale this problem arises at the pumping and reclamation of former lakes. An additional problem can be the upward seepage of brackish water in the low polders.

(d) Groundwater pollution

The most recent, but not the least important, problem concerns the pollution of soil and groundwater by a dense population and an intensive agriculture. Pollution by private and public waste dumps, by air pollution, by fertilizer, pesticides and deposing of excess manure, is easily transported by shallow groundwater to places where it can harm others.

The above types of problems led to a characteristic development of groundwater sciences, where the emphasis was more on the behaviour than on the occurrence of groundwater. Much attention was, and is, paid to analytical and numerical solutions to groundwater flow problems, rather than to a hydrogeologic description of the subsurface. Civil and agricultural engineers have made important contributions to groundwater research in the Netherlands.

3. Physiography and climate

(a) Geomorphology

The part of the subsurface relevant to the flow of groundwater largely consists of unconsolidated Quaternary sediments, deposited in a subsiding basin. The axis of the basin dips north-west, resulting in the largest thickness of the aquifers in the central and north-western part of the country. At the eastern State Boundary, Tertiary and even older sediments are at land surface. In the south, the surficial deposits are locally lower Pleistocene in age. Where the Quaternary sediments thin out, upper Tertiary layers may also form exploitable aquifers. The southernmost part of the country occupies a special position insofar that only in south Limburg are exploitable aquifers formed by consolidated sediments of Cretaceous age.

Pleistocene sedimentation was expecially important. During glacial periods the deposition of thick layers of coarse, glaciofluvial sediments gave rise to many large aquifers. Finer sediments, deposited during relatively warmer stages, locally subdivide the aquifers. In the last glacial period (Weichselian) large parts of the Netherlands were covered by eolian sand and loam. In the eastern and southern provinces these deposits form a less permeable or even a confining top layer. This covering layer hampers the recharge of the underlying aquifers; a natural drainage system of brooks discharges part of the excess rainfall.

Moreover, the presence of land ice sheets during the Pleistocene exerted a strong influence on geomorphology and, thereby, on present groundwater flow patterns. Deep valleys were scoured either by meltwater, or by the land ice itself. In the northern provinces these glacial valleys reach a maximum depth of 300 m below present land surface; they have been filled in with practically impermeable clay (so-called potclay). In the central parts of the country the valleys scoured by the ice sheet have a maximum depth of 125 m; also, these valleys have been filled in afterwards with less permeable deposits. The sediments removed by the land ice formed lateral moraines along the valleys with a maximum elevation of about 100 m. The ridges are important natural infiltration areas, generally without a surficial drainage system. Glacial impact is also present as boulder clay, covering the higher parts of the northern provinces.

Tectonic activity influenced the hydrogeologic situation of the southern part of the country. Fault zones affected the shallow subsurface and thereby the presence and the magnitude of aquifers and confining layers. In the Netherlands, many fault zones behave as a barrier to groundwater flow, owing to the clay gouge on the fault planes.

During the Holocene age, starting 10,000 years ago, fine sediments have been deposited at the relatively low parts of the Netherlands, i.e., the coastal zones and the river valleys. The fluviatile (clay and loam), peri-marine (peat and clay) and purely marine (clay and sand) layers reach a maximum depth of 20 m near the coast, but they wedge out against the higher Pleistocene terrane. Along the coast a sandy dune ridge developed which was broken in the south-west to form an estuarine landscape and also in the north, where the shallow Wadden Zee developed. Through these openings in the dune ridge the sea water could penetrate far inland, thereby salinating the shallow subsurface.

Also during the Holocene, extensive peat bogs originated both at the higher grounds at places where subsoil and drainage pattern could not fully transport the excess rainfall and at lower areas between river branches. Much of the peat has been excavated, leaving soils rich in organic material at the higher grounds and, first, lakes and then deep polders in the lower parts. Where peat is still present, the land is subject to subsidence.

The aquifer system only exceptionally consists of a single aquifer. Generally, there are a number of subaquifers divided by leaky clay and loam layers. Clearly, the covering layers are also not uniform; locally, the Holocene layers are sandier, thus losing their confining character. A diagram, compiled by Jelgersma (1977), of the hydrogeologic structure is given in figure 39.

Figure 39. Netherlands: Hydrogeologic cross-section

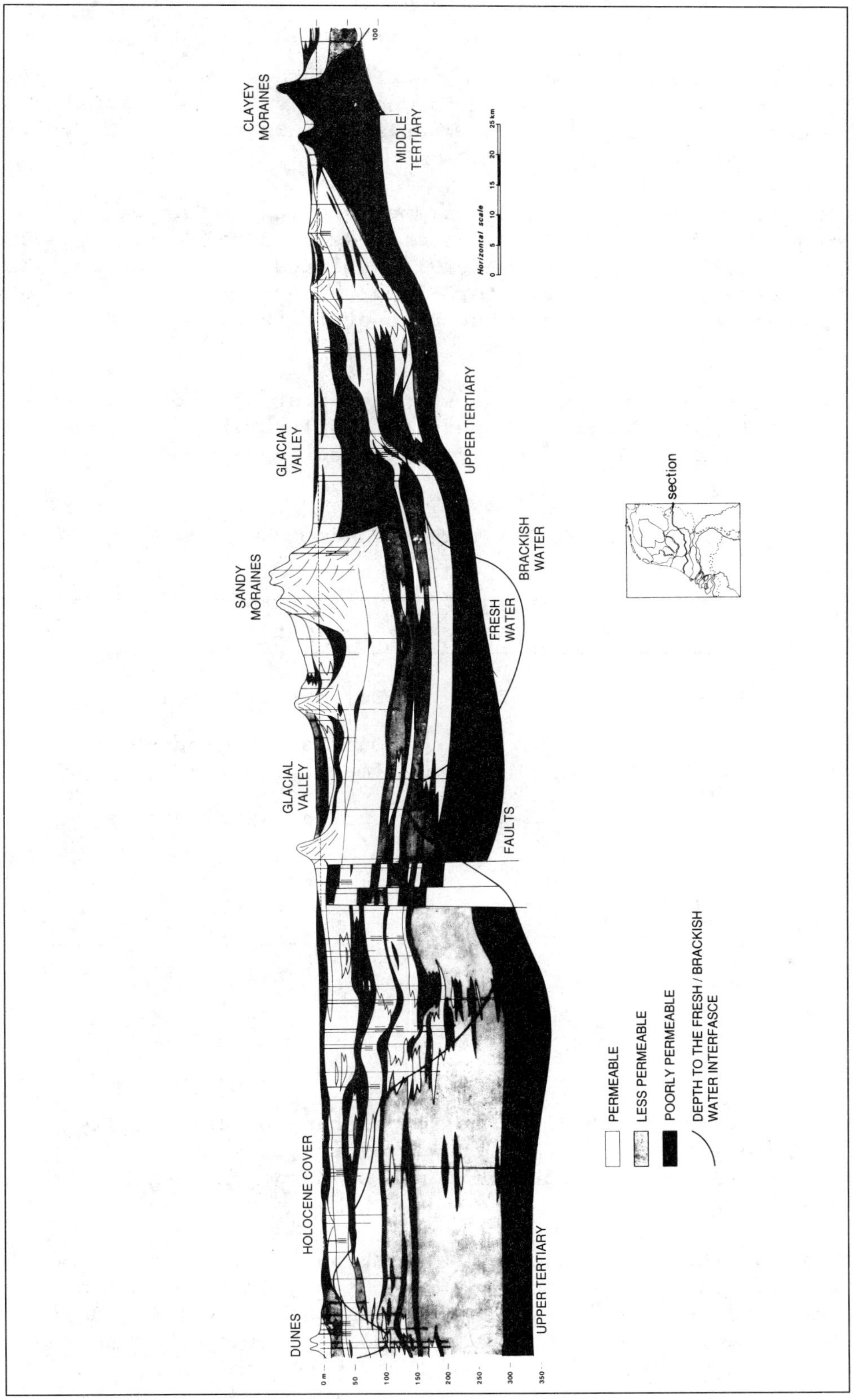

(b) <u>Water balance</u>

The Netherlands have a moderate, essentially oceanic climate. Average monthly temperatures at the De Bilt weather station vary from -2° C in January to 12° C in July.

Regional variations are small; the yearly average temperature varies from 8.5° C in the north-east to 10.5° C in the south-east. The rainfall is relatively evenly distributed over the year. The yearly amounts of rainfall vary between 500 and 1,200 mm; the long year mean is 750 mm.

Evapotranspiration is high in the summer and low in the winter. It follows that the average evapotranspiration in summer exceeds rainfall. The long year average of the actual evapotranspiration is estimated to be 465 mm. The average yearly rainfall excess thus amounts to 285 mm. The rainfall excess is potentially available for groundwater recharge, but part of it is directly discharged to open water.

In the polder areas, however, the surface water directly recharges the groundwater; notably, the Rhine branches are infiltrating. A rough water balance of all water passing through the country is given in table 41.

Table 41. Netherlands: Annual water budget

In	km^3	mm	Out	km^3	mm
Precipitation	26	750	Evapotranspiration	16	465
Inflow of groundwater (mainly from Belgium)	0.1	3	Outflow of groundwater (mainly to sea)	0.05	1
Rhine	70	2 000	Surface-water discharge	89	2 535
Meuse	8	229			
Small rivers	1	30			
Total	105	3 000	Total	105	3 000

(c) <u>Recharge</u>

Groundwater levels are shallow; in 90 per cent of the country the level is less than a metre below land surface in winter and less than 2.5 m in summer. The only areas with a deep water table are south Limburg, the central moraine ridges and part of the dune area.

Because of the shallow water tables, a strong relation holds between landscape and groundwater. Three major zones exist: the higher sandy areas inland, the also relatively high areas of the most recent coastal and fluviatile accretions and a lower transition zone, where extensive peat bogs originated. The shallow groundwater heads in the sandy areas are largely controlled by the natural drainage system, strongly adapted to the needs of agriculture. The backbone of the drainage system in the coastal sea-clay areas is the former creek and gully system. In the peaty areas, both the landscape and the flow patterns in groundwater are almost fully artificial. Peat bogs were drained by long parallel ditches. Large peat bodies were excavated, leaving lakes, later to be pumped and reclaimed.

All of the land is below sealevel, the non-excavated areas only 1 or 2 metres because of shrinkage of the peat, the reclaimed lakes to a depth of about 5 metres below mean sealevel (see figure 38). This is the land of the polders; excess water has to be pumped and groundwater flow is directed by the variations in levels of high and low polders. The low polders are the centres of groundwater discharge by a strong, upward seepage of sometimes brackish water. Present and former branches of the rivers are above the polder levels. Hence, the rivers are permanently infiltrating and even at a large rate as the river beds are mostly sand.

Groundwater is recharged by rainfall at the sandy areas and in the most recent coastal zone. In the polder areas, the higher infiltrating polders and the surface water from the river branches contribute a major part of groundwater recharge. Not all of the potential recharge reaches the groundwater in the exploitable aquifers.

The sandy areas in the northern part of the country are covered with boulder clay and brook loams. Rainfall excess is about 300 mm per year, of which about 30 mm is directly discharged by surface water, only about 200 mm reaches the upper soil zones and about 70 mm recharges the exploitable aquifers.

The central sandy areas are covered in the east with a dense drainage system of brooks and forewaters. Of the about 250 mm per year of rainfall excess, about 50 mm is directly recharged to surface water or only reaches the upper soil zone, which leaves about 200 mm for recharge of the single aquifer. At the moraine ridges the full rainfall excess of 350 mm per year recharges the mighty aquifer system underneath.

The areas south of the Meuse were affected by shallow tectonic activity. From east to west a tectonic low (Valley of the Meuse), a horst area (Peel), a deep tectonic low (Central Valley) and again a tectonic high are present. In the Peel area, the unconfined aquifers are thin and have low transmissivity. In the Central Valley, the aquifer system reaches a depth of some 500 m but it is covered with thick loamy layers. Of the 225-mm-per-year rainfall excess, about 200 mm recharges the Peel aquifer, but only about 100 mm feeds the upper aquifer at the Central Valley.

In the low areas covered with Holocene sediments most of the rainfall in excess of 250 mm per year is recharged directly, or by a short soil passage, to the surface water, except for the dunes. The natural recharge of some 300 mm per year at the dune area is locally supplemented by high amounts of artificially recharged water from the Rhine. This latter type of water will have a minimum residence time of two months before it is again treated to drinking water.

At the upstream parts of the areas, in between the river branches, an interesting situation occurs. At high river levels and low groundwater heads (in summer), the river branches are infiltrating and almost the full area shows upward seepage. In the reverse situation, the present and former river banks become recharge areas for the excess rainfall.

B. Hydrogeology

1. Aquifer structure

The whole unconsolidated subsurface acts as one interconnected aquifer system. Locally, intercalated clay forms semi-confining layers, dividing the system into a number of subaquifers. These clay layers may exert significant hydraulic resistance, but they are nowhere fully impermeable. Hence, even the deepest unconsolidated layers take part in the active hydrological cycle.

In figure 40, the top and base of the aquifer system are represented, as well as an indication of the total transmissivity. The total transmissivity has been estimated from thickness and permeability of the contributing subaquifers.

In outline, the sandy areas will have a phreatic water table in an unconfined aquifer, whereas in the other areas the shallow aquifer is confined by a less permeable covering layer. In more detail, the northern sand region is over large areas covered with surficial boulder clay and brook loam. But also many of the shallow sandy layers in the other sand areas consist of fine sand and loamy layers, not exploitable for public water supply. Commonly, an exploitable aquifer will be a deeper subaquifer, confined by less permeable layers (see figure 39). The hydraulic resistance of the above covering and confining layers generally is between 100 and 100,000 days, but is not sufficiently known as to enable mapping it.

2. Water table and flow patterns

The shallow groundwater heads in the sandy areas are largely controlled by the form and the levels of the natural drainage system, adapted to the needs of agriculture. The regional flow pattern is overshadowed by a flow directed to the brook system. Moreover, the flow pattern is not constant in time. In the summer, large parts of the brooks gradually fall dry, the flow pattern then changes to a more regular form, determined by the larger watercourses. The groundwater heads in deeper aquifers, especially if they are covered with less permeable layers, show more markedly the regional tendency. The flow patterns are largely constant in time.

In the upper reaches of the river-clay district, the shallow groundwater heads are greatly influenced by the water levels in the branches of the river Rhine. Again, the flow pattern in deeper aquifers of the river district is more regular and less time-dependent. The situation at the downstream reaches of the river branches is more similar to the rest of the polder areas in the west of the country. The river levels are relatively constant by the control of the nearby sealevel and they are permanently above the polder levels in the adjoining areas. Hence, the rivers are permanently infiltrating. The shallow groundwater levels in the clayey soil of the endiked river marshes are fully controlled by constant polder levels. The groundwater heads in the shallow exploitable aquifer are generally slightly below river level at the river beds and slightly above polder levels. The areas enclosed by river branches are predominantly groundwater discharge areas for water infiltrated at the river bed. In deeper aquifers, a groundwater flow derived from nearby sandy recharge areas can sometimes be recognized.

Figure 40. Netherlands: Top and base of the aquifer system and total transmissivity

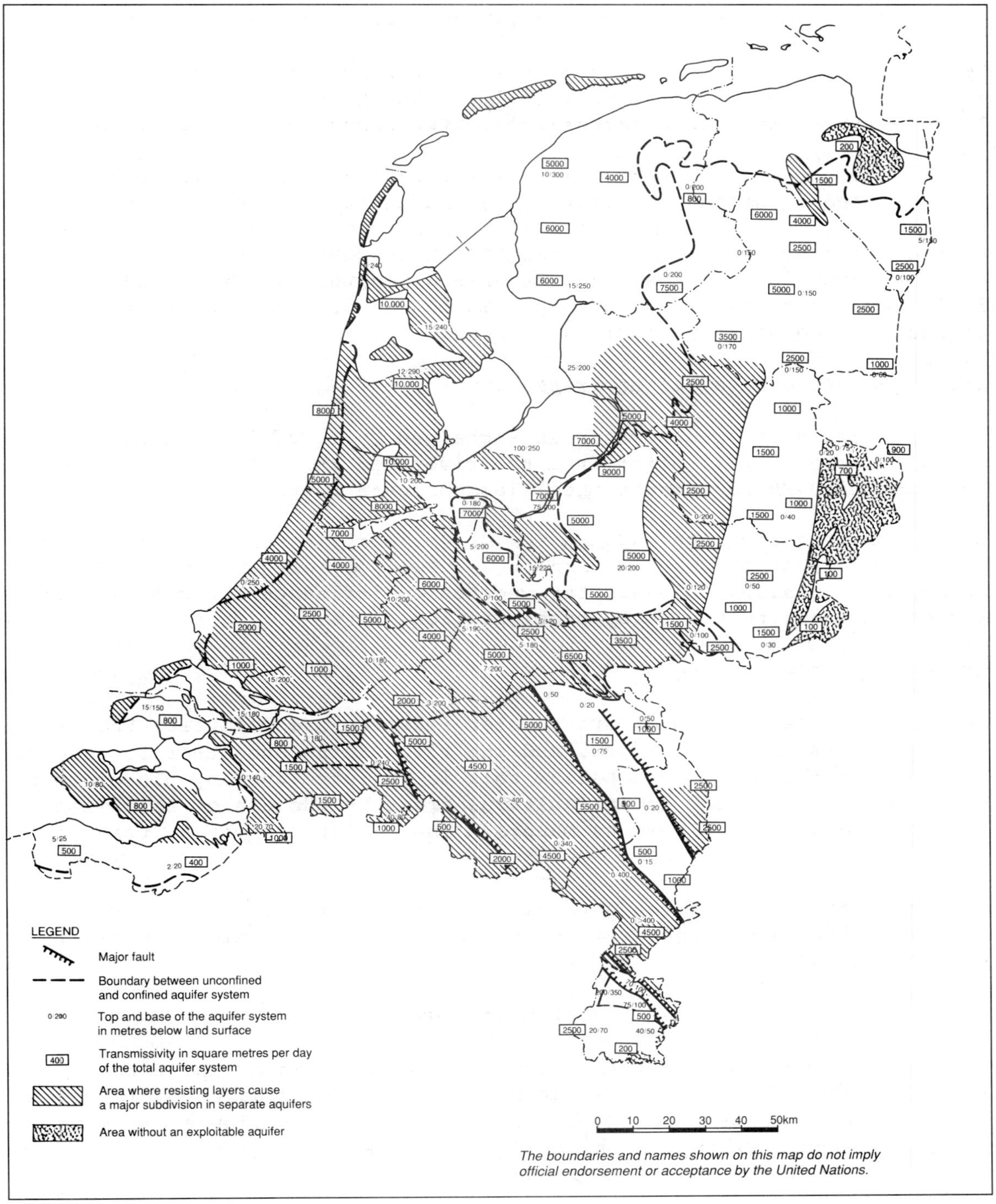

3593.33x

In the polder areas farther from the rivers, the situation may be more complex as the recharge areas are the relatively more elevated polders along former natural streams. Infiltration from the beds of the open watercourses itself is less important than at the sandy river beds. In figure 41, the general regional trends of the groundwater flow patterns in the exploitable aquifers are represented for the whole country.

3. Water quality

(a) General

In the Netherlands, four factors play important roles as affecting groundwater quality.

i. Presence of saline and brackish groundwater

Most of the shallow groundwater in the broad coastal belt is brackish due to Holocene transgressions of the sea. At greater depth, everywhere in the country a transition towards brackish groundwater can be found, which is determined by geology and hydrogeologic processes.

ii. Type of recharge of aquifers

In sandy areas the groundwater is recharged by rainfall, but notably in the river-clay district infiltration from the river bed is important. In the river valleys a modified Rhine water composition has been found.

iii. Conditions at the surface

The type of shallow soil and the vegetation influence the fresh groundwater composition of areas recharged by precipitation.

iv. Man-made pollution

At present, the sources of groundwater recharge are polluted, the rainfall by air pollution from various origins, and the river Rhine and other surface water by waste water. Furthermore, the shallow soil is polluted in many places by the presence of waste dumps and in agricultural areas by the use of fertilizer, biocides and the dumping of excess manure. The pollutants may be transported by moving groundwater.

A further detailing of the natural composition of groundwater can be based on thousands of analyses of major components in groundwater, which are available from observation and pumping wells.

(b) Distribution of fresh and brackish groundwater

The marine transgressions, flooding the present area of the Netherlands during the early Holocene age, took place during two major periods, each divided in a number of substages. The chloride content in the most shallow aquifer is represented in figure 42, which is based on a number of regional studies, including the results of geoelectrical surface soundings.

Figure 41. Netherlands: Groundwater contours at top of exploitable aquifer system

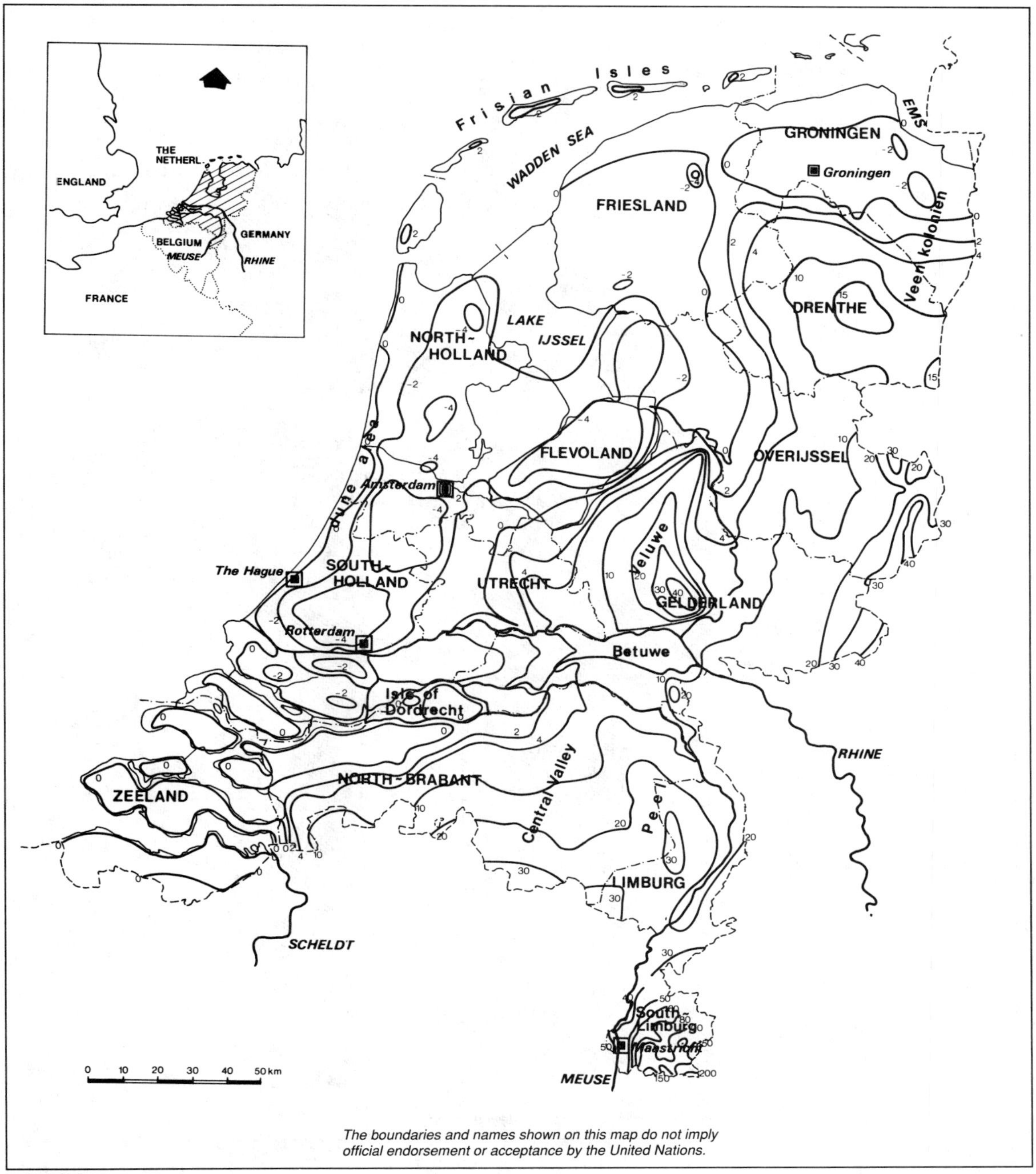

3593.34x

Figure 42. Netherlands: Chlorides at the top of the aquifer system

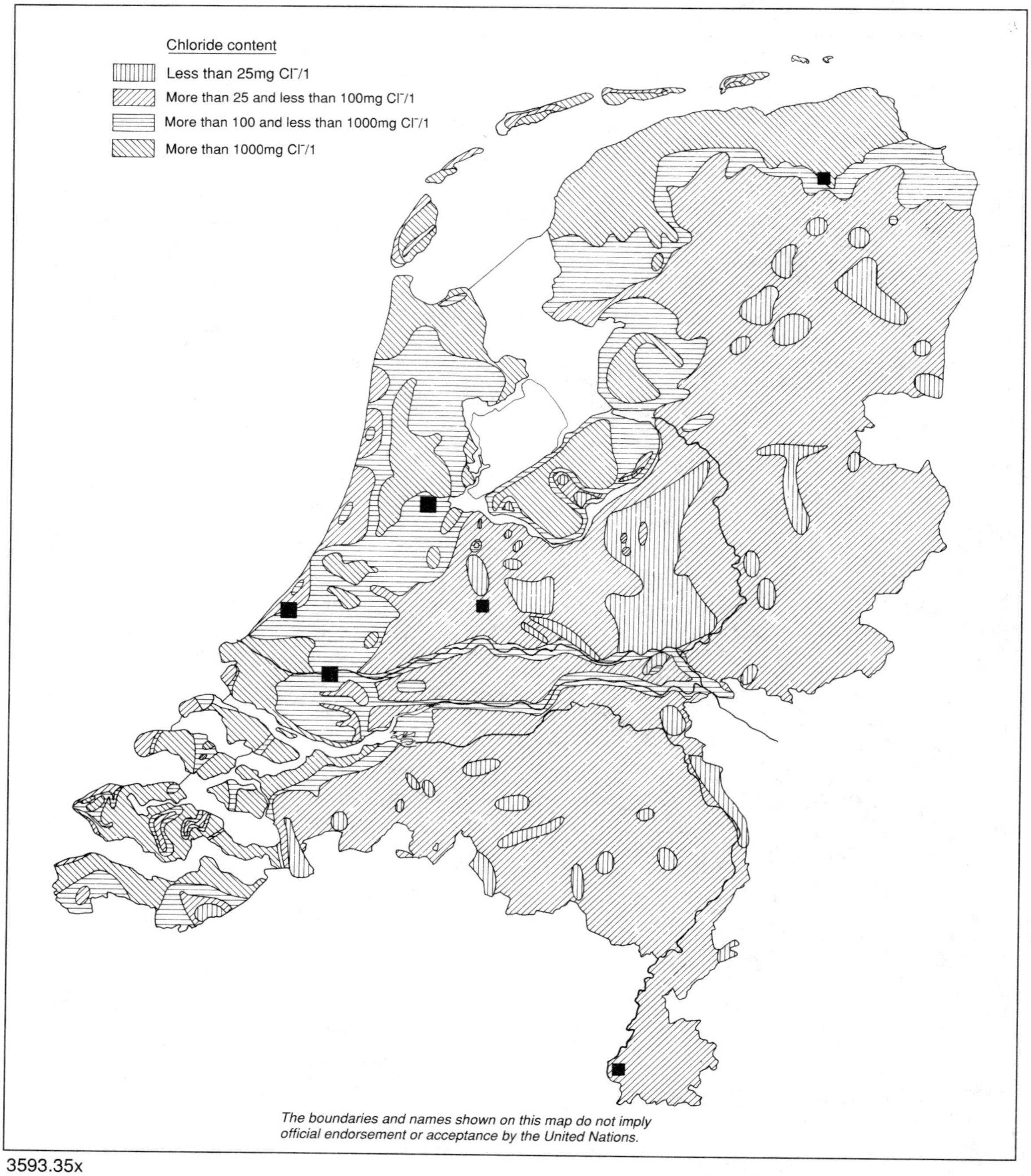

3593.35x

The area with increased chloride contents generally corresponds with the maximum reach of transgression. The contents are relatively low if compared with the salinity of sea water of 19,000 mg/l. This may be explained (Volker, 1961) by molecular diffusion being the main transport mechanism for the downward movement of chloride ions. Only at a relatively short distance from the coast (some 1 to 2 km) may it be expected that salt-water intrusion from the actual North Sea affects the groundwater. Furthermore, the differentiation in contents will be largely caused by groundwater flow patterns, which came into being mainly after the reclamation of the area. The most brackish water is found at the deepest polders. In figure 42 areas have also been mapped outside the coastal regions. Higher chloride contents in the fresh-water regions are mainly due to pollution.

In almost the whole country, groundwater at greater depth is brackish. The interface between fresh and brackish water is represented in figure 43. The origin of this salt has to be sought in older marine sediments of lower Pleistocene and Tertiary Age.

The chloride contents gradually increase with depth. The salt distribution in the inland areas can largely be explained by lateral dispersion caused by a mainly horizontal flow in the aquifer system above the marine base (Meinardi, 1983b).

The coastal regions are at the downstream end of the groundwater flow and, hence, there the upward movement of salt ions is the strongest. The result is that in the coastal areas both types of brackish water meet, such that the whole aquifer system contains brackish groundwater. An exception is the dune ridge and the nearby area of older coastal barriers where the downward percolation of fresh water was not hampered by confining layers. A lens of fresh water originated below the dune ridge (the Ghyben-Herzberg lens), with a sharp transition zone to the underlying groundwater of a relatively high salinity. Badon-Ghyben (1888) developed his theory for the situation in the dune area west of Amsterdam. Almost everywhere outside the dune areas the transition from fresh to somewhat more saline groundwater is so gradual that the Ghyben-Herzberg conditions do not apply.

(c) Fresh groundwater in sandy areas

The natural composition of fresh groundwater in sandy areas can be illustrated at the example of the Veluwe area. The Veluwe area always consisted largely of hills covered with wasteland and now forms a vast natural reserve. The subsurface is sandy to a great depth. Table 42 represents some representative analyses of groundwater samples, as well as an estimate of the average rain composition, determined 50 years ago.

From the Veluwe groundwater it may be concluded that two processes predominate, namely condensation by evapotranspiration and the dissolution of calcium carbonate by the percolation of water containing carbon dioxide. The condensation factor (rainfall divided by actual recharge) is about 2 to 3. The deeper strata still contain calcium carbonate, the more shallow layers are already leached and there the pH value and the Ca and HCO_3 contents are lower.

Figure 43. Netherlands: Depth to interface between fresh and brackish water

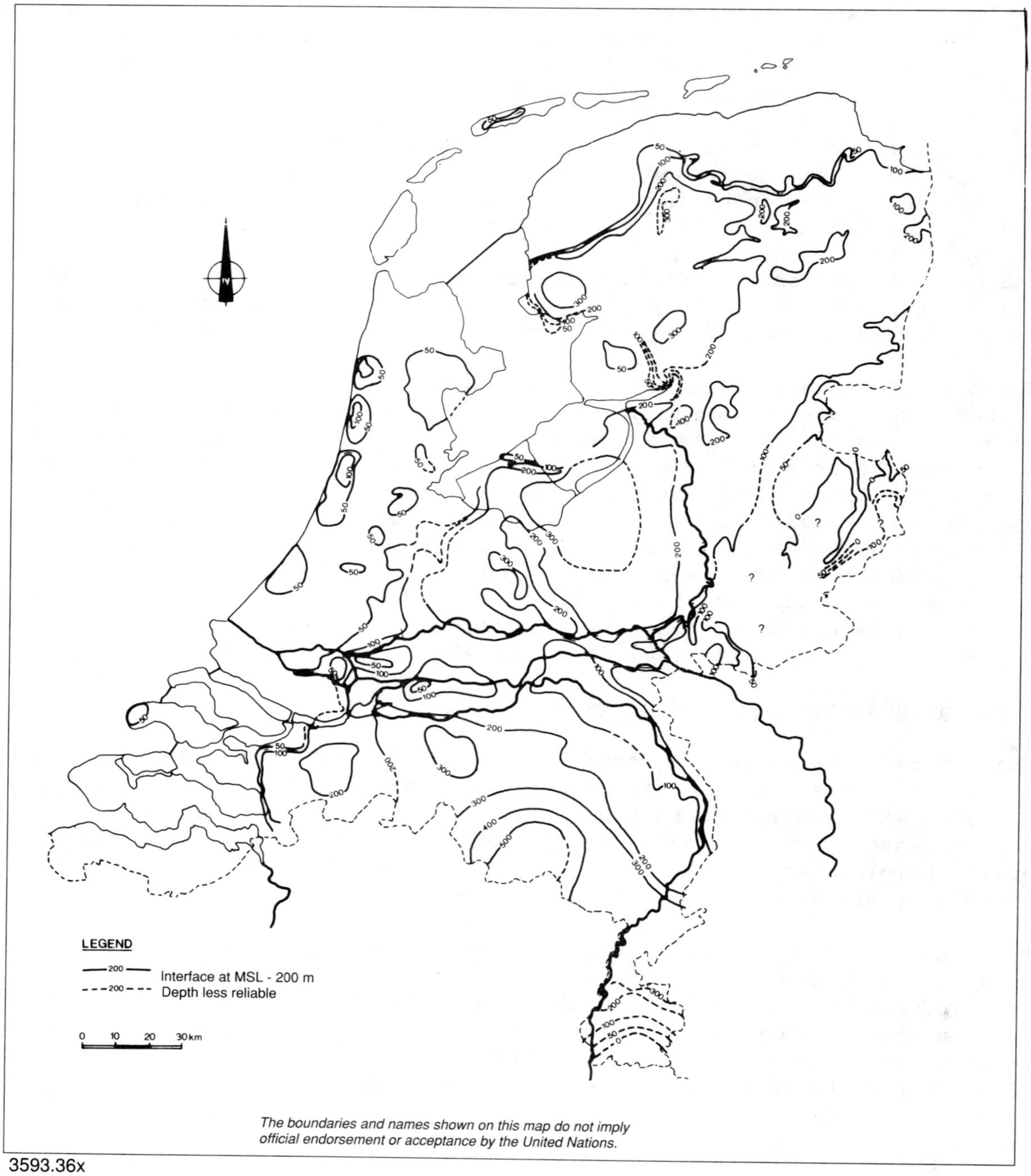

3593.36x

Table 42. Netherlands: Chemistry of water from aquifer at Veluwe

Borehole No.	Screen depth (m)	Month of sampling	Ca^{2+}	Mg^{2+}	$Na^{+}+K^{+}$	NH_4^{+}	Cl^{-}	SO_4^{2-}	NO_3^{-}	HCO_3^{-}	pH	Fe	Mn
						(contents in milligrams per litre)							
26H145	149	Dec. 1967	26.6	1.1	12	0.23	11	11.5	--	88	7.6	1.3	0.22
27B155	132	Sept. 1968	26.0	2.4	14	0.3	11	11.1	--	98	7.1	0.13	--
32G137	30	June 1973	31.5	1.3	8.0	0.24	7	6	0.1	107	8.1	0.25	0.04
32G137	100	June 1973	36.5	4.0	8.0	0.20	8	9	0.1	134	7.9	0.45	0.04
33A 20	108	Mar. 1953	9.9	1.7	14.8	--	18.7	8	4.3	31	6.6	0.07	--
40A323	84/117	Mar. 1964	19	2.2	10	0.10	12	4.9	--	67	7.7	0.02	--
Natural rainfall		1933/1937	1.7	1.0	2	0.5	4	4.5	0.1	25	..	..	..

Table 43. Netherlands: Chemistry of water from bog at Veenkoloniën

Borehole No.	Screen depth (m)	Year of sampling	Ca^{2+}	Mg^{2+}	$Na^{+}+K^{+}$	NH_4^{+}	Cl^{-}	SO_4^{2-}	NO_3^{-}	HCO_3^{-}	pH	Fe	Mn
						(contents in milligrams per litre)							
12F385	73	1928	122	7	21	5	21	--	--	445	7.1	26	0.8
12F385	128	1928	114	5	16	4	18	2	--	440	7.1	9	0.4
12F687	24/39	1973	116	11	15	5	35	10	..	390	6.7	29	0.5
12F957	44	1976	77	7	25	4	48	25	--	256	6.8	15	0.08
12F960	54	1976	107	10	24	1	29	20	--	418	7.0	..	..
12F973	47	1976	111	13	24	6.5	24	28	--	440	6.7	32	0.26

(d) Fresh groundwater under former raised bogs

Large parts of the higher grounds were covered with raised bogs up to some 100 to 200 years ago. These raised bogs generally originated where the transport capacity of open water and the subsurface was insufficient to discharge excess rainfall. At the raised bogs, increased evaporation took care of the discharge of the remaining water. Currently, the peat has been excavated; only soils rich in organic material remained. The largest raised bog was present in south-east Groningen (Veenkoloniën). Examples of the deeper groundwater composition at the centre of this area are represented in table 43.

At the Veenkoloniën example, some major ions contents may be explained from rain composition multiplied by a condensation factor of about 5 (morass evaporation). The groundwater contains relatively large amounts of ammonia, iron and manganese. The Veluwe and the Veenkoloniën examples are the two extremes in between which the natural composition of unpolluted groundwater varies at higher ground.

(e) Groundwater in the upstream river valleys

River valleys alongside the present branches of the river Rhine are recipients of groundwater. Only some minor local recharge areas exist in the nearby higher sandy areas and river beds.

An example is the Over-Betuwe area (see figure 41). A most remarkable type of groundwater in that area is derived from Rhine water, but affected by evaporation. This type of groundwater is present over large areas in the shallow aquifer. An interesting explanation is that this water is the remnant of floodings which occurred in the period before reclamation. The flood water filled the lower basins enclosed by the river banks and was further discharged by evaporation and percolation into the soil at a subsequent period of lower river levels.

(f) Fresh groundwater in the lower parts of the country

The situation at the downstream reaches of the Rhine and the lower polder areas in the west and north of the country is exemplified by the shallow aquifer of the island of Dordrecht. Without withdrawal, almost the entire area experiences upward seepage. The groundwater flows from the surrounding river beds, except at the north-east side where unreclaimed river marshes form an area of natural recharge.

After installation of a well field drawing about 4 million m^3 per year the polder area around the well field became an infiltration area. In that situation the northern wells almost exclusively attracted infiltrated water from sandy river beds and the southern wells water which had infiltrated at the river marshes with a peaty soil and at the polders. Both types of groundwater are represented in table 44.

Wells 18, 20 and 30 in table 44 contain water almost fully derived from the river. Well 6 yields water infiltrated locally. Wells 13 and 27 represent mixtures. The river-bed infiltrated water is characterized by relatively high salt contents, corresponding with high contents in the river, but relatively low contents of NH_4, Fe and Mn. The groundwater derived from the local polders has low salt content, indicating a local origin of the recharge. The contents of NH_4, Fe and Mn and also the hardness are high. The groundwater is heavily anaerobic.

Table 44. Netherlands: Chemistry of Dordrecht groundwater supply (Screen depth, 15-25 m; all samples collected April 1974)

Number	Ca^{2+}	Mg^{2+}	$Na^{+}+K^{+}$	NH_4^{+}	Cl^{-}	SO_4^{2-}	NO_3^{-}	HCO_3^{-}	pH	Fe	Mn
	(contents in milligrams per litre)										
6	118	14.4	32	24.2	64	1.4	--	476	6.84	15.0	0.70
13	152	17.8	30	26.0	41	2.1	--	641	6.76	19.4	0.80
18	96	13.4	84	8.2	169	64.7	--	226	7.34	3.0	1.4
20	96	13.4	71	7.8	132	38.1	--	287	7.17	4.0	0.80
27	167	21.1	36	31.2	54	1.0	--	689	6.81	14.4	0.75
30	94	15.8	74	11.9	141	52.1	--	287	7.09	6.0	0.90

The river-bed infiltrated groundwater of table 44 is largely representative for a relatively narrow zone along the lower reaches of the Rhine; outside that zone the infiltrated river water is older and has lower content of dissolved compounds. The type of groundwater derived from recharge of local water percolating through peaty layers (well 6) can be found throughout the coastal zone of the Netherlands, as far as the groundwater is fresh.

(g) Pollution from man-made activities

The groundwater of the island of Dordrecht, discussed above, already bears the marks of a man-made pollution as the river-bed infiltrated water is modern Rhine water, being polluted by the upstream industry and population. But also in the areas where rainfall recharges the local groundwater the effects of pollution are present almost everywhere in the shallow groundwater. The sources of pollution are manifold. A distinction as to the areal effects is the division into point, line and diffuse sources. Line sources are the infiltrating surface waters like the river Rhine.

Important point sources of pollution are sanitary landfills and industrial waste dumps. The sanitary landfill at Noordwijk at the western dune ridge is an example.

Although the effluent of the landfill towards the groundwater is heavily polluted, the areal extend of the polluted zone after 20 years of landfill presence is still limited to a few hundreds of metres mainly downstream of the landfill (figure 44).

In the long run, however, sanitary landfill and other waste dumps form a serious threat to groundwater quality in general; the more as an inventory yielded a large number of sites over the whole country. The sites of industrial waste dumps are not always known. Many of the minor dumps are situated at industrial zones. Groundwater may be polluted by a variety of compounds. Pollution with halogenated carbons caused a threat to a number of well fields for public water supply.

Figure 44. Netherlands: Extension of polluted effluent from Noordwijk sanitary landfill

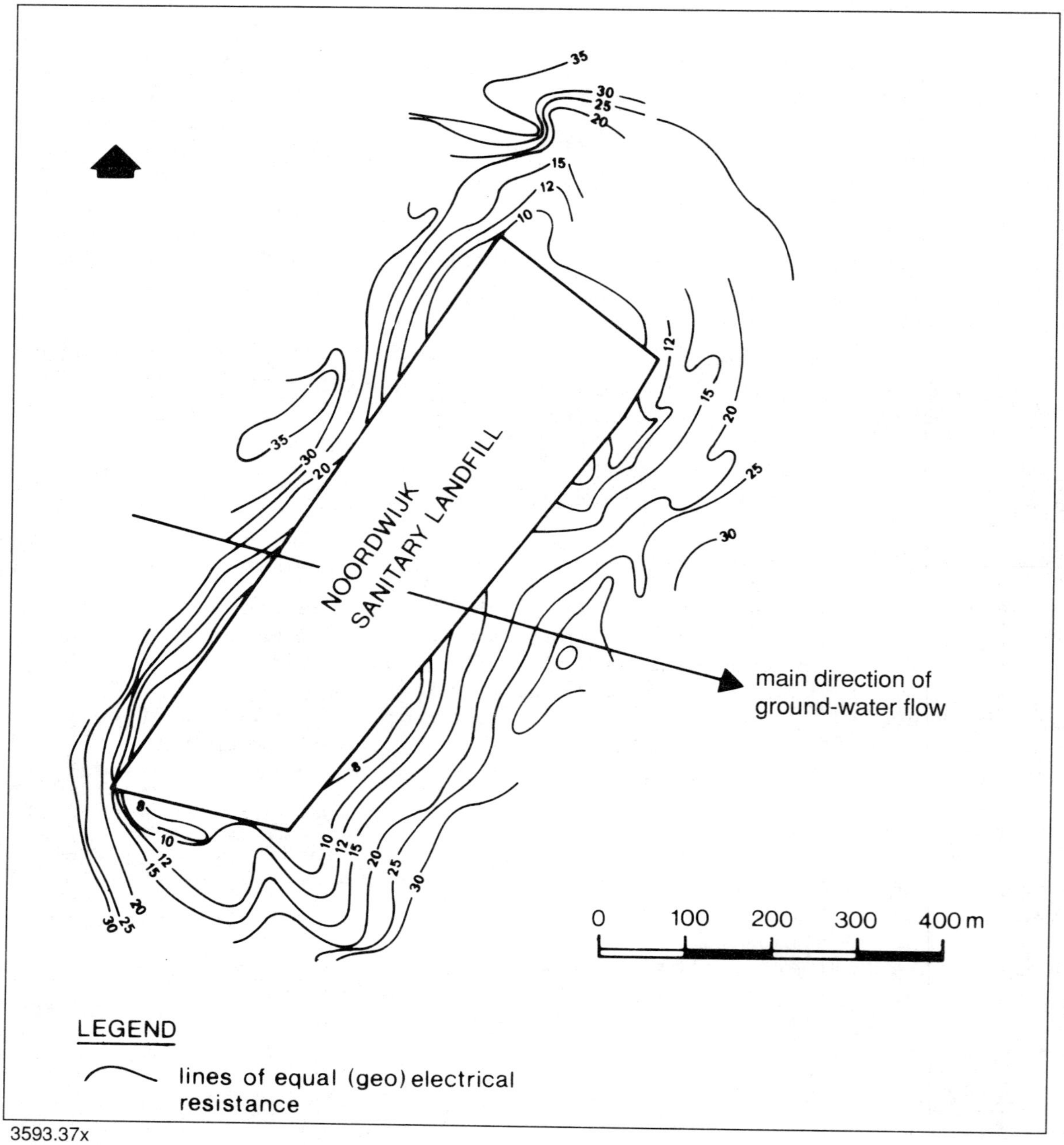

3593.37x

The main sources of diffuse pollution are the use of fertilizer and other chemicals in the intensive agriculture, disposing of excess manure in areas of intensive cattle breeding, and air pollution by industrial and agricultural sources.

The use of fertilizer proper will generally lead to an increase in the inorganic compounds in groundwater, of which, notably, the nitrate content may surpass acceptable limits.

The use of chemicals in agriculture started some 100 years ago in the form of fertilizers, and was extended some 40 years ago by the use of herbicides. Notwithstanding this relatively short period, cases have already been reported where well fields are threatened by nuisible organic compounds from agricultural sources.

Intensive animal husbandry developed from 1960 onwards. It led to manure production far beyond the need for fertilizer in the areas concerned. Yet, the excess manure is disposed of by dumping it on arable land. The resulting pollution of the shallow groundwater is severe, mainly with organic compounds but also with dangerous trace constituents. A vertical profile (Boumans, 1985b) in the shallow groundwater of the Peel area shows the effects of a dumping of excess manure on some compounds (figure 45). The situation is alarming, especially because the areas concerned are largely composed of sandy soil.

The effect of air pollution is felt throughout the country in the form of an increased deposition of several major and minor compounds, especially by means of acid rain. The vertical profile in the centre of a vast forest area of figure 46 shows the effects of recent air pollution on shallow groundwater (Boumans, 1985b).

4. Age of groundwater

The determination of the age of groundwater is of practical interest in several ways. For example, the age of shallow groundwater in the upper soil layers (to a depth of several metres) will express the downward movement of recent pollution. Also, the age of water in aquifers, exploitable for public water supply (depth from 10 m to 400 m in the Netherlands), mostly antedates the period of industrial pollution. A large reservoir of good quality groundwater is present. However, pumping may attract younger water. Lastly, deeper, less permeable, layers have a certain potential for the storage of dangerous waste. An age determination of groundwater will give a clue about the rate of circulation in these layers.

A relatively direct way to determine the age of groundwater is by means of the content of radioactive isotopes. For shallow groundwater the tritium content will give reliable results up to an age of about 40 years. Hence, this method can only be used to date shallow groundwater. A more suitable isotope for groundwater in deeper aquifers is carbon-14, to be used for ages up to about 40,000 years. Unfortunately, the age determination of groundwater by means of carbon-14 is unreliable for a number of reasons. Nevertheless, a first estimate of the age of the groundwater in exploitable aquifers can be given (figure 47).

Figure 45. Netherlands: Vertical profiles of pollutants at site of excess manure dumping in Peel area

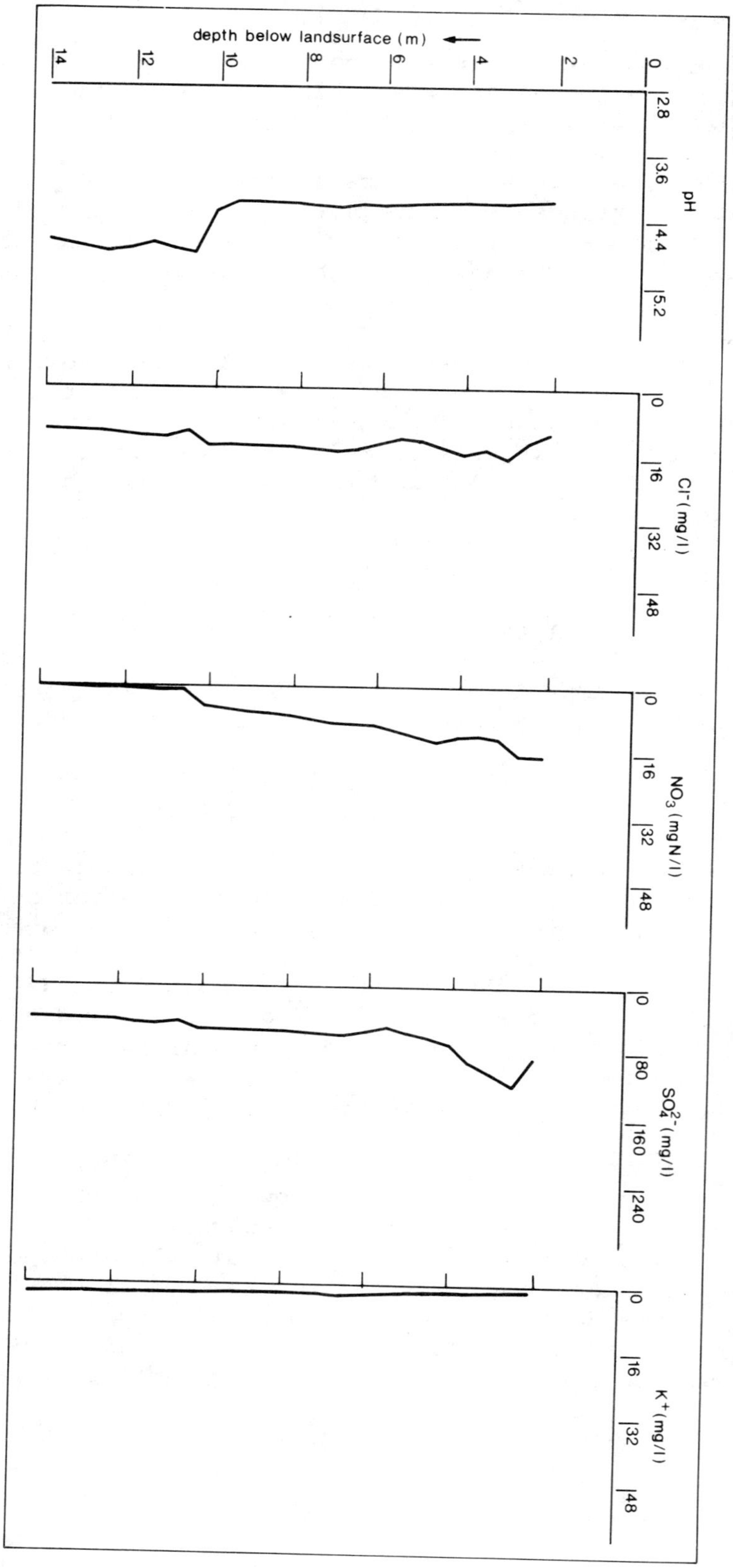

Figure 46. Netherlands: Vertical profiles of pollutants from acid rain within a natural reserve in Peel area

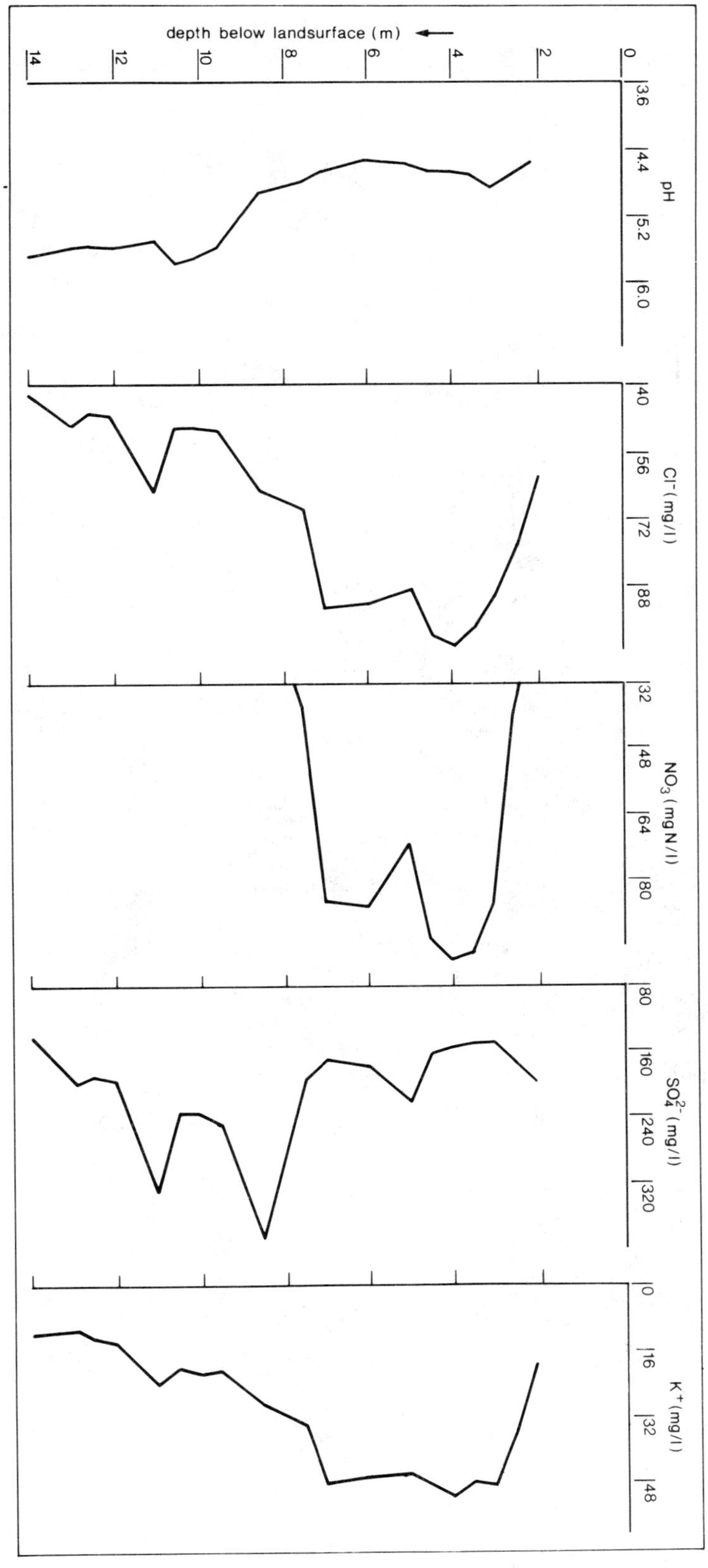

Figure 47. Netherlands: Age of groundwater in exploitable aquifers

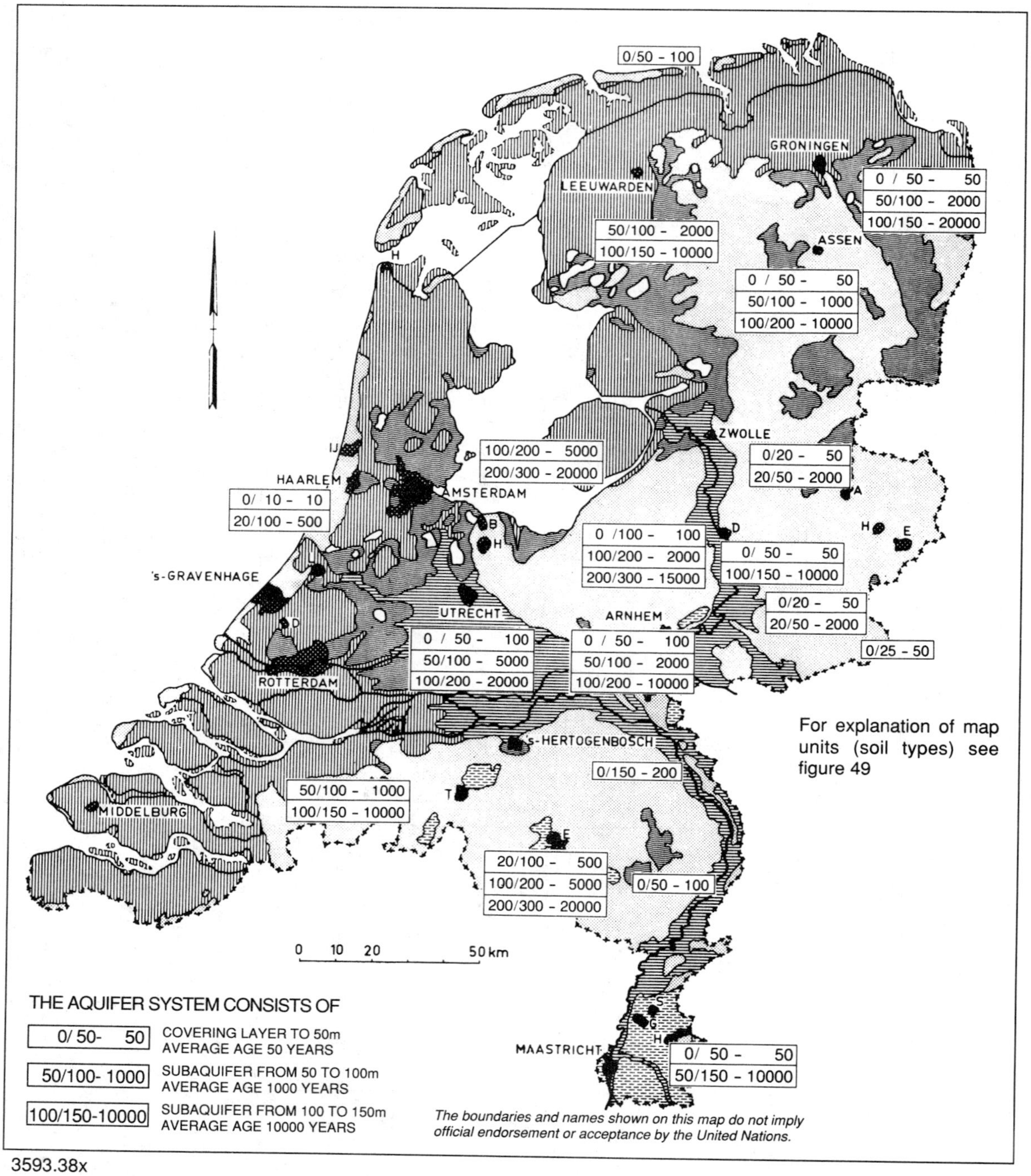

For explanation of map units (soil types) see figure 49

The boundaries and names shown on this map do not imply official endorsement or acceptance by the United Nations.

3593.38x

C. Groundwater development

1. Current status

In the Netherlands groundwater development generally includes not only active withdrawal by means of wells or other intakes, but also maintenance of a shallow water table on behalf of agricultural crop production and the management of natural reserves and recreational grounds based on the mere existence of shallow groundwater. In fact, many other economic activities are related to groundwater, owing to the particular situation that water tables are shallow and exist in phreatic or slightly confined aquifers. Consequently, the withdrawal of groundwater for a particular purpose always requires care with regard to other parties involved.

Active withdrawal of groundwater serves three major purposes:

(a) Public water supply

The total amount of groundwater used for public water supply amounted to 696 million m^3 in 1980. Groundwater is withdrawn at 244 well fields of which about 100 fields produce more than 2.5 million m^3 per year. Only 10 per cent of the well fields have yields of less than 25,000 m^3 per year (see figure 48).

Table 45 represents the withdrawal per province; the amounts withdrawn in the western provinces are relatively small because of the absence of fresh groundwater over large areas. At the dune ridge protecting these provinces from the sea, an additional 150 million m^3 per year of water is produced by artificial recharge of the dune groundwater; loations are indicated in figure 48. Sources of the recharge water are the river Rhine (Amsterdam, Provincial Water Works of North-Holland), the river Meuse (The Hague) and local polder water (Leiden).

(b) Industrial and other private use

The withdrawal of fresh groundwater by industry and other private users in 1980 amounted to 325 million m^3 of which 164 million m^3 of water was used for cooling purposes. The industrial abstractions are relatively small; of the total 2,700 abstraction points only 28 well fields produce more than 2.5 million m^3 per year of water. A number of industrial abstractions pump brackish groundwater to a total amount of about 25 million m^3 per year, most of which is used for cooling purposes.

(c) Agriculture

In parts of the summer, the soil water falls short to sustain crop growth (actual evapotranspiration is less than potential). This situation is partly remedied by sprinkling, although studies indicate that the economical basis for sprinkling is small or even lacking (Romijn, 1985). In the polder areas mostly surface water is used for this purpose, but in the sandy areas with a less developed drainage system the sprinkler installations are fed by groundwater. The individual abstractions are relatively small, but the total amount of groundwater withdrawn in the sandy areas is considerable (table 45 and figure 48).

Figure 48. Netherlands: Abstractions of groundwater, 1980

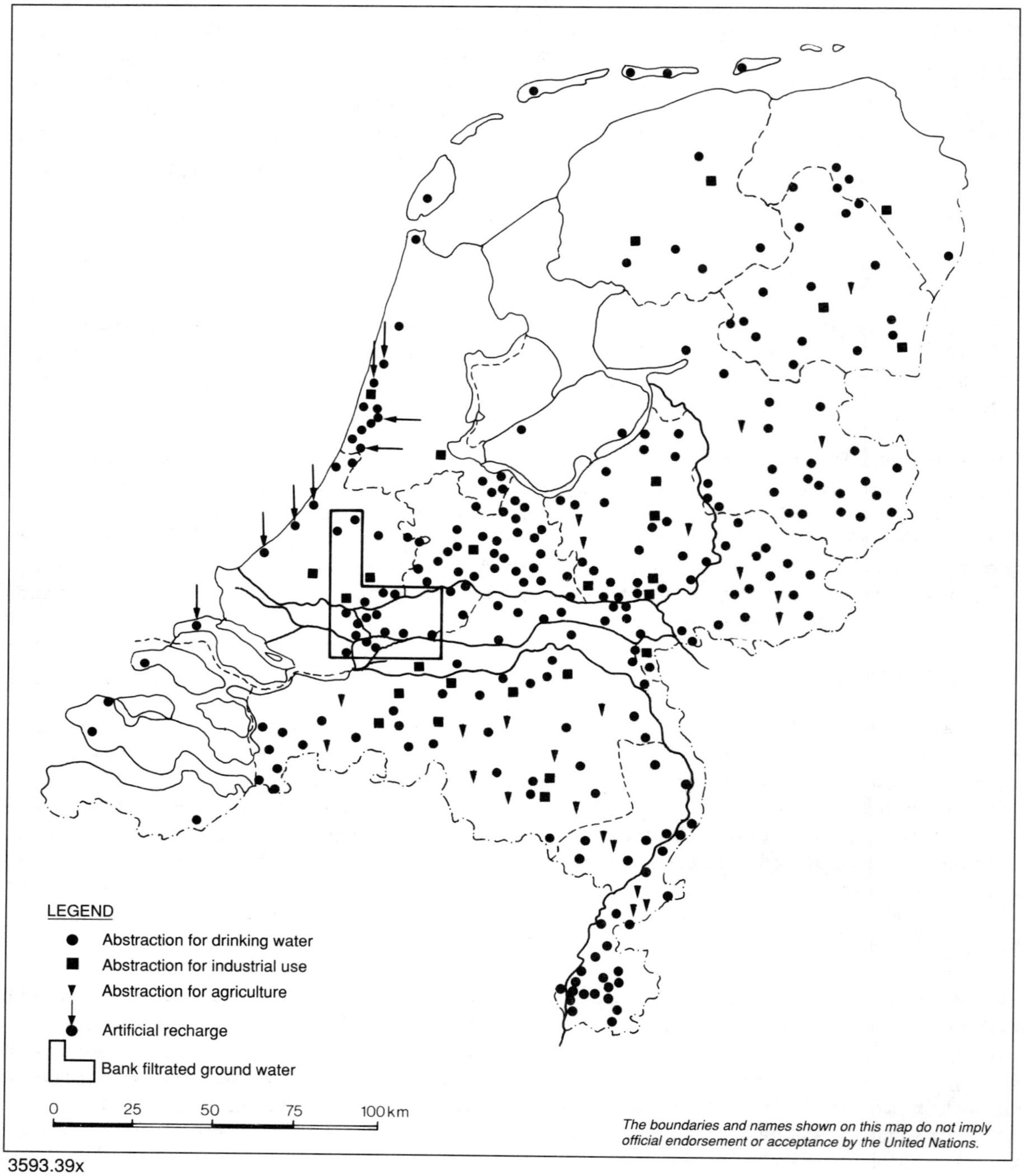

3593.39x

Table 45. Netherlands: Groundwater withdrawals, 1980

Province	Water supply companies	Industry and other private well owners: Cooling water	Industry and other private well owners: Other uses	Sprinkling (estimate)	Total
	(in million cubic metres)				
Groningen	21.5	3.2	16.6	0.1	41.4
Friesland	43.3	16.5	0.8	0.2	60.8
Drenthe	51.2	21.6	4.7	0.6	69.1
Overijssel	65.4	16.6	10.9	5.9	98.8
Gelderland	113.2	35.1	49.0	18.8	216.1
Flevoland	7.3	..	..	0.1	7.4
Utrecht	62.3	12.0	5.6	1.7	87.6
North-Holland	33.2	2.9	2.9	0.5	39.5
South-Holland	52.4	9.3	6.0	0.1	67.8
Zealand	4.1	..	0.3	..	4.4
North-Brabant	172.1	39.6	36.5	31.3	279.5
Limburg	69.6	16.3	24.9	14.7	125.5
The Netherlands in total	695.7	164.4	160.7	74.0	1 091.9

2. Well design

Out of a wide variety of drilling methods, well construction and pumping practice originally used, the public water supply branch has developed a practical standard. A representative exploitation well (see figure 49) at present has been drilled with the rotary method at reversed circulation, sometimes with air lifting of the drilling fluid. The well-screen consists of vertically slotted PVC pipe, surrounded by a gravel packing; the pump is of the submersible type.

Depending upon the desired capacity, the well diameter ranges from 0.2 to 1 metre. Directly after reaching the final depth, a well logging is executed consisting of spontaneous potential (SP), long normal (LN) and short normal (SN) resistivity, natural gamma radiation and caliper determination. Results of the drilling master's report and the well logging determine the sections of the casing either to be screened or to be kept blind.

Well yields depend on the desired quantity and on hydrogeological conditions, they may vary between less than 10 m^3 per hour at thin subaquifers in the river-clay district, to about 700 m^3 per hour (once abstracted during half a year from a test well) at the mighty aquifers in the central Netherlands. No strict rules have been developed to determine the maximum yield of individual wells.

Figure 49. Netherlands: Parts of a typical well

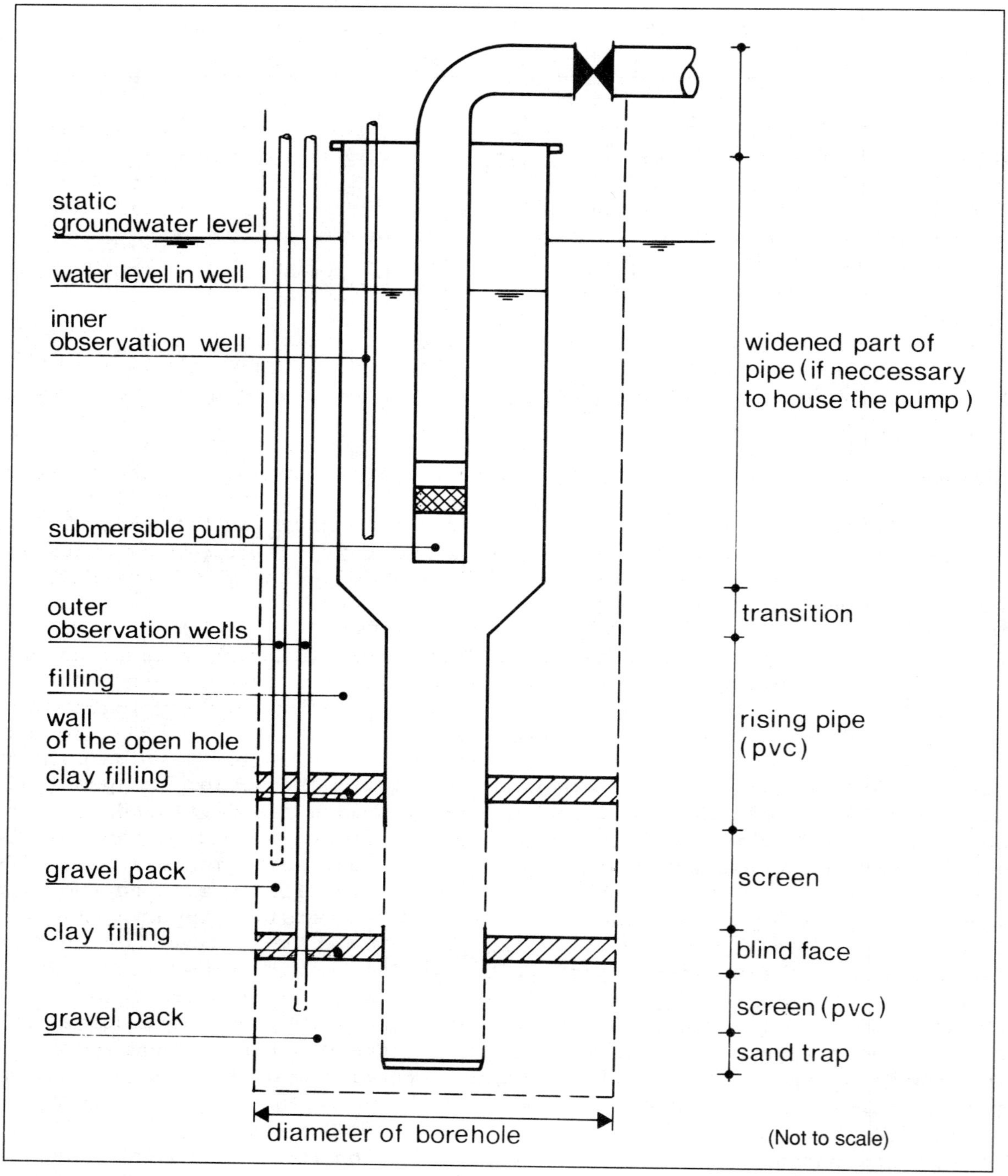

A relatively common problem in the Netherlands is the clogging of wells during exploitation. Two main types occur. At a mixing of aerobic and anaerobic ground water the precipitation of iron or manganese together with bacterial growth may clog the slots; wells can be rehabilitated with strong acids. In anaerobic groundwater, a not yet fully understood mechanism can clog the aquifer outside the gravel pack; this type of clogging can be remedied by applying chloride of soda (Beek and Kooper, 1980).

3. Exploration

Technical possibilities of withdrawal are not commonly the critical factor in the design of well fields. Almost everywhere in the country suitable aquifers can be found. Normally the availability of fresh groundwater will be sufficient, due to good recharge possibilities of the aquifers. Water quality, however, may constitute a technical constraint. Especially in the peaty areas the natural groundwater composition may pose difficulties to treatment. Another threat to quality of shallow groundwater is modern pollution of various origins.

Most important are the environmental constraints, such as damage to natural reserves, reduction of agricultural production, land subsidence, changes in surface-water discharge. The impact of the environmental effects varies over the country. In sandy areas, the reduction in crop growth and damage to natural reserves are limiting factors. In river-clay areas, damage to natural or agricultural vegetation plays a less dominant role owing to a more clayey subsoil. The danger of land subsidence, however, has to be feared, especially if the subsurface contains peat. In peaty areas, damage to vegetation due to hydrological changes may be considered negligible as water levels are controlled by polders. Possible land subsidence puts a limit to withdrawal, especially the unevenly occurring subsidence due to an irregular geologic structure.

The monitoring of large withdrawals in sandy areas ultimately requires the design of models simulating simultaneously the flow in the saturated and the unsaturated zones. Future groundwater levels and actual evapotranspiration have to be predicted. Such models need an extensive and costly collection of field data, consisting not only of a full knowledge of the parameters determining saturated groundwater flow, but also those of the unsaturated zone, evapotranspiration and the interaction with surface water. In the river-clay and polder areas the numerical modelling may be based on the superposition principle, such that only changes in water levels are predicted.

In the Netherlands, where groundwater is easily polluted from above, due attention must be given to protection of the surroundings of a well field. The following practice has developed.

Around a well field, zones are distinguished of residence times of 60 days, 10 years and 25 years. The zoning is based on travel times at horizontal flow in the pumped aquifer, residence times at vertical flow in confining layers are neglected, which provides an additional safety factor. Water companies are recommended to own and exploit the 60-day zone (to prevent bacteriological contamination). In the 10-year and 25-year zones, regulating measures can be taken as to land use (fertilizer, biocide) and other potential sources of pollution.

4. Well fields for recharge

Induced recharge in the Netherlands takes the form of the withdrawal of bank water along the branches of the major rivers. In the past, these withdrawals were considered from the viewpoint of groundwater abstractions. Investigations involved land subsidence, invasion of brackish water, clogging of wells and treatment of poor-quality water. At present also the quality transformation between parent river and wells is investigated. Travel times for each particular well field are computed using numerical modelling of the hydrogeologic situation (Meinardi and Grakist, 1985).

After the installation in the western dune ridge of artificial recharge works using Rhine water, the need for hydrogeologic research diminished. The upper soil layers, through which the river water percolates, are considered to function as a naturally slow sand filter. At present the interest in the hydrogeology of the dunes increases again, owing to their large recreational and natural values, which may be conflicting with water supply development activities.

D. Administration and management

1. Provincial and national interests

Until now, groundwater resources have not actively been managed at a national level. Yet, there is an indirect management, in that users who want to withdraw a significant amount of water need permission. From 1952 onwards only withdrawal on behalf of public water supply needed governmental approval, by force of the Grondwater Wet Waterleidingbedrijven. Since 1983 all abstractions are regulated by the Grondwater Wet (Groundwater Act). This act explicitly opens the possibility for an active management to be executed by the provincial administrations.

Passive management up to now consisted of an evaluation of plans, submitted by potential users, in the light of the effects on other interested parties like agriculture. A future active management has to be based on Provincial Groundwater Plans, to be composed by each province. The plans will be brought in line at the national level, also considering other water resources and other water users.

A number of investigations have already been carried out as a first step towards direct management. The most important are investigations by the Provincial Water Board of Gelderland and investigations on the future water supply of the province of South-Holland. Rijkswaterstaat made an ambitious start in 1984 with the Policy Analysis of Water Management (PAWN) in the Netherlands.

2. Monitoring networks

Groundwater levels, quality and other hydrogeologic parameters have been intensively investigated through various types of study. National monitoring networks make use of these local data collections, or the agency concerned has installed a specific network. The following national networks are particularly significant:

a. The archive of groundwater levels, operated by Groundwater Survey TNO, collects observations from 9,700 observation wells, with 19,000 screens at various depths. Generally, the level observations are done every fortnight by a corps of volunteers.

b. About 25 per cent of these wells are observed only four times a year. Abstractions are registered by the Ministry of Welfare, Health and Cultural Affairs (WVC). The REGWAT system of the Ministry of WVC collects and administers the data on all withdrawals, sufficiently large (more than about 10,000 m^3 per year) to obligate registry by the provincial authorities.

c. The Royal Meterorological Institute (KNMI) operates different types of weather stations, where weather data are collected on a regular basis.

d. The State Water Board (Rijkswaterstaat) maintains a national network on surface-water levels and discharges. Either continuously, or at intervals, the level or discharge are measured at a great number of points along rivers, lakes and the sea.

e. KNMI and the National Institute of Public Health and Environmental Hygiene (RIVM) collect data on rain water quality at 21 sampling points on a monthly basis.

f. The RIVM started in 1980 a national groundwater quality monitoring network. In 1984, 370 observation wells had been installed (see figure 50) with observation screens generally at a depth of 10, 15 and 25 metres below land surface.

3. Data bases

Apart from the data bases belonging to the monitoring networks, a number of other data bases are operated by national agencies. These data bases have been combined in a national retrieval system of hydrogeologic data, which in itself does not contain data, but indicates where the data can be found. The data concerned consist of:

a. Logs of about 60,000 boreholes with a description of the soil samples and often a geologic interpretation,

b. About 40,000 chemical analyses from 14,000 wells,

c. About 15,000 wells with data about the technical construction,

d. A few hundred wells where a geophysical well log, or a cable with electrode pairs (salt/fresh interface) or a temperature profile is available,

e. Results of about 1,000 pumping tests,

f. Data on groundwater levels recorded by the Groundwater Survey TNO.

Figure 50. Netherlands: Groundwater quality monitoring network, location of observation wells

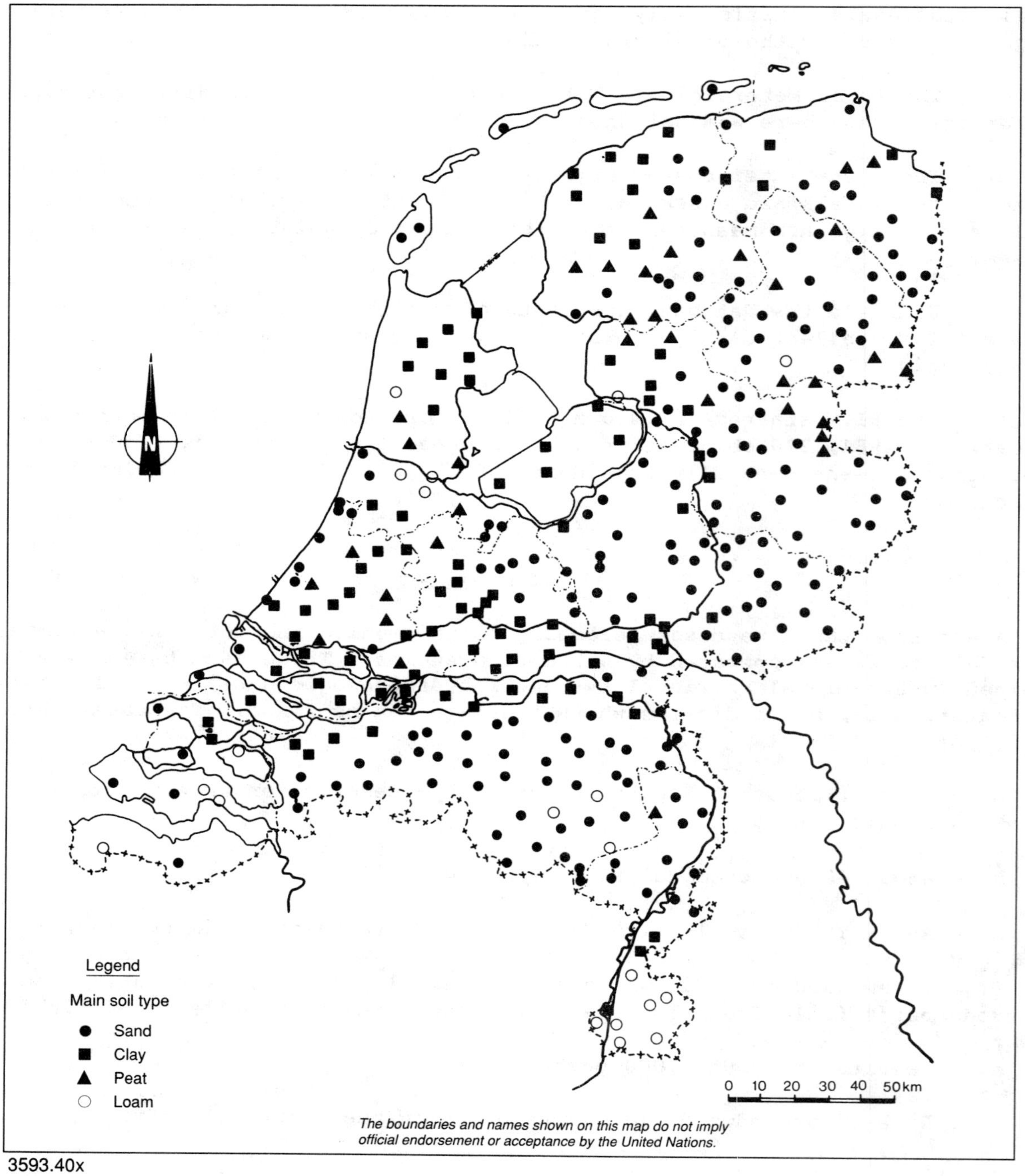

3593.40x

4. Mapping

The following mapping activities are groundwater related.

The Geological Survey prepares geologic maps at the scale 1:50,000. Three levels are distinguished, namely the shallow level (depth less than 40 m), a moderately deep level (between 40 m and about 500 m below land surface) and all deeper levels. The shallow level especially is represented on the 1:50,000-scale map and additional maps at scale 1:100,000. The moderately deep and the deeper strata are only described. Additionally, the Geological Survey has published general maps to the scale 1:600,000.

The Groundwater Survey TNO edits groundwater maps to a scale 1:50,000, which contain the locations of boreholes and geoelectrical surface soundings, the geometrical situation of aquifers and confining layers, sections showing aquifers and confining layers, groundwater contours at different levels, the transmissivity of aquifers, the major chemical groundwater components, and the distribution of fresh and brackish groundwater.

5. Training

The three starting points for groundwater research outlined before, namely, agriculture, public water supply and civil engineering, can still be recognized in the research institutes and even the educational institutions, dealing with groundwater investigations. The geologic approach never played a dominant role, but functioned rather as support to the engineers. The result is that no centralized institute for groundwater studies exists, but that a number of organizations working in the above fields have important branches for groundwater research.

The same holds for training facilities. No specific centre for training in groundwater is present, neither at an academic nor at lower levels of education. Groundwater studies are part of the curriculum of Delft Technical University, of the Agricultural University at Wageningen and of the Geological Department of the Free University in Amsterdam. At the secondary level, some institutes of professional training provide introductions to groundwater research.

The main agencies at the national level dealing with groundwater studies are:

a. Rijkswaterstaat (RWS)

b. National Institute of Public Health and Environmental Hygiene (RIVM)

c. Institute for Land and Water Management Research (ICW)

d. Soil Survey Institute (Stiboka)

e. Netherlands Geological Survey (RGD)

f. Groundwater Survey TNO (DGV-TNO)

g. Delft Hydraulics Laboratory (WL)

h. Testing and Research Institute of the Waterworks Association (KIWA)

i. A number of consulting firms vending expertise in groundwater.

6. International cooperation

The international cooperation of Netherlands groundwater researchers with other countries takes different forms. In the first place there is the exchange of knowledge, realized at international symposia, by publications and lectures. Another type of cooperation is the support of international organizations by contributing to their groundwater publications and other activities in the field, notably the International Hydrogeological Programme of UNESCO.

In bilateral activity, the Netherlands Development Cooperation water supply and sanitation projects are an important item, as well as irrigation and drainage projects. A number of governmental agencies and private firms bring in their knowledge in the field of groundwater studies to execute projects in various countries.

Training of staff from developing countries in the field of groundwater is provided by the International Institute for Hydraulic and Environmental Engineering (IHE) at Delft and the International Institute for Aerial Survey and Earth Sciences (ITC) at Enschede. Both institutes receive each year a considerable number of professionals from developing countries for postgraduate training in, amongst others, groundwater investigations.

Overseas activities have been undertaken in such countries as the United Republic of Tanzania, Burkina Faso, Egypt, and the Sudan through projects dealing with water balance calculations, data resources assessment, hydrogeologic maps, computerized data banks, exploration, drilling condition of wells, and the drilling of boreholes for rural water supply.

References

Appelo, C. A. J., G. J. W. Krajenbrink, C. C. D. F. Van Ree and L. Vasak. Beïnvloeding van de grondwaterkwaliteit in het infiltratiegebied van de noordwestelijke Veluwe. VU report, Amsterdam, 1982.

Bakker, H. de. Soils and their geography, Proc. symp. on peat lands below sea level. Edited by De Bakker and Van den Berg, ILRI publ. 30 : 85-97, Wageningen, 1982.

Beek, C. E. G. M. van and W. F. Kooper. Clogging of shallow discharge wells in the Netherlands river region. Groundwater 18 : p. 578-586, 1980.

Bon, J. Afvoer en berging in verband met beekverbetering, toegelicht aan het stroomgebied van de Lunterse Beek. ICW mededeling 107, Wageningen, 1968.

Boswinkel, J. A. Spatial salinity variations in the Netherlands. Seventh Salt Water Intrusion Meeting, Uppsala, 1981.

Bots, WCPM, P. C. Jansen and C. J. Noordewier Fysisch-chemische samenstelling van oppervlakte- en grondwater in het noorden des lands. ICW regionale studies 13, Wageningen, 1978.

Boumans, L. J. M. Kwaliteitsvariaties van het grondwater naar de diepte en de tijd. RIVM report, The Hague, 1986.

Bruin, H. A. R. de. Neerslag, openwaterverdamping en potentieel neerslagoverschot in Nederland. Frequentieverdelingen in het groeiseizoen; WR 79-4; KNMI, 1979.

Commission of The European Communities. Grondwater in Nederland, behoeften en beschikbare hoeveelheden. CEC, Brussels, 1986.

__________. Kwetsbaarheid en kwaliteit van het grondwater in Nederland. To be published by CEC, Brussels.

CoGroWa. Aanbevelingen voor het geohydrologisch onderzoek ter onderbouwing van vergunningsaanvragen voor grondwateronttrekkingen. CoGRoWA, Utrecht, 1983.

Cramer, W. Development of groundwater resources in the Netherlands. Proc. of the Exeter Symp., IAHS publ. No. 136, 1982.

Drabbe, J. and W. Badon-Ghijben. Nota in verband met de voorgenomen putboring nabij Amsterdam. Tijdschr. Kon. Inst. v. Ing. Verh., p. 8-22, 1888.

Duijvenbooden, W. van. Groundwater quality in the Netherlands, collection and interpretation of data. Sci. of the Tot. Envir. 21, p. 221-232, 1981.

__________. Effects of local sources of pollution on groundwater quality in the Netherlands. In Ward, C. H. and others (Eds.). Groundwater Quality (John Wiley, New York), 1981.

Geirnaert, W. Hydrogeology and hydrochemistry of the lower Rhine fluvial plain. Thesis, Leiden, 1973.

Groundwater Survey TNO. Groundwater mapping of the Netherlands. Various sheets and various supporting reports, n.d.

Herweijer, J. C., A. van Luijn and C. A. J. Appelo. Calibration of a mass transport model using environmental tritium. J. Hydrol, 78 p. 1-7, 1985.

Jelgersma, S. Holocene sealevel changes in the Netherlands. Ned. Geol. St. Ser C-VI-7, Haarlem, 1961.

__________. De geologische grondslag van het geohydrologisch onderzoek in Nederland. H_2O, 10, p. 543-550, 1977.

Meinardi, C. R. Characteristic examples of the natural groundwater composition in the Netherlands. Versl. Meded. CHO/TNO'21, p. 12-33, 1976.

__________. Fresh and brackish groundwater under coastal areas and islands, Geojournal 7.5, p. 413-425, 1983(a).

__________. Groundwater recharge in the Rhine fluvial plain, Zt dt geol. Ges. 134, p. 581-611, 1983(b).

Meinardi, C. R. and G. J. Hey. Samenstelling en stroming van zoet grondwater in het West-Nederlands poldergebied (Eiland van Dordrecht). RID med 78-2, 1978.

Meinardi, C. R. and G. Grakist. Hydrology of bank groundwater withdrawal in the Netherlands. J. of Hydrol., 78, p. 151-163, 1985.

RAND Corporation. Policy analysis of water management for the Netherlands. Report R-2500/1- NETH, 1983.

Rolf, H. L. M. Modeltoepassing voor de Over-Betuwe. Basisrapport Comm. Best, Waterhuish. Gelderland (in preparation), 1985.

Romijn. Economic importance of groundwater abstraction in the Netherlands. Memoirs, 18th congress International Association of Hydrogeologists, Cambridge, 1985.

Rijks Geologische Dienst. Toelichting bij geologische overzichtskaarten van Nederland. Eds. Zagwijn en Van Staalduinen RGD, Haarlem, 1975.

__________. De Geologische Kaart van Nederland. Various sheets with explanatory notes, n.d.

Stichting voor Bodemkartering. De bodem van Nederland. Stiboka, Wageningen, 1960.

Stichting Wetenschappelijke Atlas van Nederland. De Atlas van Nederland, Staatsuitgeverij, Den Haag, 1975.

Volker, A. Source of brackish groundwater in Pleistocene formations beneath the Dutch polderland. Economic Geology, vol. 56, 1961.

Vries, J. J. de. Inleiding tot de Hydrologie van Nederland. Rodopi, Amsterdam.

__________. Anderhalve eeuw hydrologisch onderzoek in Nederland. Rodopi, Amsterdam, 1982.

Werkgroep in de provincie Drenthe. Eindrapport 1969-1975.

Werkgroep Midden West-Nederland. Hydrologie en Waterkwaliteit van Midden West-Nederland. ICW Regionale Studies 9, Wageningen, 1976.

Werkgroep Noord-Holland. Grond- en Oppervlaktewater Noord-Holland benoorden het IJ. ICW Regionale Studies 16, Wageningen, 1982.

Zagwijn, W. H. Palaeographic evolution of the Netherlands during the Quaternary. Geol. en Mijnb., 53, p. 369-385, 1974.

XII. PORTUGAL

Area: 92,082 km^2 Population: 10.4 million

A. Physiography and climate

The Portuguese Republic on the continent is roughly the shape of a rectangle. Its maximum length is 561 km and its maximum width is 218 km. To the west and south it is bounded by the Atlantic Ocean with coastlines of 660 and 170 km, respectively. To the north and west, Portugal borders on Spain with frontiers of 400 and 800 km, respectively.

The archipelago of Madeira (population 260,000) is situated in the North Atlantic Ocean, 978 km south-east of Lisbon. It is made up of the volcanic islands of Madeira (741 km^2) and Porto Santo (43 km^2) and the uninhabited islands of Desertas and Selvagens.

The archipelago of the Azores (population 250,000) is situated in the North Atlantic 1,400 km west of Lisbon. It consists of nine volcanic islands. They form three groups: (1) the eastern group made up of the islands of Sao Miguel, the largest of the archipelago with an area of 742 km^2, and Santa Maria (97 km^2); (2) the central group made up of the islands of Terceira (400 km^2), Graciosa (62 km^2), Sao Jorge (246 km^2), Pico (433 km^2) and Faial (172 km^2); and (3) the western group made up of the islands of Flores (143 km^2) and Corvo (17.5 km^2).

Continental Portugal may be divided into three zones of relief: (1) the lower zone, below 50 m, which occupies about 11.8 per cent of the territory; (2) the intermediate zone 50-1,200 m, which occupies 87.7 per cent; and (3) the subalpine zone, 1,200-2,000 m, which occupies barely 0.5 per cent.

The river Tagus (Tejo) divides continental Portugal into two distinct orographic zones: it represents the approximate boundary between the lower areas of the south and the higher areas of the north. The average altitude of the country is 240 m: to the north of the Tagus it is 370 m and to the south 160 m. The country's principal river basins are shown in figure 51.

There are two main forms of relief: (1) the plateaus and (2) the mountains of moderate height. The most important mountain range consists of the sierras of Estrela, Montejunto and Sintra, which trend roughly north-east. The highest plateau is Tras-os-Montes in the north-eastern part of the country, which is about 1,000 m. To the south there is the plateau or peneplain of Alentejo. It forms an inclined plane with heights ranging from 200 m to 800 m near the Spanish frontier. On the coast are a number of important alluvial plains such as those of the rivers Vouga, Mondego, Tagus and Sados.

The archipelago of the Azores is mountainous. The mountain known as Pico on the island of Pico is the highest point (2,351 m). The island of Madeira is also mountainous: the highest point, Pico Ruivo, is 1,861 m. The island of Porto Santo is less mountainous: the highest point, Pico do Facho, is only 517 m.

Figure 51. Portugal: Principal river basins

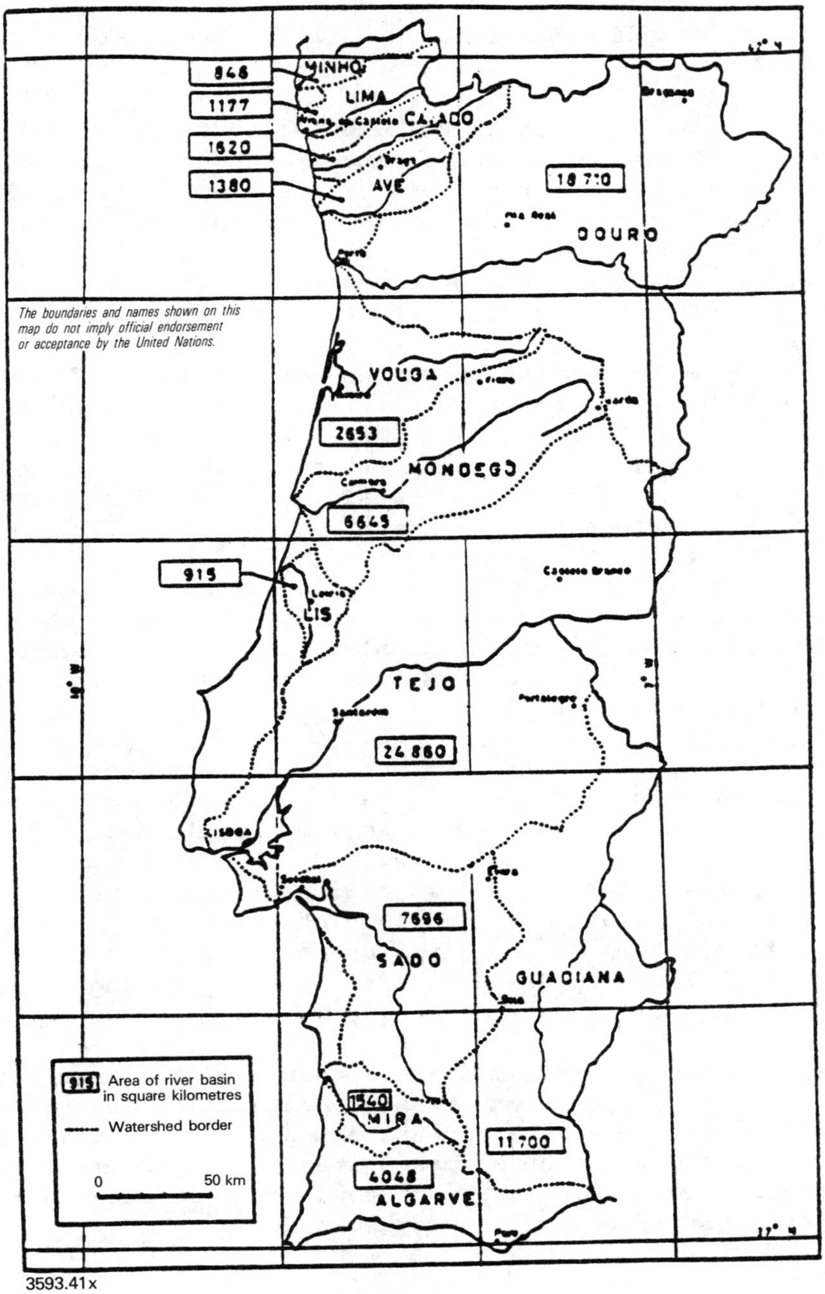

3593.41x

On the basis of the relief, the rainfall and the Atlantic influence to the north and the Mediterranean influence to the south, Cunha and others (1980) have divided continental Portugal into three large natural geographic regions: (1) the North Atlantic zone which includes the region of Minho, Douro Litoral, Beira Litoral and the western part of Beira Alta and Beira Baixa, with altitudes over 1,000 m, high precipitation and subject to oceanic influence; (2) the north-eastern zone including the regions of Trasos-Montes and Beira Transmontana, with high plateaus separated by deep valleys and with precipitation lower than that of the Atlantic zone to the north; (3) the Mediterranean zone to the south, which consists of the south-east part of Beira Baixa, Estremadura, Ribatejo, Alentejo and the Algarve, with low relief and moderate precipitation much below that of the other two zones.

The mean annual precipitation in continental Portugal is 900 mm. In the northern regions, the annual mean precipitation attains about 3,000 mm in the sierras of Geres and Peneda. To the south of the Tagus, the rainiest zone is in the sierra of Monchique in the Algarve, with an annual mean precipitation of 1,260 mm. In the drier areas north of the Tagus the annual mean precipitation may drop to 400 mm in the valley of the Douro. South of the Tagus, the lowest mean annual precipitation is on the coast of the Algarve, with 340 mm.

Throughout the year precipitation varies considerably. It is concentrated in the wet season (October-March) and this winter rainfall is some 80 per cent of annual precipitation. In the dry season (April-September), four months may pass without a single day of rain in the southern regions. Over the years there are also marked variations in precipitation totals. These variations may be of the order of 1 to 3 or even 1 to 4.

The annual mean potential evapotranspiration is about 880 mm; real average annual evapotranspiration is estimated at 500 mm. The regions with the highest potential evapotranspiration are in Baixo Alentejo, with readings of 900 mm. The regions with the lowest potential evapotranspiration are in the mountain zones, where they attain mean values of 570 mm.

The mean annual temperature for the whole continental territory is about 15° C. Mean daily temperature varies regularly throughout the year attaining normally minimum values in January and maximum values in August.

The climate of the Azores is temperate with an annual mean temperature of 17° C. Annual mean precipitation is fairly high, ranging on the island of Sao Miguel from 1,000 to 3,000 mm in the lowlands to 6,000 mm at the summit of Pico. Precipitation occurs throughout the year with maximum readings in January and minimum readings in August.

The Madeira climate is also temperate (oceanic). The mean summer temperature is 21° C and for winter 16° C. Mean annual precipitation on the island of Madeira varies between 600 mm on the southern coast and 2,840 mm in the mountain areas. On the island of Porto Santo mean annual precipitation is 260 mm. The maximum monthly precipitation occurs between November and February, depending on the locality. The minimum monthly precipitation usually occurs in July.

B. Geology

Continental Portugal may be divided into four geologic units (figure 52): (1) the western massif (also known as the ancient massif or Iberian Meseta); (2) the western Middle Cenozoic fringe; (3) the southern Middle Cenozoic fringe; and (4) the sedimentary basin of the Tagus and the Sado.

The western massif occupies an area of about three quarters of continental Portugal. It is made up of pre-Mesozoic rocks. These formations were affected by metamorphic phenomena associated with various tectonic phases of the Hercynian orogeny. There are many types of metamorphic and igneous rocks: the former occur most generally south of the sierra of Estrela and the latter occur mostly in the north. The most representative types of rock are schist, metagreywacke, quartzite, marble, granite, granodiorite and porphyry. In certain zones these formations are covered by more recent sedimentary deposits of continental origin. Late Hercynian orogeny gave rise, in the western part of the massif, to a series of tectonic events. These included the sinking of an extensive tectonic trench containing a thick series of sediments which today constitute the western Middle Cenozoic fringe.

The formations of the western massif were sharply folded and fractured by the Hercynian orogeny and some of the events were reactivated during the Alpine orogeny. In various localities, the results of these orogenies are superimposed on the remnants of former orogenies, in particular during the Caledonian period. The Hercynian orogeny gave rise to the striking geomorphologic features of the territory, namely, the north-west to south-east direction of the folds and the general pattern of fracturing. The main granite intrusions date also from this period.

From the geomorphologic point of view the eastern massif differs from the northern and southern regions. The northern region is mountainous and many of its valleys are enclosed. This is the region of heaviest rainfall and it has the densest and most important drainage system. This region also contains Portugal's highest peaks. The southern region consists largely of lowlands with occasional gentle hills. There is less rainfall here, the drainage system is less dense and the indigenous rivers are smaller. However, the Tagus and the Guadiana, which traverse this region, are two large waterways which drain extensive parts of the surrounding countryside.

The Middle Cenozoic fringes occupy the coastal areas between Espinho and Sines on the west, and between Sagres and Vila Real de Santo Antonio on the south. These geologic units contain a highly diverse mixture of lithologic types, with ages ranging from the Triassic to Recent. They are almost exclusively sedimentary. The formations are basically carbonates (limestone, marl and dolomite), clay and sand with some intercalations of igneous rock, both intrusive and extrusive.

The western fringe is made up of a series of sediments, more than 3,000 m thick. They were deposited in an extensive tectonic trench caused by the sinking of the western massif. The formations consist of bedded limestone, marl, sand, siltite and argillite. Over these formations is an evaporite covering which in various localities gives rise to diapiric uplifts. Some of these are heavily eroded with outcrops of the more ancient formations of the diapiric nucleus. From the end of the Cretaceous to the Eocene, there were tectonic phenomena related to the movement of the salt-bearing massifs. At the same altitude there is an uplift

Figure 52. Portugal: Principal geologic units

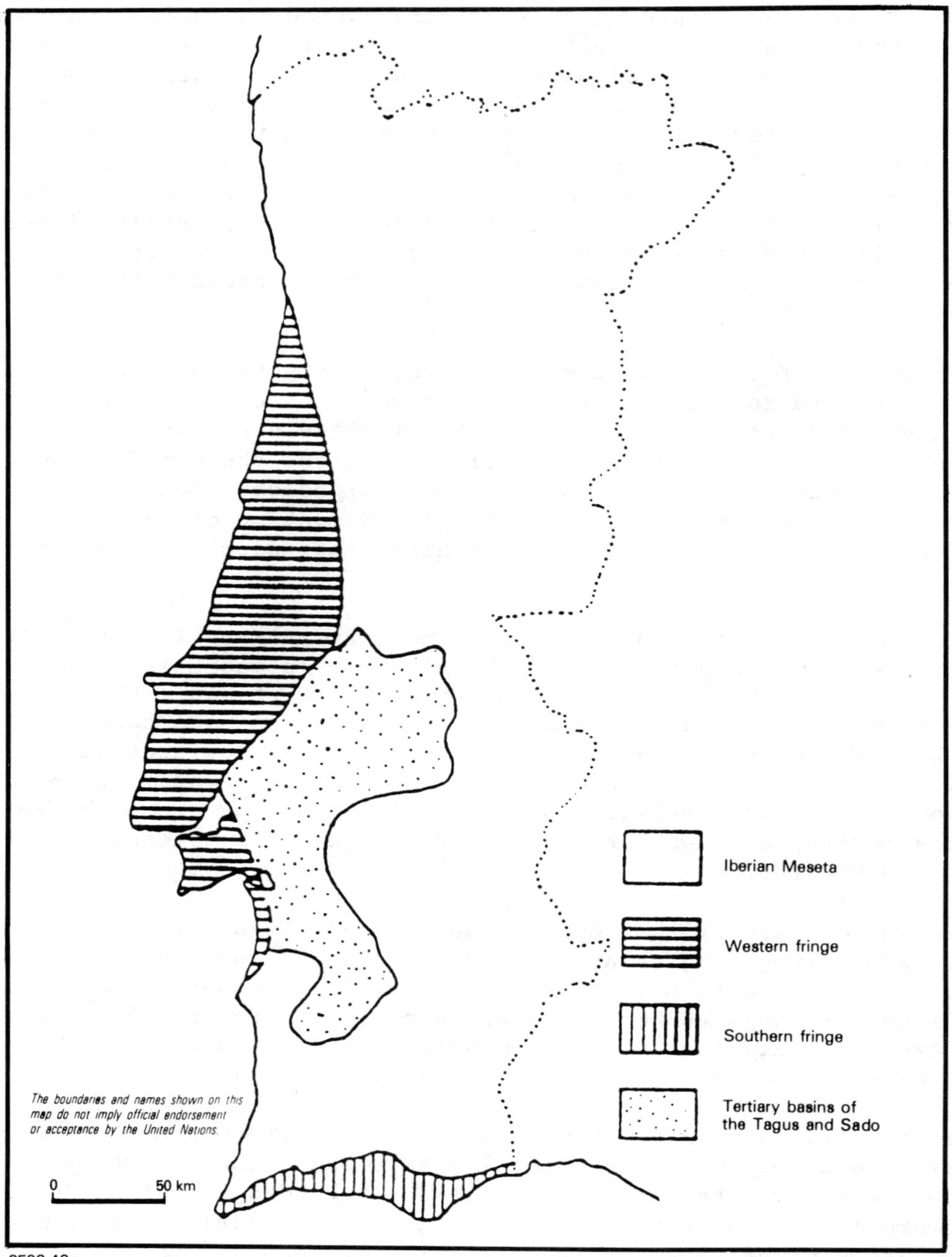

3593.42x

of basaltic magma, especially in the Lisbon region. The sub-volcanic massifs of Sintra and Sines are also formed in this fringe. The sedimentary formations are deformed and highly fractured. Some folds gave rise to the jagged relief that is a feature of the geomorphology in this region.

The karstified formations of the Middle Jurassic constitute the best aquifers in the western fringe. They contain important reservoirs of water for the Lisbon area. The Lower Cretaceous, consisting of detrital rocks, constitutes a good aquifer especially in the southern part of the fringe. Extensive areas where beds of marl and argillaceous sand occur are poor in terms of groundwater.

The alluvial formations of certain major rivers are of great hydrogeologic significance, but the sandy coverings, the fluvial terraces and the deposits of ancient beaches are of only local interest.

From the lithologic standpoint, the southern fringe is similar to the western fringe. Beds of limestone, dolomite, sandstone, argillite and marl are abundant. Near the western massif there are volcanic rocks including basalt, pyroclastic rocks, dykes and sills. There are also a number of diapiric structures.

The southern fringe was subject to considerable tectonic activity which gave rise to extensive folds and significant fractures. With the gradual increase in thickness of the sedimentary series there occurred various folds trending roughly east-west. These have played an important role in the number and operation of the aquifers in the region.

The formations of the Lower Jurassic were subject to intense karstification and contain the main reserve of groundwater in southern Portugal.

The sedimentary basins of the Tagus and the Sado are made up of formations which range in age from the Paleogene to Recent. They occur in the terminal part of the hydrographic basins of these two rivers. The Cenozoic formations consist mainly of limestone, sandstone, siltite, argillite and marl, whereas the Recent formations consist of sandy and sandy-argillaceous material of alluvial origin.

The formations of this region are practically horizontal since they have been subject to little tectonic activity. The occurrence of certain faults along the limits of the basin suggest that the structure is a graben. The detrital levels are of great hydrogeologic significance. They contain the best water reservoirs in Portugal with depths attaining 500 m.

The Madeira and Azores islands are made up almost exclusively of volcanic rocks. From the lithologic point of view, they are mainly basaltic or similar, although trachitic and rhyolitic rocks are often found. Collapsed rock and pyroclastic materials are the most common but there are abundant sills and dykes, especially in the more ancient massifs. Generally speaking, the most recent rocks are more porous and permeable, while the older rocks are more compact. Also, in general, the abraded basaltic formations, especially the most recent, are those with the greatest water-bearing potential. Certain pyroclastic formations, with abrasions, flakes and lapilli, when there are favourable intakes, may provide important reservoirs.

The relief of the islands is of great importance for groundwater supplies since it influences the precipitation and hence the recharge of the aquifers. For this reason, the islands of Madeira and Porto Santo, which are barely 40 km apart, have quite different hydrogeological conditions. The island of Madeira is mountainous with altitudes over 1,800 m, heavy rainfall and consequently abundant water. The island of Porto Santo has altitudes which do not exceed 517 m, the precipitation is light and, as a result, the island is relatively dry with a marked lack of water.

The same may be said of the Azores; the two extremes are represented by the islands of Sao Miguel and Graciosa.

C. Hydrogeology

1. Main features

The large geotectonic divisions of Portugal described above also correspond to the large hydrogeologic divisions. On the one hand, the western massif contains a vast series of igneous and metamorphic rocks; here there are hardly any porous acquifers. On the other hand, the western and eastern Middle Cenozoic fringes and the Tertiary basin of the Tagus and the Sado are made up almost exclusively of sedimentary rocks, which contain a number of porous and karstified aquifers.

Figure 53 shows a lithologic map prepared for hydrogeologic purposes. The lithologic types are grouped in the light of the hydrogeologic characteristics considered most relevant. This map is based on the geologic map of Portugal produced by the Portuguese geological services on a scale of 1:1,000,000 (SGP, 1968). In relation to the basic cartography, the map contains a number of simplifications and omissions due mainly to the small scale of the map and, in certain cases, the fact that the hydrogeologic significance of certain areas is undoubtedly much less than their geologic importance.

2. Western massif

The western massif is made up essentially of crystalline rocks, both igneous and metamorphic. In several areas it is covered by sedimentary formations of continental origin but usually small in area and shallow. These sedimentary formations nowhere give rise to aquifers of any significance.

Metamorphic rocks predominate south of the Coimbra parallel. The geologic maps of Portugal refer to the occurrence of formations with ages ranging from Precambrian to Carboniferous. The continuity of these formations is interrupted at various points by igneous formations which include the eruptive massifs of Fundao and Castelo Branco in Beira Baixa, the eruptive complexes of Beja and Evora, the granitic massifs of Nisa, Portalegre and Monforte in Alentejo and the eruptive massif of Monchique in the Algarve.

To the north igneous rocks predominate also. In several places they are interrupted by important outcrops of schist. These occur along the border with the western fringe, in the metasedimentary belts between Esposende and Castro de Aire, between Mesao Frio and Barca d'Alva and the great schistose area of Tras-os-Montes.

Figure 53. Portugal: Lithology as affecting groundwater

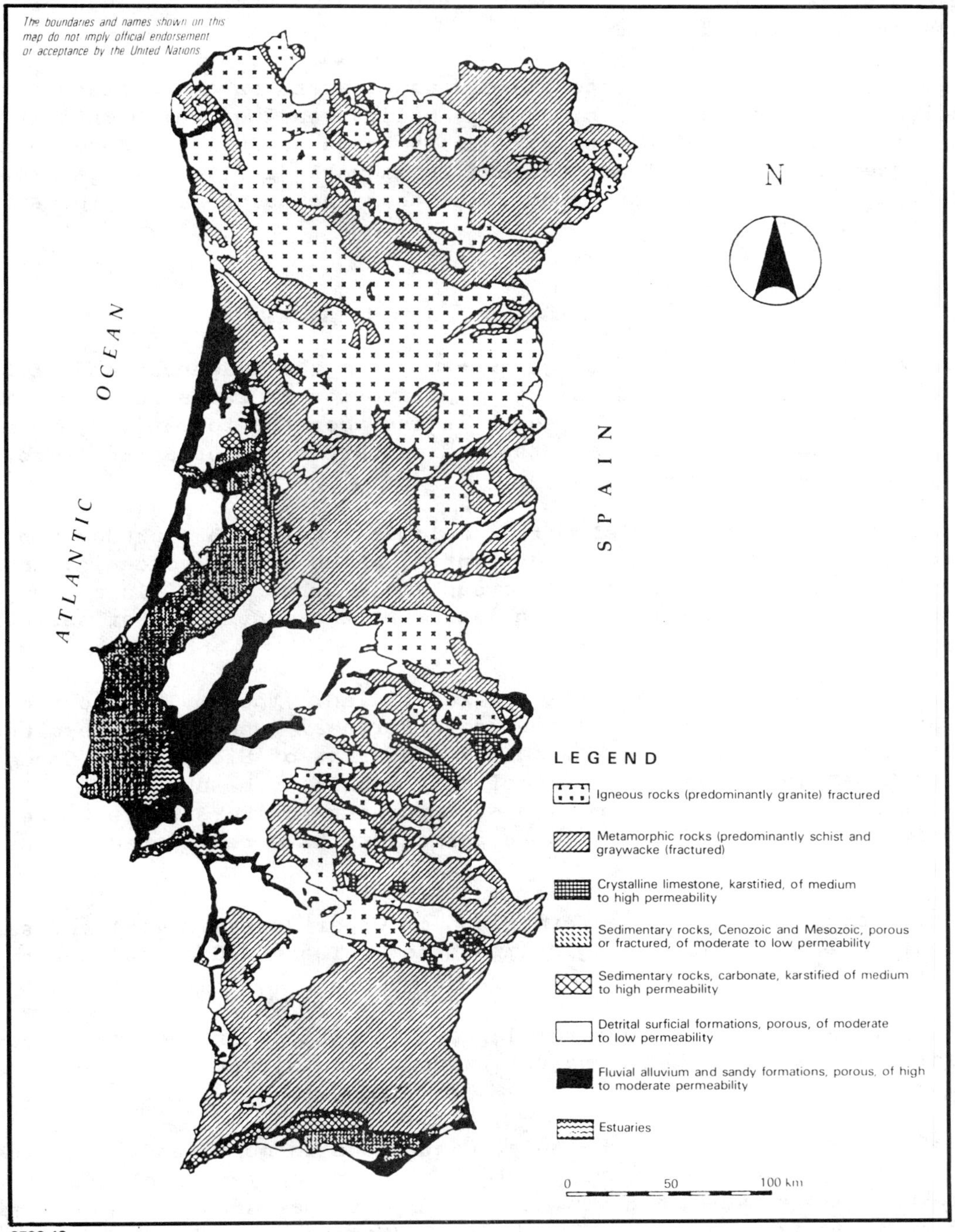

3593.43x

The detrital formations covering the crystalline basement are generally thin. Most are relatively undeveloped or even skeletal. In various areas of the country, erosion has reached severe levels. This is related to the indiscriminate use of agricultural and forestry methods which are not favourable to soil conservation.

The high precipitation observed in the north-west makes this region more interesting for the use of groundwater. Yields with capacities between 1 and 3 l/s are common. In exceptional cases, in areas of large fractures, dykes or sills, these values may exceed 10 l/s. In the regions more to the south, the productivity and the reserves are markedly lower.

The limestone massifs of Alentejo are unusual areas in the western massif. Karstification is responsible for high permeability readings which make it possible to install waterworks with high yields, some as large as 80 l/s. Some specific yields are higher than 70 l/s (Paradela and Zbyszewski, 1971). The gabbro-diorite complex of Beja is considered more productive than the neighbouring formations and has yields which reach 8 l/s.

3. Western Middle Cenozoic fringe

The western Middle Cenozoic fringe extends from the neighbourhood of Espinho to the north as far as Lisbon and then to the interior as far as Coimbra, Tomar, the Maior River and Alenuer. To the south it includes the formations of the Arrabida in the Setubal peninsula and the sedimentary basin of Santo Andre, between Sines, Santiago do Cacem and Melides.

This fringe has a great variety of lithologic types with a predominance of carbonate rocks. The eruptive massif of Sintra, which is sub-volcanic, the basaltic complex of Lisbon-Mafra and numerous small dykes and sills are manifestations of igneous activity which interrupt the continuity of sedimentary formations.

The main belts of carbonate rocks are karstified. They belong to the Lower and Middle Jurassic and occur in the centre of Portugal, where they form the so-called limestone massif of Estremenho. The sierras of Sico, Aire, Candeeiros, Montejunto and Ota are an integral part of this massif. The abundant karstification has allowed the formation of important aquifers where there are numerous springs with high yields. These formations have registered yields of more than 200 l/s.

North of Lisbon are extensive formations of Upper Jurassic marl and sandy clay, in general without hydrogeologic importance. The yield of these formations is low.

The marly-limestone and sandy-argillaceous formations of the Cretaceous between Cascais and Torres Vedras, do not produce significant aquifers. They are of merely local importance.

North of the Mondego River, the Lower Cretaceous is more sandy and with its greater permeability has water-bearing potential. In the region of Aveiro there are well fields which can extract some tens of litres per second. Transmissivity varies between 10^{-2} and 10^{-3} m^2/s. The Upper Cretaceous, which is clayey, serves as an aquitard.

Pliocene and Quaternary rocks are abundant north of Nazare. They are predominantly sandy-argillaceous and do not give rise to aquifers of regional significance. North of Figueira da Foz, the alluvial and sandy formations of the Quaternary are more continuous and their hydrogeologic importance greater. In certain areas, they give rise to yields above 15 l/s. The transmissivity may attain 6 X 10^{-3} m^2/s.

South of Lisbon Mesozoic limestone formations outcrop in the sierra of Arrabida, although over a reduced area and in geomorphologic conditions unfavourable for aquifers.

West of Santiago do Cacem is the small sedimentary basin of Santo Andra, which forms the southern edge of the western Middle Cenozoic fringe. These formations date from the Triassic to the Quaternary. Various tectonic events have given rise to a number of blocks that are suitable for the development of aquifers.

The most productive aquifers are the karstified limestone of the Middle and Upper Jurassic. The highest yields from this aquifer are about 125 l/s. However, karstification is only just beginning and the productivity is variable and uncertain. The transmissivity is also highly variable and ranges from 10^{-4} to 10^{-2} m^2/s.

The Miocene formations superimposed on the carbonate layer are essentially sandy and form an aquifer of some importance. Waterworks installed have produced 20 l/s. The transmissivity is about 1.5 X 10^{-3} m^2/s.

4. Southern Middle Cenozoic fringe

The southern Middle Cenozoic fringe is almost exclusively made up of sedimentary formations. They range from the Triassic to the Quaternary with a predominance of Mesozoic formations. On the edge of this fringe throughout the extension of contact with the western massif, there are volcanic lavas, pyroclastic rocks, and some dykes and sills. In close association with these volcano-sedimentary formations is a formation of Silves sandstone, predominantly sandy-argillaceous, of little hydrogeologic interest.

The carbonate formations of the Lias, many dolomitic, occur in various regions where they constitute two important aquifers. The main series occurs north of a line between Silves and Querenca and constitutes the most important aquifer of this fringe.

The other Jurassic formations contain diverse lithologic types and therefore do not form extensive aquifers. However, the Dogger formations are of some significance as aquifers in certain localities.

Cretaceous rocks are more marly and do not produce good aquifers. In most of the fringe, these less permeable formations are intercalated between the more productive Jurassic formations and the sea, thus serving as barriers to marine intrusion.

The Miocene is represented by limestone and sandy formations which usually constitute good aquifers but their proximity to the sea makes them highly vulnerable to marine intrusion.

Of the Quaternary formations, the most important are the dunes of Monte Gordo in the extreme east of the fringe, where there is an aquifer of good potential but highly vulnerable to sea-water intrusion.

In this region, considerable use is made of groundwater. It is (1988) the only source of domestic supply for the whole region.

The Lias aquifer (including certain levels of Dogger with which it is in continuous contact) is of the karstic type and can produce yields of 100 l/s. Its extent and the fact that it is crossed by important rivers make it subject to a significant annual recharge. Although rare, there have been transmissivity readings of about 2×10^{-2} m^2/s.

Although they do not form extensive aquifers, the Upper Jurassic and Cretaceous formations, where karstified, can provide yields of 50 l/s. But because they are in small units, their yields always decrease considerably during dry periods.

Miocene formations, which occur in the most densely populated areas of the coast, have been extensively explored. In most cases, they may give yields of 20-30 l/s per well. The aquifer in the Faro region, for example, has been overexploited and shows a continuing lowering in phreatic levels.

The transmissivity of the Miocene formations ranges from 7×10^{-2} to 6×10^{-4} m^2/s.

5. Sedimentary basin of the Tagus and Sado

The Tertiary formations of the Tagus and Sado basin are bounded on the north and west by Jurassic formations and on the east and south by the formations of the western massif.

The area occupied by the Tertiary sediments was subjected to large-scale orogenic movements which caused their collapse along lines of structural weakness in the basement rock. This collapse along a line trending north-east was accompanied by a tilt in the substratum towards the west. This zone of collapse which already contained Mesozoic sediments was covered by Oligocene continental deposits. On these were superimposed the material deposited during the Aquitanian-Burdigalian transgression. At the same time, the sea advanced deeply into the depression and covered about two thirds of the present basin. Subsequently, during the Helvetian, the regression began and was prolonged until the upper Miocene. While these marine formations were deposited in the zone covered by the sea, fluvio-continental and lacustrine sediments formed in the continental part of the zone. Thus, in this basin, sediments of the same age have distinct facies and consequently different hydrogeological characteristics.

Although the Jurassic limestone formations underlie the Tertiary sediments, at least in the western part of the basin, their depth makes them of little interest for groundwater purposes. In the Setubal peninsula, where the sedimentary series is thicker, these formations have been detected by drillers in search of oil at depths greater than 1,000 m. North of the Tagus, they were detected at about 500 m. At the north-west boundary of the basin, there is a limestone massif of the Jurassic age with important springs. Although some of these springs are drained by the Tagus, the limestone massif is enclosed in the Middle Cenozoic fringe.

From the hydraulic point of view, the Jurassic aquifer may be considered to be independent of the Miocene aquifer which lies above it; the thickness and low permeability of the Oligocene-Cretaceous formations which separate them make the circulation of water between them difficult.

The basin of the Tagus and the Sado also contains a complex of aquifers in the Miocene and Quaternary formations. These are interrelated from the hydraulic point of view and for practical purposes may be considered as a single aquifer system.

The lower level of this system is limited by the argillaceous-marly formations of the Aquitanian-Burdigalian, in the western part of the basin and by the Oligocene-Cretaceous formations in the other zones.

The aquifers which occur in the limestone-detrital formations of the Burdigalian and lower Helvetian and in the clay-detrital complex of the middle and upper Miocene are confined. The aquifers in the Pliocene and Quaternary formations (alluvium and fluvial terraces) are unconfined.

In the upper part of the limestone-detrital complex, there is a more marly zone which separates it from the clay-detrital aquifer and which acts as an aquitard. The piezometric difference between the two aquifers is not more than 2 m. Nevertheless, this aquitard plays an important role in the operation of the system; it prevents saline intrusion into the existing reservoirs in the neighbourhood of the estuaries.

The upper aquitard, which limits the clay-detrital aquifer of the middle and upper Miocene and establishes the hydraulic relationship between this aquifer and the overlying unconfined aquifer, belongs to the Pliocene although locally it may be part of the uppermost Miocene. Although it is lenticular, this formation serves hydraulically as an aquitard and there are piezometric differences about 10 metres between the two aquifers.

The hydraulic parameters of the confined aquifers are high: there have been specific capacities of 3-8 l/s per metre and some reservoirs have yields of 40 to 100 l/s. The transmissivity values are about 10^{-2} m^2/s and the storage coefficients vary between 3×10^{-4} and 7×10^{-4}.

The unconfined aquifers cover practically all the basin and are found in the Quaternary alluvial formations of the Tagus and Sado and their tributaries, in the fluvial terraces of the Tagus and in the Pliocene formations. Their base is made up of the lenticular aquitard referred to above and the aquifers are drained through the alluviums and the terraces into the surface-water system.

The thicknesses of these aquifers differ widely and in some cases are difficult to define; normally they do not exceed 120 m. The hydraulic parameters vary considerably. The alluvium of the Tagus range from 1.5×10^{-2} to 3.0×10^{-2} m^2/s for transmissivity. Storage coefficients range from 10^{-2} to 10^{-3}.

The hydrology of the aquifer system of the Tagus-Sado is similar to other basins of the same type. The intake area is in the higher parts where the deep aquifers are fed from the overlying layers. The lower part of the basin constitutes the discharge zone of the system and the deeper aquifers are drained through the unconfined layers into the surface streams.

D. Groundwater development

1. Potential

Figure 54 contains a map showing the aquifer productivity in the different geologic formations, according to Paradela (1975). The zones of higher productivity lie in the sedimentary formations of the western and southern Middle Cenozoic fringes and in the basin of the Tagus and Sado.

Table 46 and figure 55, extracted from MPAT (1987a and 1987b), make it possible to compare the average volumes extracted annually from the two main aquifers and the volumes of the annual recharge. The conclusions are indicated in each case in one of the columns of the table.

Groundwater in Portugal plays a basic role in meeting human, industrial and agricultural needs. According to the most recent agricultural census (INE, 1979), there were in Portugal in 1979 about 620,000 irrigation units of which 550,000 were in operation, at least in part, using spring, well or tubewell water representing about 90 per cent of the total.

In the case of drinking water, it is also confirmed, according to MPAT (1987a and 1987b), that of all the communes of Portugal only nine are supplied from surface water. For the others, the water comes either exclusively from underground reservoirs or from a combination of ground and surface water (figure 56).

2. Water quality

In terms of the quality of groundwater, the highest amounts of chlorates, sulphates and total dissolved solids are found in the south. In the Algarve region, considerable amounts of chlorates have been observed. In this region, agricultural needs are great and the volumes used are equivalent to more than 80 per cent of the total supply. At the same time, the balance between resources and consumption in this region is not favourable. For these reasons, there has been considerable interest in studying it.

Such studies, according to MPAT (1987a), reveal the gradual salting of water intakes, especially in the regions of Faro, Tavira, Alvor and Portimao. Studies made so far indicate that geologic activities (dissolution of salts forming the basis of sedimentary formations in the Algarve), marine intrusion and intensive use of groundwater for agriculture are the causes.

Figure 57 shows the distribution and characteristics of certain mineral water springs. It may be observed that in the western massif sulphurous and bicarbonate waters predominate. In the middle Cenozoic fringes the water is more highly mineralized (chlorates). The less mineralized water (used for table and thermal purposes) is to be found in the western massif, which is mainly composed of granite and schist in the northern regions. The mineralization is greater south of Portalegre, attaining the highest readings south of Beja.

In continental Portugal there are 194 mineral water springs of which 50 are being exploited, 41 have ceased to operate and 11 have been abandoned or are without concessions. Of the total number of springs, three quarters are situated north of the Tagus.

Figure 54. Portugal: Productivity of principal aquifers

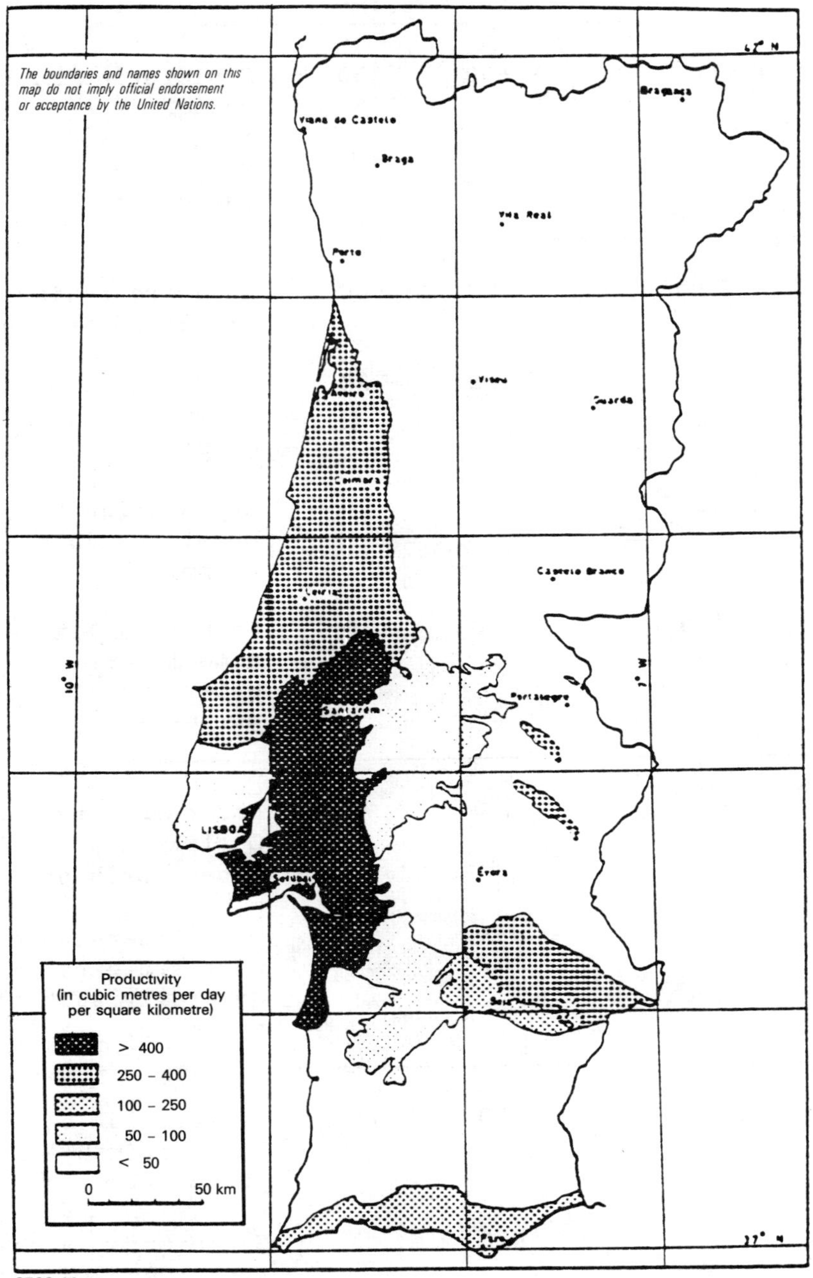

3593.44x

Table 46. Portugal: Characteristics of the principal aquifers

Aquifer subsystem	Annual average: Recharge (million cubic metres per year)	Annual average: Withdrawal (million cubic metres per year)	Suitability
Middle Pliocene of the Tagus and Sado	1 600	150	Exploration much below resources, though it contains an intensive zone of exploration in Setubal Peninsula. Water suitable for various purposes.
Cretaceous - Baixo Vouga	100	60	In Avero region, withdrawals are within advisable limits.
Cretaceous - Mondego	100	25	Exploration much below resources. Water suitable for various purposes.
Cretaceous - Vila Nova de Ourem	80	20	Exploration below resources. Water suitable for various purposes.
Tertiary of the Lourical	70	10	Generally little exploration with some intensive exploration zones. Water quality suitable for various purposes.
Pliocene-Quaternary of Baixo Vouga	250	70	Under-explored aquifer. Qualitative problems in coastal zones north of Vouga.
Pliocene-Quaternary of Figueira da Foz	170	20	More intensive exploration in northern part of subsystem. Water generally suitable for various purposes.
Jurassic of Baixo Vouga	50	20	Exploration below resources. Water suitable for different purposes.
Jurassic of Orla Estremenha	470	40	Exploration much below available resources. Water suitable for different uses.

Table 46 (continued)

Aquifer subsystem	Annual average: Recharge (million cubic metres per year)	Annual average: Withdrawal (million cubic metres per year)	Suitability
Northern Estremadura, Alcobaca and Torres Vedras	70	..	Satisfactory, small local withdrawals, water suitable for different uses.
Coastal - Algarve Central	25	10	Withdrawals below resources. Generally, water is of acceptable quality except for that withdrawn in zones of Vilamoura and Lagoa.
Coastal - Algarve Eastern	50	50	Intensive exploration. Locally, withdrawals above resources. Water quality: (a) Campina de Faro: upper aquifer, salts and nitrates. Lower aquifer, medium quality; (b) Elsewhere: medium to poor.
Dunes of Monte Gordo	4	1	Exploration below resources. Good quality water.
Jurassic - Algarve Eastern	35	10	Withdrawals below resources. Generally water is of good quality. Locally there are health problems because of deficient installation of waterworks (near coast or in diapiric zones).
Jurassic - Northern Algarve	80	40	
Upper Jurassic - Algarve	90	35	Withdrawals below resources, except for area between Tavira and Praia Verde. Water of good quality except for local diapiric outcrops (chlorates).

Source: MIAT (1987a).

Figure 55. Portugal: Comparison between extraction and reserves of groundwater

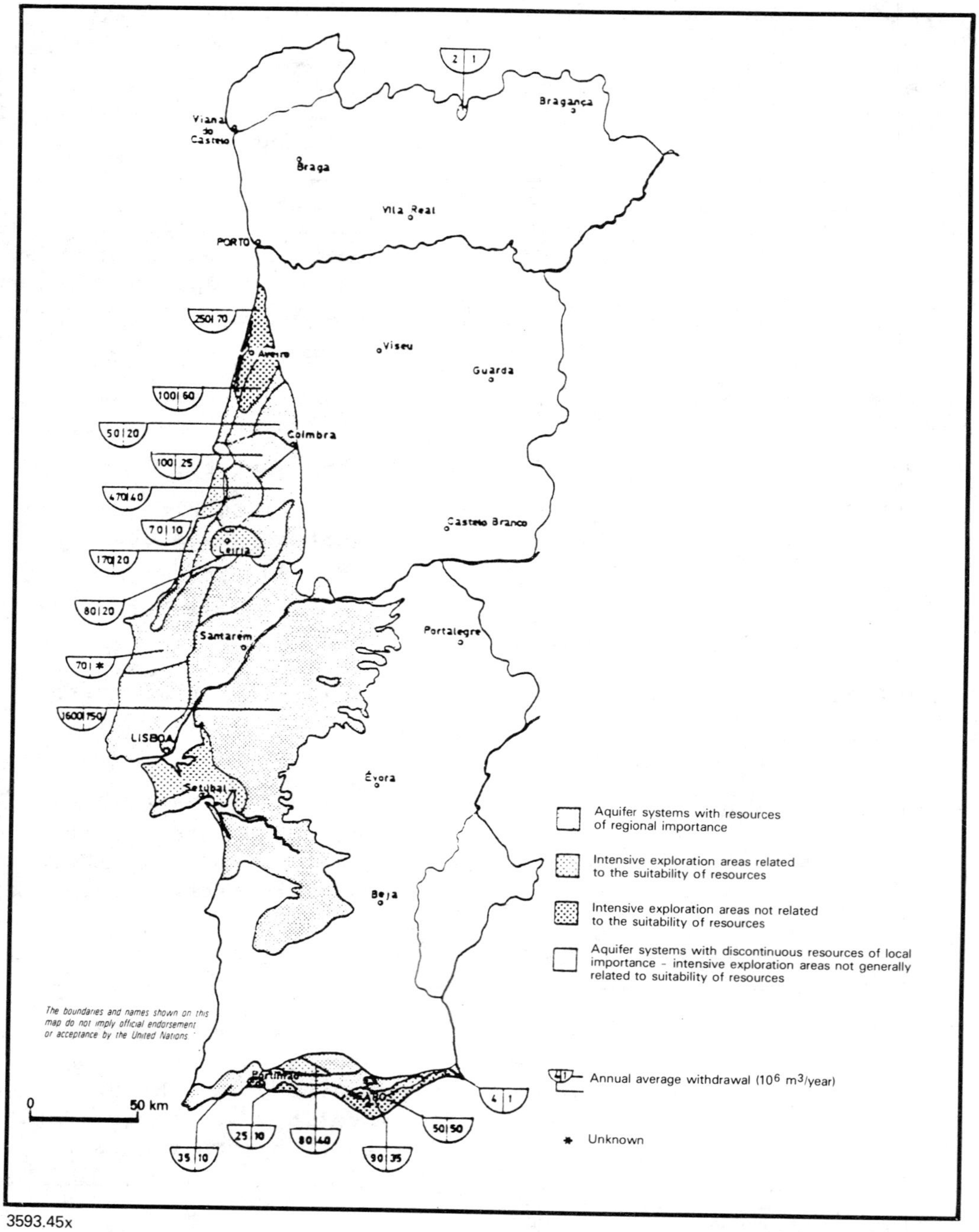

3593.45x

Figure 56. Portugal: Sources of water for public supply

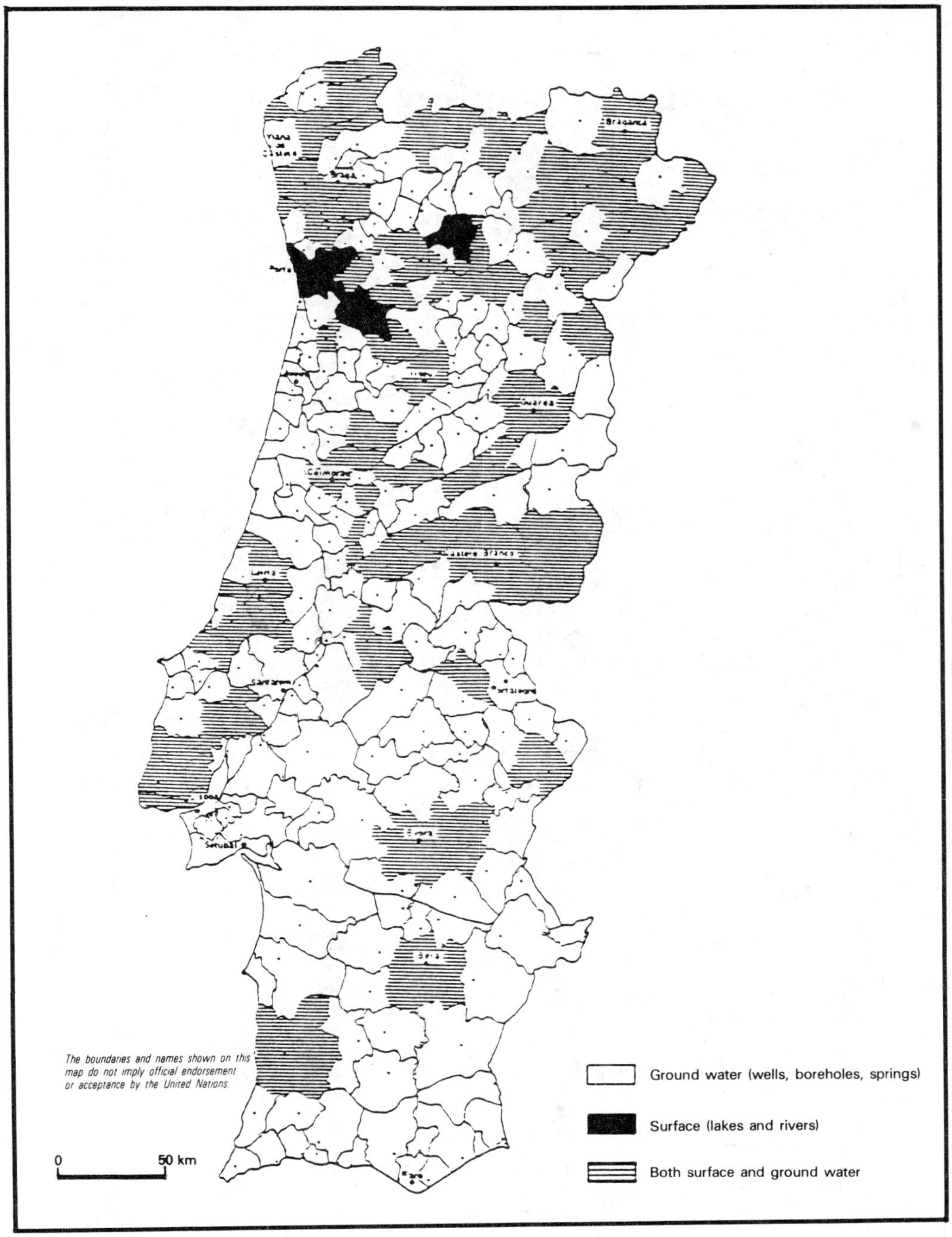

3593.46x

Figure 57. Portugal: Mineral water

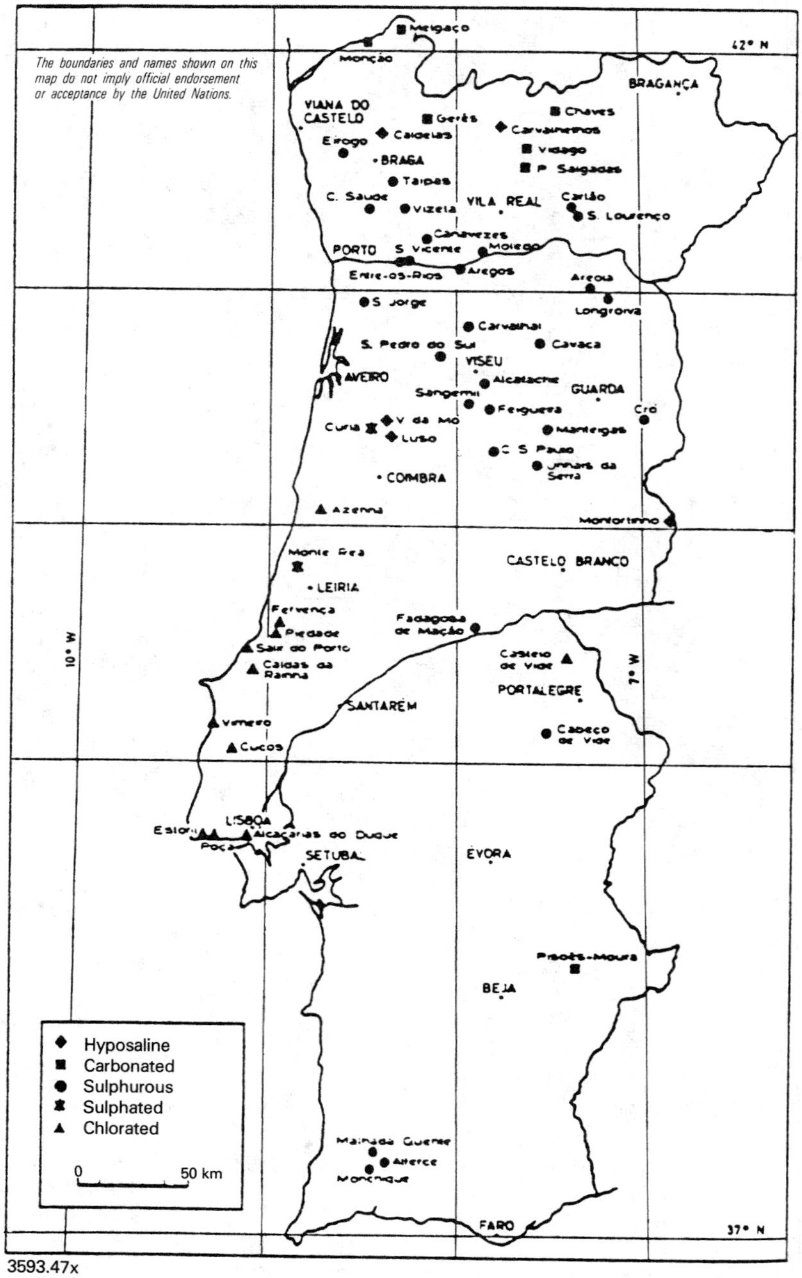

3593.47x

E. Legislation

The main law dealing with the management and utilization of water in Portugal dates from 1919 (CNA, 1981). It says little about the control of water quality. It merely states that industrial or agricultural effluents which have to be returned must not make the water insalubrious and useless or prejudicial for those who also are entitled to use it and it cannot be defiled by substances that are harmful to agriculture, health or fish. This principle concerning water pollution is also embodied in later legislation but no provisions were made for the imposition of penalties or for the control of water quality.

It was at the end of the 1960s that concern began to be expressed about these aspects of groundwater. This concern resulted in the adoption of legislation which so far has formed the main basis for the control of water quality in Portugal.

In addition to these laws, there are a number of sectoral laws which include provisions designed to diminish the adverse effects of activities which affect the environment and the quality of the water.

The whole legal system governing water administration, which so far has been subject to the Water Law of 1919, is to be overhauled and new laws known as the Water Law and the Water Quality Law will be adopted.

The groundwater in Portugal in private hands belongs legally to the owners; the Water Law of 1919 makes no provision for public action at this level.

An inventory of groundwater resources carried out in the 1960s showed that there has been a marked increase in the use of these resources. The importance of groundwater in the Portuguese water system makes it essential to adopt additional legislation to control its exploitation.

The draft bills for the Water Law and the Water Quality Law, referred to above, will deal with public ownership of water, the role of the State in the management of water and the establishment of organizations with water responsibilities at the central, regional and local levels. The law will also make clear the link between the water authorities and the users, both in industrial and institutional. These draft laws also make clear the principle that the user pays.

<u>References</u>

CNA. Legislaçao sobre Lisboa, Lisboa, Comissao Nacional do Ambiente (CNA), 1981.

Cunha, L. V. <u>and others</u>. A Gestäo da Agua. Principios Fundamentais e sua Aplicaçao em Portugal. Lisboa, Fundaçao Caloust Gulbenkian, 1980.

INE. Recenseamento Agricola do Continente. Lisboa, Instituto Nacional de Estatistica (INE), 1979.

MPAT. Relatório do Estado do Ambiente e Ordenamento do Territorio.

Lisboa, Ministerio do Planeamento e da Administraçao do Territorio (MPAT), 1987(a).

__________. Compêndio (Experimental) de Estatisticas do Ambiente. Lisboa, Ministerio do Planeamento e da Administraçao do Territorio, 1987(b).

Paradela, P. L. Recursos Aquiferos Subterraneos. Produtividade. Lisboa, Comissao Nacional do Ambiente, 1975.

Paradela, P. L. <u>and</u> Zbyszewski. Hidrogeologia Geral do Central e Sul de Portugal. Livro-Guia da Excursao No. 9, in "Actas do I Congresso Hispano-Luso-Americano de Geologia Economica", Lisboa e Madrid, 1975.

Quintela, A. C. Recursos de Aguas Superficiais em Portugal Continental. Lisboa, 1967.

SGP. Carta Geologica de Portugal (a escala 1:1,000,000). Lisboa, Servicos Geologicos de Portugal (SGP), 1968.

SMN. Atlas Climatologico de Portugal Continental. Lisboa, Servico Meteorologico Nacional (SMN), 1974.

XIII. SPAIN

Area: 504,750 km^2 Population: 39.2 million

A. General geography

1. Physiography

The Kingdom of Spain is situated in the south-western corner of Europe and is separated from Africa only by the Strait of Gibraltar. It occupies most of the Iberian Peninsula and includes four non-coterminous entities: the Balearic Islands, the Canary Islands, Ceuta, Melilla and some smaller islands.

Although Spain has no high peaks - the maximum altitude in the peninsula is slightly over 3,000 m, and the highest point in the entire territory of Spain is a little over 3,700 m, on the island of Tenerife - it has the largest number of steep gradients in Europe. The central and western parts constitute an elevated shield, known as the Meseta Ibérica; adjacent to this and crossing it are several mountain ranges that divide the country into sectors, and in earlier times communication between these sectors was difficult. Figure 58 shows the principal topographic features and the hydrographic network in simplified form.

2. Climate

Climatic features are sharply contrasted due to the varied geomorphologic features, the influence of the Atlantic Ocean and the Mediterranean, the presence of the high Pyrenees to the north and the African continent to the south. In humid and temperate areas of the Iberian Peninsula, annual average rainfall may exceed 1,500 mm while in the most arid areas of the south-east and some islands, especially Lanzarote, it may barely reach 100 mm.

In the Canary Islands it is possible within a distance of a few kilometres, to travel from forest or humid pastureland to desert. Precipitation in the Mediterranean areas is highly irregular, and severe rainstorms are frequent. Sudden catastrophic rises in the water level of the rivers and extensive floods are not unusual. In some cases more than 300 mm of rain have been recorded within a few hours, and it has even happened that almost 1,000 mm of rain fell in one day.

Much of the rain in the peninsula comes from moisture originating from the Atlantic and entering from the north and north-west, and less frequently from the south-west also from the Atlantic. The amount of moisture held by the clouds decreases rapidly as they travel eastward, so that the Mediterranean regions would be a desert if it were not for rainfall of local origin, generally produced by the warm and moist Mediterranean air masses interacting with cold fronts. These rains do not penetrate far into the continent, but they keep the eastern and north-eastern coast from being semi-desert.

There are, in general, two dry periods per year - sharply marked summer and winter seasons, and two humid periods - a stormy autumn and a spring. Differences in rainfall are sharper in the Mediterranean area. The sequence of dry years and rainy years is not always the same in the north-east as in the rest of the peninsula.

Figure 58. Spain: Physiography

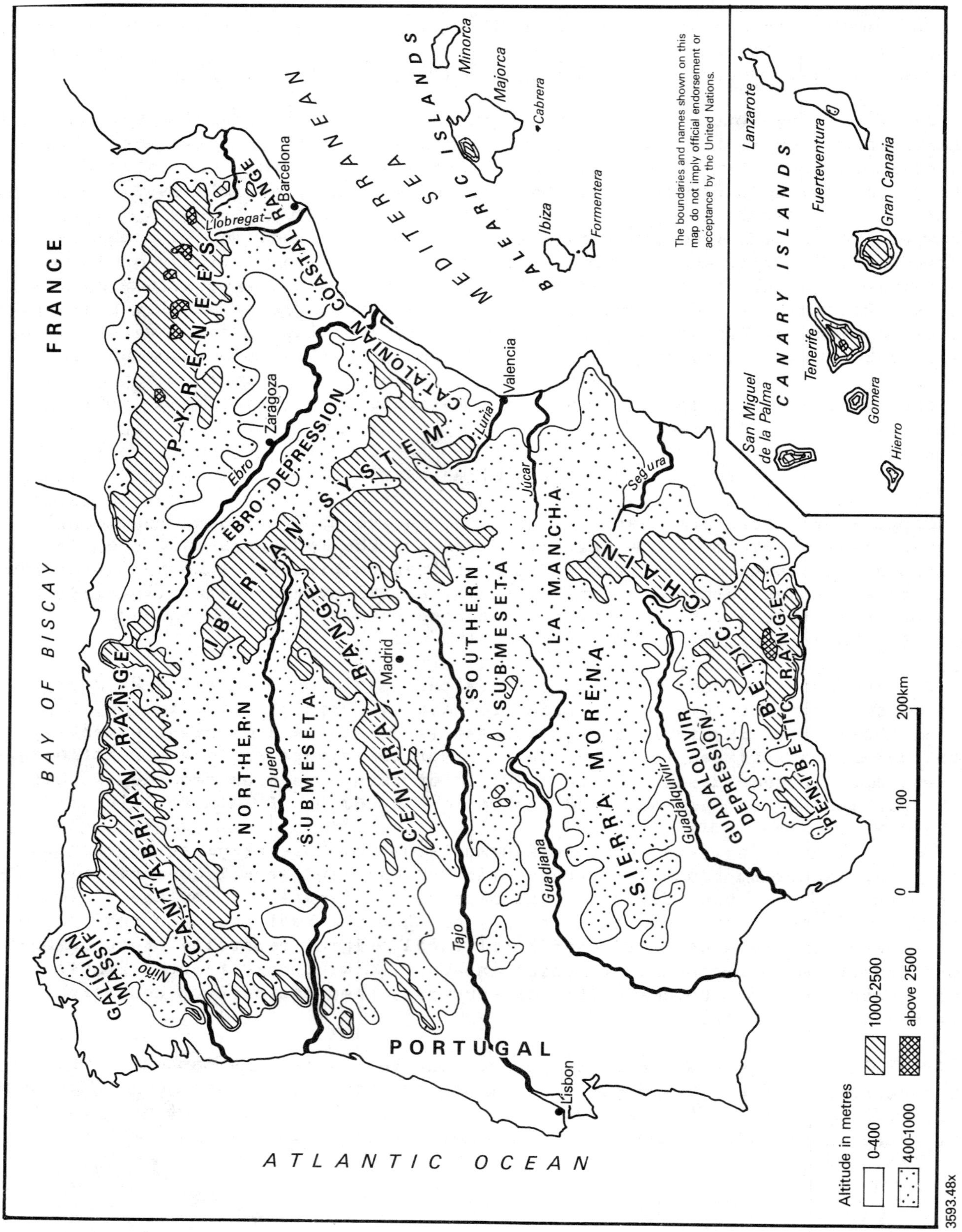

In the mountains, especially in the northern and central areas, much of the precipitation takes the form of snow, so that the rivers of the northern half of the country have a snow component at the end of the spring season.

The annual average isotherms range from 10° C in the mountainous parts of the northern and central areas to 18° C in the south, the Mediterranean coast and the Balearic Islands, and as much as 20° C in parts of the Canary Islands.

In general terms, the temperature is lowest in January and February and highest in July and August, with the greatest differences found in the interior of the peninsula subjected to a continental type climate, ranging up to 20° C as a monthly average, and only 14° C in the Levant area. In the interior areas of the peninsula there may be summer temperatures as high as 45° C and winter temperatures as low as -20° C. In the Mediterranean and south-western areas it seldom freezes. In the Canary Islands the climate is subtropical, with no winter season in the coastal areas, although snow falls on the peaks; the presence of the trade winds greatly moderates the climate with regard to maximum temperatures, and only the areas sheltered from these winds may have high temperatures (50° C) on those few days when there is a very hot and dusty wind blowing from the Sahara.

The natural vegetation ranges from Atlantic forests (oak [roble], chestnut and beech; pine and fir in the Pyrenees) in the north and the north-west to Mediterranean forests (holm oak [encina], and secondarily pine, mostly degraded to matorral) in most of the rest of the peninsula and the Balearic Islands, with pines and high-mountain vegetation (matorral, mosses, lichens) in the mountain ranges, and steppe areas (esparto grass, low bushes) in part of the Levant and other isolated areas in Aragon and La Mancha. In the Canary Islands the vegetation is of the steppe type, with cactus and euphorbia, evolving towards the peaks into Canary Island pines and remnants of laurisilva, and finally high-mountain vegetation in the highest portions.

3. River systems

Sixty-three per cent of the Spanish territory drains towards the Atlantic Ocean, either directly to the Spanish shores, or through Portugal, and 57 per cent of the population lives here.

The 37 per cent remaining and which flow to the Mediterranean Sea include the Ebro basin with 18 per cent of the territory and 9 per cent of the population, and small basins totalling 19 per cent of the territory with 34 per cent of the population.

Table 47 indicates the principal characteristics of the river basins, based on a 1980 report of the National Hydrological Plan, published by the Ministry of Public Works and Urban Development [Ministerio de Obras Publicas y Urbanismo (MOPU)].

Table 47. Spain: Principal river basins

Basin	Area (km^2)	Average precipitation		Average total accumulation (10^6 m^3/ year)	Total average runoff %	Large enterprises		Uniform total accumulation (10^6 m^3/ year)	1980 population (10^6)	Per capita resources (m^3/year)	
		(mm/ year)	(10^6 m^3/ year)			Volume (10^6 m^3)	No.			Natural	Regulated
Northern (*) (.)	53 804	1 353	72 797	44 876	62	3 227	48	7 448	6.7	6 698	1 112
Duero (.)	78 972	587	46 357	12 170	26	6 503	31	8 355	2.2	5 532	3 798
Tajo (.)	54 769	650	35 600	10 185	29	9 873	49	7 372	5.3	1 922	1 391
Guadiana (.)	59 873	534	31 972	5 110	16	4 037	17	1 897	1.6	3 194	1 186
Guadalquivir	63 085	615	38 797	9 234	24	4 834	37	2 152	4.4	2 099	489
Southern (*)	18 391	557	10 244	2 830	28	412	5	689	1.9	1 489	363
Segura	18 254	380	6 936	960	14	907	11	1 243	1.3	738	956
Júcar	42 904	520	22 310	4 366	20	2 656	23	3 104	4.0	1 091	776
Ebro	86 098	591	50 884	18 341	36	6 446	67	11 271	2.7	6 793	4 174
Eastern Pyrenees (*)	16 493	747	12 320	3 250	26	661	8	1 654	5.5	591	301
Balearic Islands (**)	4 834	590	2 852	690	24	1	2	290	0.6	1 150	483
Canary Islands (**)	7 273	350	2 545	965	38	10	15	496	1.4	689	354
Total or average value	504 750	661	333 614	112 977	34	39 667	313	45 971	37.6	3 005	1 223

(*) Combination of small basins.

(**) Islands.

(.) Up to the Portuguese frontier or excluding the coast in Portugal.

The total values include the underground part which leaves the basin and enters Portugal or which flows to the sea.

It can be seen from the foregoing table that a large amount of water has been impounded in reservoirs, many of them behind hydroelectric dams, although there are also reservoirs established for irrigation and community water. If it were not for this reservoir damming capacity and its proper operation, many of the major rivers would have very low base flow rates in the summer, and in some cases in the winter as well. The Ebro River, with an average flow rate of 600 cubic metres per second, and some peaks reaching as much as 4,000 cubic metres per second, had in their natural state base values as low as 80 cubic metres per second, which are currently maintained at over 250 cubic metres per second thanks to regulation by means of reservoirs. The only navigable river is the Guadalquivir up to Seville.

4. Geology

Spain includes every age of geologic formation since the Cambrian, affected by numerous deforming processes, as a result of Spain's position at the junction of the African and European plates. The successive periods of orogeny have left clear traces, and in many places there is superposition of the Hercynian with the Alpine.

Figure 59 shows the distribution of rocks, grouped into siliceous rocks (granite, gneiss, schist and quartzite, mainly Paleozoic), calcareous rocks (limestone, dolomite and calcareous marl) and argillaceous rocks (marl, lutites), mainly in the Triassic and in the large Tertiary sedimentary basins, which include considerable amounts of gypsum and anhydrite, and some salt).

The Canary Islands are entirely constituted of Miocene, Pliocene and Quaternary volcanic formations which also occur locally in the Peninsula (Olot, Campo de Calatrava, Cabo de Gata) and in the Mediterranean (the Columbrete Islands). Volcanism is still active in the Canary Islands, and in Olot (Gerona) there is a volcano dating from historic times.

5. Population and land use

In the past, Spain has been primarily an agricultural country, with a great tradition of irrigation. Cultivated land occupies a significant portion of the entire territory with 2.8 million hectares (28,000 km^3) being irrigated.

Favourable Mediterranean climatic conditions render it possible to harvest several crops a year, so that intensive vegetable gardening is widespread, and so are fruit trees (chiefly orange) and some semi-tropical crops. Bananas, tomatoes and a large variety of other fruits grow in the Canary Islands. There are large areas of concentrated irrigation and winter irrigation, which have done much to raise local standards of living. Figure 60 shows the principal irrigated areas.

The population is concentrated along the coast and on the islands, while the interior is sparsely populated, with the exception of Madrid and some cities on major rivers. The great concentration of population in the Mediterranean and island areas does not match the areas of greater availability of water in the Atlantic basins, and this creates problems of water management.

Figure 59. Spain: Principal rock types

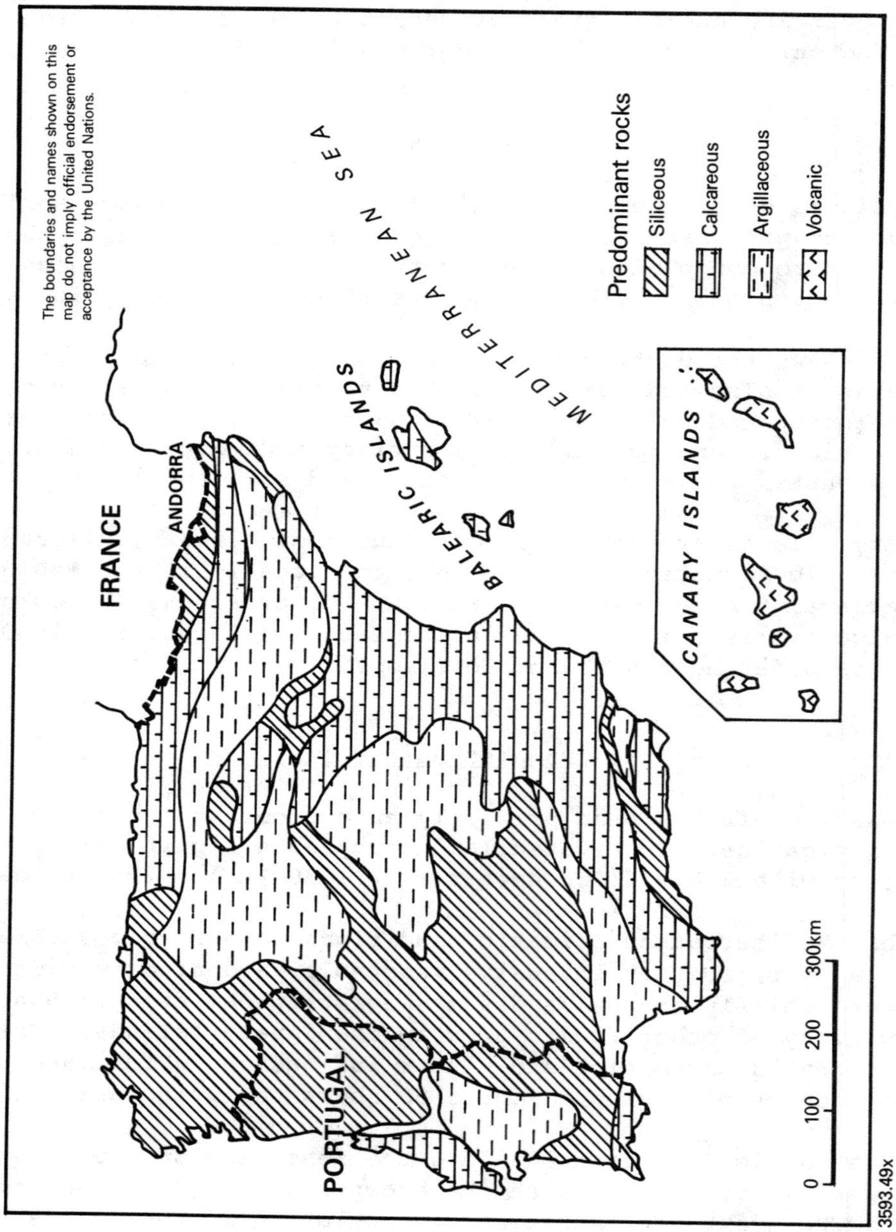

Figure 60. Spain: Irrigation zones

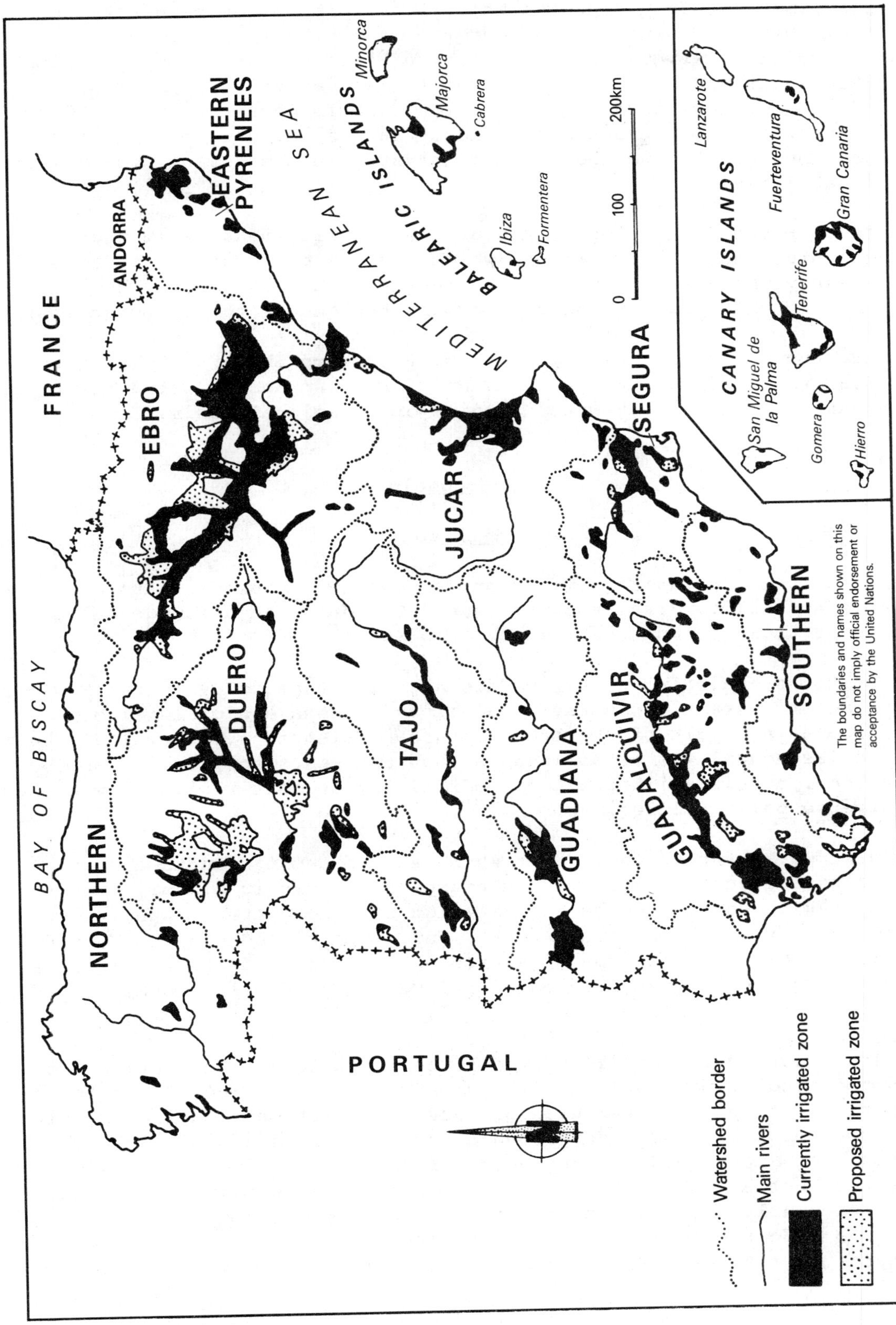

Industry is widely distributed with high concentrations in Catalonia (north-east), the Basque Provinces (north) and Asturias (north), and more recently in Madrid (centre), Zaragoza (north-east), Valencia (Levant), the northern region and somewhat in the Seville-Jaén area (south), with all types of manufacturing present. From the standpoint of high water demand, the following are to be mentioned: textile manufacturing in Catalonia, iron and steel and other metals in the Basque Provinces and Catalonia, automobile manufacturing in Madrid, Barcelona, Zaragoza and the north-western region, agriculture in Navarre, Aragon, Catalonia and Murcia, chemical and petrochemical industries in the Basque provinces, Catalonia and the Levant, and the leather industry in the Levant, the Balearic Islands and Catalonia.

Also to be noted is the importance of the tourist industry in the Mediterranean area and in the islands, which creates high points of seasonal demand, precisely where and when the resources are the scarcest. There are towns which sometimes double their population during the two or three summer months.

From the above considerations, it clearly appears that Spain has great geographic and seasonal variations in demand for water and in water resources which make it necessary to carry out water studies on a regional basis.

B. Hydrogeology

1. Background

During the years of Roman domination, almost two thousand years ago, a number of remarkable hydraulic works, such as aqueducts, dams and drainage galleries, were built. Some of these structures are still in service.

Later, at the beginning of the Middle Ages and especially in the Mediterranean areas, irrigation systems and water regulations reached a high level of sophistication as a result of the Arab culture. Valencia's Tribunal of the Waters [Tribunal de las Aguas] was established a thousand years ago, during the years of Arab domination, and it still functions officially in that region, having survived the most diverse political regimes.

Hydraulic works for surface water such as dams and aqueducts and legal institutions have unquestionably made a conspicuous mark in history, but there were also many hydraulic works for the exploitation of groundwater, especially infiltration galleries of the qanat type. For example, the city of Madrid was supplied exclusively by this type of source until 1958, and eventually an extensive network of such galleries was developed in Barcelona. Similar works may be found in many other places in Spain.

The water sources for Madrid proved to be insufficient to meet the demand created by the growing population, which amounted to about 200,000 at the end of the eighteenth century. The failures of successive attempts during the nineteenth century to obtain water from underground sources by means of drilled wells in the Tertiary basin of Madrid led the engineers of that time to the conclusion that groundwater was not a usable resource in Madrid. This, combined with other factors, particularly the rigid centralism of the administration, had the result that those who made the decisions in State water policy applied a similar conclusion to the rest of Spain, although the reality was quite different.

For this reason the Government's interest in the development and study of groundwater was limited until recent times. In fact Spain seems to make less use of groundwater for the public water supply than any other Western European country except Norway. Nevertheless, private and local public initiatives gradually increased the availability of groundwater, in many cases with great success, in many peripheral areas, especially the Mediterranean region and the islands.

During the past two decades the picture has changed considerably, owing to the introduction of modern methods of water-well drilling and the widespread use of turbine pumps, to the conduct of important regional hydrogeologic studies with modern criteria since 1965 and to the training of about one thousand professionals in hydrogeology in half-year courses which have been held every year since 1967 in Madrid and Barcelona.

Water legislation has given preferential attention to surface water, and it was only recently that a new law which regards groundwater as an essential part of total water resources was adopted.

The first hydrologic studies, dating from the eighteenth century, were those of Father Feijoo and, especially, Teodoro Ardemans (1724). The second landmark came near the end of the nineteenth century, in connection with the National Geological Map Commission [Comisión del Mapa Geológico Nacional] and the extraordinary work done by Moragas (1896), which made use of the latest knowledge available at that time.

After a long interval with little activity, modern studies began in 1964, with the Besós-Llobregat Study (Ministry of Public Works, 1966) and the Guadalquivir Study (UNDP-FAO), which were followed by the Canary Islands Study (UNDP-UNESCO), the Eastern Pyrenees Study (REPO, 1971), several Compilation and Synthesis studies carried out by the Geological Service of the Ministry of Public Works, the Hydrogeological Studies of the Geological and Mining Institute of Spain [Instituto Geológico y Minero de España (IGME)] within the framework of the Ground Water Research Plan [Plan de Investigación de Aguas Subterráneas (PIAS)], and studies conducted by the National Hydrological Plan.

2. Principal aquifers

Practically all types of hydrogeological conditions occur in Spain when almost every kind of large and small sedimentary basin, porous and fissured rock, consolidated and unconsolidated rocks are found.

At the beginning of the 1970s the Geological and Mining Institute of Spain, with a view to preparation of the Ground Water Research Plan, published the Hydrogeological Map of Spain on a scale of 1:1,000,000, which first identified the major aquifer systems in a general although preliminary manner. At the same time, it prepared a synthesis map on a scale of 1:500,000, which presented an estimate of the average annual renewable resources of each aquifer system. Subsequent studies have led to the publication starting in 1972 of additional hydrogeologic maps of various kinds and on various scales by the Geological and Mining Institute of Spain working mainly on the scales of 1:50,000 and 1:200,000.

On figure 61 the major Spanish aquifer formations of Spain are presented. Within the porous formations one may distinguish two groups, depending on whether

Figure 61. Spain: Principal aquifer systems

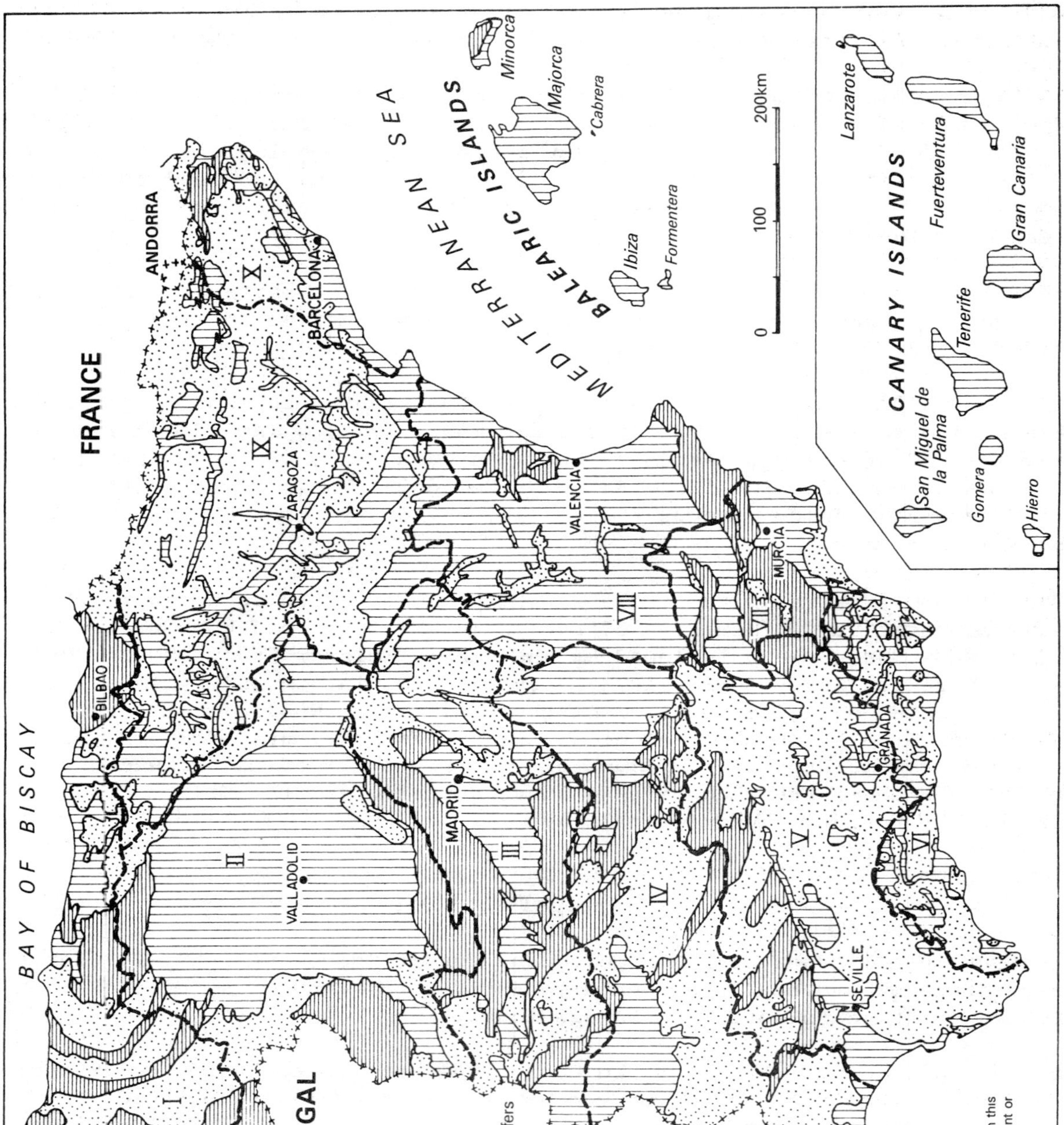

they are more permeable or less permeable. The first group includes essentially the alluvium or Quaternary terraces of the main watercourses, as well as the delta plains. In general these aquifers have been exploited earliest and most intensively, for readily understandable reasons. Because of their close connection with major watercourses, they constitute general zones of discharge of groundwater at the local or regional level, in addition to which they may induce recharges from rivers; both of these factors tend to guarantee that they will remain exploitable and that there will not be any significant decreases of water level in the wells. Because of the flat relief, they are zones well suited to irrigation. Indeed, most of the irrigated areas in Spain using surface water are situated in regions close to major rivers (see figure 60). As a result of this and of a State water policy that has economically favoured irrigation with surface water, there has been little use of groundwater in alluvial aquifers. An exception is the alluvial waters of the many torrential watercourses in the Mediterranean area, where the collection of groundwater is the only way to ensure availability. There seems to be a growing awareness that good management will require a programmed and combined use of the aquifer-and-river systems in much of Spain. The small delta of the Llobregat River in the Barcelona area (about 80 km^2) is the place where such combined use has been most developed; extractions in recent years have reached 150 hm^3 per year, and at present they still remain in the general vicinity of 100 hm^3 per year, preferably for industrial uses and urban water supply.

Almost all the coastal plains of the Mediterranean are underlain by semi-consolidated Tertiary or Quaternary materials, with variable but generally acceptable or high permeability. This factor, combined with the high demand for water for tourist and agricultural use - and in some cases also for industrial use, as in Tarragona - has led to the intensive exploitation of groundwater, which has caused problems, especially with marine intrusion and excessive drops in water level.

Approximately one fifth of Spain's area is occupied by tectonic troughs filled with Tertiary materials of continental or mixed continental and marine origin. Continental Tertiary materials, with thicknesses which may reach several kilometres, contain in their detrital margins some sandy lenses included in a more or less discontinuous mass of limestone and sandstone; in their central formations the Tertiary deposits are formed chiefly by pelitic and evaporite rocks, some crowned by a table of lacustrine limestone. The four main continental Tertiary basins are the Duero basin, the Madrid basin, the La Mancha basin and the Ebro basin. Other smaller basins are of considerable interest, such as that of the Vallés-Penedés in Catalonia. The exploitation of groundwater in these aquifers became relatively intensive only about two decades ago. Those which are most intensively exploited for irrigation purposes today are the Duero and La Mancha basins. The aquifers chiefly exploited in the latter are the Tertiary limestones. Exploitation in the Madrid basin, which is also relatively intensive today, is carried on mainly for urban water supply and industrial use. In the detrital Tertiary formations of the Ebro basin, groundwater development is still low but increasing; this is due partly to the low permeability of those Tertiary deposits, as well as to the extensive programme of irrigation using surface water which has taken place since the turn of the century. Considerable advances in the knowledge of these large sedimentary basins have been made during the past decade, and many research studies are still being conducted on it. In some areas, such as the Guadalquivir trough, the Tertiary is of marine origin, and lutites of low permeability are predominant. However, in other areas, or at certain levels, marine Tertiary sediments hold more or less consolidated reef or calcarenite formations which may constitute productive aquifers. These include, for example, those of the coastal plains of various islands of the Balearic archipelago, or the calcarenites of Seville and Huelva or those of the Penedés-Garraf in Catalonia.

Among the rocks which became permeable by fissuring, the most prominent in Spain are calcareous and dolomitic formations such as the Tertiary lacustrine limestones or the calcarenites of Seville or Majorca. On the whole, most of the carbonate formations are Mesozoic. They occupy approximately 20 per cent of the area of Spain and are found especially in the mountainous regions: the Cantabrian Range and the Pyrenees in the north, the Betic Chain in the south, and the Iberian System and the Coastal-Catalonian range linking the first two. Karstic phenomena are common in all of these formations. There is generally less exploitation of these aquifer systems, because in the mountainous areas where they predominate there is little demand for water. Nevertheless, in the areas in which limestone beds form or underlie large flat areas, as in La Mancha, where they are close to important centres of demand for water, as occurs along much of the Mediterranean coast, they have been and still are exploited, and this has led to local occurrences of marine intrusion or of spectacular drops in water level, amounting to as much as 10 metres per year, as is happening in Murcia.

The group of fissured rocks includes volcanic rocks, which are poorly represented within the peninsula; on the other hand, they constitute practically all of the Canary Islands (7,000 km^2 and about 2 million inhabitants). This population density, combined with a benign climate which promotes the cultivation of tropical crops, has led to intensive exploitation of groundwater in the archipelago. The crucial importance of groundwater for the economy of the Canary Islands and the numerous and varied types of problems that resulted from it created pressure for a joint hydrogeologic study, and this was carried out by UNESCO and the Spanish Government during the early 1970s. Since then, various organizations and individuals have continued to conduct many hydrogeologic studies in the Canary Islands, which constitutes a natural laboratory for this type of research.

3. Productivity of aquifers

Alluvial aquifers are the ones which normally present the most predictable characteristics. Formations with high permeability are commonly found in areas with rivers that have steep gradients, with materials originating chiefly in calcareous zones, as is common on the Mediterranean slope. Permeability values of 25 to 100 metres per day are common, and near the coast it is not unusual to find some formations whose permeability coefficient exceeds 500 metres per day. Under such conditions, and with saturated thicknesses which may reach several dozen metres, it is not surprising that drilled wells may succeed in extracting more than 50 litres per second, and in some cases close to 500 litres per second.

In coastal areas of the delta type it is normal to find two alluvial aquifers superposed, separated by fine semi-permeable or slightly permeable materials (from 1 m per day to less than 10^{-5} m per day). The captive deep aquifer comes to have a connection with the sea, and under natural conditions it contains fresh water in many areas where the land has a steep gradient, for example in the Llobregat delta, while in areas with shallow gradients it contains only marine salt water, as in the delta of the Ebro.

The conditions in the large sedimentary basins are highly variable, but in general the Tertiary filling materials have low permeability, with typical values ranging from 0.1 to 1 m per day.

The flow rates of the wells, with rare exceptions, range from less than 1 litre per second to 10 litres per second, and in exceptional cases up to 20 litres per second, generally with large drawdown. In many of these areas, wells with low flow rates are being exploited because it is clearly preferable to use such sources for community and industrial use than to bring water from other areas by means of lengthy and expensive piping.

The behaviour of the calcareous materials is highly variable. Both Mesozoic and Tertiary carbonate rocks are productive in the eastern and south-eastern areas of the Iberian Peninsula and the Balearic Islands, the most outstanding among them being calcarenite. In general, Triassic and Jurassic rocks are highly fractured, and may form excellent captive aquifers. It is not unusual to find wells with flow rates of over 50 litres per second, much more in some cases, although the number of unproductive wells is also considerable. In the north-eastern region, however, the limestone beds are much less productive. Best conditions are usually found in Triassic rocks. Cretaceous and Miocene formations constitute acceptable aquifers only occasionally or only near the sea, where there is considerable karstification, possibly due to the action of the area where fresh water and salt water mix. In many cases, severe salinization problems occur after prolonged exploitation.

Volcanic rocks display a broad range of characteristics, depending on their nature and their age. Recent volcanic rocks are usually highly permeable, although in the Canary Islands they are usually found above the saturation level except on the coast itself, where they contain salt water. Permeability may reach values of up to 500 metres per day and wells with yields ranging up to several hundred litres per second have been drilled. At Olot (Gerona) these volcanic materials are less permeable than the alluvium with which they are intercalated, and while the former may yield more than 50 litres per second, the productivity is usually no more than 10 litres per second, where only basalts of lava flow or pyroclastic origin are found. The older - Pliocene or Miocene - volcanic materials, both in the Canary Islands and in the occasional formations on the peninsula, have low or almost no permeability. Miocene basalts may have permeability values ranging from less than 0.01 m per day to 0.1 m per day, while deep wells ranging down to several hundred metres and galleries may produce yields from practically nothing to a few dozen litres per second, especially if there are lavas of the pahoehoe type. Phonolite has low permeability and yields, at best, a few litres per second. Ignimbrite is even less productive, but occasionally such rock intersects fissures or comes into contact with other materials, such as basaltic basement works.

4. Water quality

Inland, the main problems relating to the natural quality of groundwater are related to the presence of gypsum in many formations such as those of the Triassic and the middle Cenozoic at various places in the country. These problems are also reflected in the quality of the water in the rivers, and they are particularly serious in the Ebro basin and at various places in Catalonia, where there have also been terrain collapses caused by dissolution of the gypsum. They are also found in the Madrid basin and in some areas of the Levant. In the highly permeable coastal areas it is common to find considerable salt-water intrusion extending up to several kilometres, some captive aquifers being salinized. In some areas it is not unusual to find highly saline springs of water, which sometimes can even damage surface waters and reservoirs (as was the case for Meliones in Malaga).

Except for local problems with iron and manganese, captive aquifers rarely exhibit any other problems related to abnormal amounts of minor elements or heavy metals, although some carbonated waters may contain as much as 7 mg/l of fluorine.

In the Canary Islands highly mineralized waters may be found in old volcanic materials, mainly with bicarbonates of sodium, calcium and magnesium. In some areas they are highly encrusting, and in others they are corrosive because of an unbalanced excess of CO_2.

C. Groundwater development

1. Design of wells and pumps

Today there still are many wells dug by manual methods and lined with stone or brick, some of which may have galleries at the bottom, of varied lengths, extending as far as several hundred metres. Depths range from a few metres to more than 50 metres, and the oldest wells usually hold no more than 1 or 2 metres of water, while those constructed more recently and equipped with a sump pump may have up to 10 metres. Water was mostly lifted by means of a draw-wheel (noria) powered by horses, mules or donkeys, and some of these norias were fairly ingenious in design. The well diameter of 1.5 to 3 metres was just large enough to house the noria (some of them were elliptical in cross section), and in the case of low permeability formations, large enough to store the amount of water needed (including possible galleries to permit several hours of extraction, often associated with an irrigation pond. At present these wells are equipped with electrical pumps or pumps driven by internal-combustion engines, and only in a few locations, such as the plains of Palma and Fuerteventura, were they equipped with wind-driven pumps; these were abandoned, and efforts are being made to reinstall them.

To meet large demands for community water supply, some large diameter wells (up to 10 metres) were dug by the turn of the century, mostly in thick alluvial beds, down to 30 metres), within which there were installed hermetically sealed metal chambers housing the heavy and bulky piston pumps, powered by a steam engine on the surface. A few of these wells still remain operating today. Water is now lifted, however, by means of electrically driven centrifugal pumps.

For the past several decades, most wells have been constructed by means of conventional drilling machines. All of common methods are used, although cable percussion is by far the predominant one. This predominance may be explained by the fact that drilling companies are very small, usually owning only one machine or a few machines, and by the great diversity of the geologic terranes and circumstances in which they must drill, necessitating the use of equipment which is as diversified as possible and requires a minimum of special tools and relatively unskilled personnel. Drilling is frequently done with diameters ranging from 300 to 500 mm, the well being lined with welded steel 3 to 6 mm in thickness and 250 to 450 mm in diameter. In alluvial land, in order to obtain large yields with reduced losses of load, drilling is sometimes done with diameters as high as 800 to 1,000 mm. The most frequent drilling depths range up to 100 m, although much deeper wells, ranging up to 500 metres or more in depth, are drilled in increasing numbers. These too are often drilled by percussion with heavy drills, sometimes with glass spheres for limestone and basalt, and low-extension cables.

For the last 10 years, large numbers of small-diameter boreholes (120 to 150 mm) have been drilled by means of air-powered and down-the-hole hammers, with or without foam mostly for domestic supplies and small industries or farms, in limestone, calcarenite, granite and schist, with depths ranging to 100 metres. In the Canary Islands larger rigs are coming into use, for drilling with diameters up to 300 mm and depths of as much as 300 metres, in highly varied volcanic materials, although the problems that arise are in some cases very difficult.

Even though rotary drilling methods are used, both with direct circulation and with reverse circulation, these cannot be regarded as generalized techniques, except in the case of a few large borehole-drilling companies, which devote most of their attention to large public projects, or in combination with mineral and hydrocarbon prospecting.

The equipment used is highly diverse in origin, although much of it was manufactured in Spain, in small workshops or even in well-established industries with good technology. Cable drilling machines are generally predominant, although Spain also manufactures air drilling machines or highly versatile mixed-type machines capable of combining several drilling methods, which are adopted in accordance with the lithologic sequence encountered. Some combination rigs are now being exported.

Except for some firms that have highly qualified professionals, the quality of drilling work is not up to the level of sophistication of the equipment available nor to the needs of the country, and in many cases it can be described as bad or very bad, owing to the existence of family firms using obsolete equipment, and clients who are interested only in the lowest price without attention to the real water economy that they will obtain. The situation has grown worse in the past few years because of economic crises, the increasing cost of money and the fact that Spain has neither technical associations of drillers nor a system for providing information to those who conclude contracts for wells or grant credits for helping to drill wells.

The result of this is not only that in many cases drilling data and test data are not available but also that the casings are often of poor quality. Also, there is a lack of cementing and separation between aquifers, and the active parts of the well are limited to slotted casings, or at most to screens which are inadequate and short-lived. It is not unusual to find very inefficient wells, many of which are not deep enough even at the time they are constructed or after being operated for some time, or even wells that collapse during pumping, so that equipment is trapped and lost. In contrast, there are some firms with full capability which assume responsibility for the construction they do and which work to specifications, although in recent years they have suffered from the competition of bargain-rate drillers with little knowledge of technology or desire to provide a good product, which go into the market with very low prices and take advantage of the lack of knowledge on the part of those who order the wells, whether they are private individuals, industrial firms or even public agencies. They offer to do work for total prices ranging from 5,000 to 10,000 pesetas per metre ($US 30-60 per metre), as opposed to the 20,000 to 30,000 pesetas per metre ($US 120-180 per metre) charged for a drilling operation with a guarantee, good execution and longer well life.

The recent establishment of a Drillers' Association [Asociación de Perforadores], together with the need for State authorization for extraction of groundwater, will probably help to improve this situation.

In the Canary Islands the techniques for extracting water are different, designed to be adapted to the characteristics of steep gradients and very low permeability found in many of the volcanic materials. Conventional methods of drilling by machinery have been introduced only recently in the Canary Islands, but there are limitations with regard to drillable areas. On Gran Canaria the predominant type is a well dug by mining methods, 2.5 to 3 metres in diameter, with lining of the well - reinforced, where the terrain makes it necessary - and with machinery for horizontal drilling installed at the bottom. This machinery is used to drill radial boreholes ranging up to a length of several hundred metres each. In the past such drains consisted of dug galleries. The continuing drop in water levels makes it necessary to deepen the wells periodically and to construct new radial boreholes. The inventories conducted indicate that in an area of 1,500 km^2 there are more than 1,000 of these wells, with an average depth of 100 metres (some as deep as 500 metres), and in many cases with more than 1,000 metres of drains in the part that is active at a particular moment; this may be associated with the abandonment of several similar but inactive levels above the currently active one.

On the island of Tenerife, because public lands are more extensive and gradients are steeper, people have opted in favour of galleries or "water mines", also drilled by mining methods, and with cross sections of the order of 1 to 1.2 m^2, in some cases with several branches. Of the island's area of 2,000 km^2, in the approximately 1,000 km^2 with steep gradients there are almost 2,000 galleries with an average length of 1,000 metres and maximum lengths of 6,500 metres. As on the island of Gran Canaria, the continuing drop in water levels makes it necessary to lengthen the gallery periodically, sometimes under difficult conditions due to high temperatures (up to 55° C) or the presence of gases (chiefly CO_2). When the centre of the island is reached, the productivity will no longer be sustainable; in some areas this situation is already developing or is close to coming into being.

The output of the gallery may temporarily be increased by drilling drains from its interior, as has been done on Lanzarote and Gran Canaria.

In special areas of the delta type, as for example in the Empordan (Gerona), or previously in a sector of the Llobregat delta, there has been exploitation of semi-confined moderately deep aquifers by sinking pipes through which the water is actually spouting or is captured by suction with pumps which are installed only when irrigation becomes necessary.

The construction of special wells, such as radial-drain wells, has been tried occasionally in alluvial areas, although their cost is so high that they have been tried only for large community or industrial water supplies. In the alluvial areas of the Llobregat valley there are wells capable of providing 750 litres per second.

The pumping machinery most commonly used is the submersible centrifugal pump. Motors installed at the top and equipped with a transmission shaft were widely used some time ago, especially for large yields, but at present most pumps are equipped with a submersible motor. They are usually driven by power from the electrical supply lines, although in rural areas there are still some pumps operated by motor-generator sets using an internal-combustion engine. Direct driving of the pump by means of a motor or by using the power of farm tractors is only occasional. Pumping by means of air energy remains quite limited.

The scarcity of water in many areas and its high economic value makes it possible in many Mediterranean areas to operate very deep wells; it is not unusual to find wells more than 400 metres deep, with water levels that are also very deep. Submersible motor-pump sets for these conditions exist on the Spanish market and are also manufactured in Spain; in some cases, they are capable of raising the water to considerable heights as to reach the area of demand or the irrigation area. There are cases in which the direct total lifting is more than 500 metres, at rates of several dozen litres per second.

2. Pollution and salt-water intrusion

No general inventories on the contamination of groundwater have as yet been made but it is clear that such contamination is significant in several areas, particularly Catalonia, the Levant, La Mancha and some of the islands. A recent compilation of problems gives an overall view of the principal situations that exist. In their broad outlines, these are the following:

a. Marine intrusion into coastal aquifers as a result of pumping. This situation is very frequent and affects all types of formations. Many community water supply systems cannot avoid delivering salty water for lack of alternative sources, with the result that for drinking water the towns in question must use imported bottled water. In some cases the process has been a slow one, developed over many years, while in other cases - as for example, those of small coastal alluvial areas - it is seasonal in nature and much of the salinity is eliminated during the winter.

b. Accumulation of nitrates under irrigated areas, generally because of the use of large quantities of inorganic fertilizer, although the utilization of used water for irrigation may also play a significant role. In many areas of Catalonia, the Levant and the Balearic Islands the amount found is more than 100 mg/l, and in some areas of the Maresme, a coastal fringe north of Barcelona, 600 mg/l may be found. In the volcanic aquifers of the Canary Islands there is a similar tendency, which is already appreciable today at considerable depths. The accumulation of nitrates is usually accompanied by an increase in sulphates that results from the chemical form of the fertilizer used. Agricultural contamination by potassium does not seem to be severe, and the same is true of contamination by phosphates, although some exceptions exist.

Other contamination problems are more localized and occasional, and among them the following have been studied or documented:

a. The creation of a reducing medium by the infiltration of water with organic material, river water, used water or water that has passed through rubbish and mud. There are high concentrations of iron and sometimes of manganese;

b. The occurrence of heavy metals, chiefly chromium, of industrial origin;

c. The accumulation of hydrocarbons through dumping or leakage over alluvial areas;

d. The generation of methane near a large rubbish dump or through the injection of spoiled wine (vinaza) which diffuses through a karstic medium (the case of Garraf and La Mancha);

e. The accumulation of boron originating in residues of industrial detergents;

f. Effects related to mining, although in general there is not enough information available. In the mining of potassium salts, there are cases in which discharges of salt-laden water into rivers affect the aquifers; the best-documented case is that of the Llobregat delta, where chlorine values as high as 400 mg per litre are found in areas recharged by the river.

3. Overexploitation, recharge and subsidence

The development of groundwater in various parts of Spanish territory is producing effects of various kinds, which generically may be grouped under the heading of overexploitation, although the aspects involved are quite diverse. The drying up of springs and marshes is notable in the central areas of La Mancha, where large areas of previously dry-farmed land have been put under irrigation, affecting the water yields of the springs of the Middle Guadiana and consequently that of the river, in addition to affecting the humid areas of the natural reserve of the Daimiel. Implementation, already in progress to some extent, of new extensive irrigation projects in the southern part of Spain may constitute a serious threat to the humid areas of ecological interest in the Coto de Doñana, in the lower Guadalquivir, which depend in large measure on the discharge of groundwater. In the salt marshes (aiguamolls) of the Empordan (Gerona) the exploitation of groundwater seems to have affected the salt balance of some of the pond areas. On Gran Canaria most of the existing springs have been permanently dried up for many decades owing to the intensive exploitation of groundwater, and the deep gullies no longer have a continuous flow of water, a situation that has existed for so long that the resident population can no longer remember a time when it was otherwise.

In some areas the water level continues to drop even though total extraction does not exceed recharge. The transit times are long and, as a result, extraction becomes more and more expensive, and even starts to decrease, before a new balance is reached. This situation is typical of the islands of Gran Canaria and Tenerife, where there are accumulated drops of as much as 300 metres in the central portions of the islands, at a current rate of 10 metres per year, while the discharges to the sea still remain relatively high. Similar situations are found in the Campo de Tarragona and Levant area, in connection with discharges to rivers or to the sea. There are situations in which the drop in water levels, although severe, is only temporary and the available water is sufficient to meet the demand arising during dry periods. This happens in some parts of the island of Majorca, where water levels may drop by more than 100 metres but subsequently return to their previous value during the more humid season.

Fresh water may be replaced by sea water or by salt water of other origin. This happens in certain coastal areas, which have been mentioned in connection with problems of marine intrusion. Characteristic examples are found in the delta of the Llobregat River and in the Levant area of Spain.

Overexploitation in the true sense, in which extraction exceeds recharge occurs only in a relatively limited part of Spanish territory, but it affects areas where there is a large demand from the local population, the tourist industry and highly productive agriculture, especially those of Murcia, Almería and Alicante. In areas like these there is an ongoing search for water reserves, which are exploited like a mineral; when one deposit is exhausted, people set out to look for another. This mostly involves voluminous limestone massifs whose surface area is limited owing to the notable tectonic complexity of the terrane.

In many areas, for a better and more rational use of limited water resources, work is being done on the joint use of surface and groundwater, with a view to integrating them into Basin Hydrological Plans, which are regarded as a basis for the management of State water (both at the general level and at the level of autonomous regions). Such schemes for integrated development of surface and groundwater include artificial recharge as a water-management tool.

There already exists in Spain a certain tradition for artificial recharge, especially on the outskirts of Barcelona where the Barcelona General Water Company [Sociedad General de Aguas de Barcelona] has been using wells for the past several decades, one of them since 1953, in addition to promoting river infiltration through careful scarification since 1948. In this same area there have been brief experiments with recharge of treated urban used water through a well, as well as experiments in the filtration of highly contaminated river water by means of reservoirs. On the island of Majorca there have been experiments with the recharge of treated used water through wells drilled in calcarenite, and on the island of Gran Canaria runoffs from gullies are being introduced into deep wells. Recently, work has been done on the recharge of mine water in the Granada area, and there are ambitious projects aimed at artificial recharge through ponds in the area of the Duero depression and by means of ponds in the deeply karstified limestone of La Mancha.

The scarcity of irrigation water and the high agricultural productivity of many areas has led to the reuse of water for irrigation, under highly diverse circumstances and with little technical administrative supervision; in addition, there are also completed projects and projects still in the official planning stage in several areas of the islands of Majorca, Gran Canaria and Tenerife. These actions all bring an increase in the recharge of aquifers, to the point where exploitation is maintained as a result of irrigation combining river water and used water. In general, there is an increase in the mineralization of the water and of its nitrate content, but no detailed studies of the positive or negative effects have been carried out. Particularly noteworthy is the amount of water infiltrated from rivers with flow rates maintained by the discharge of used water (as is the case with several tributaries of the Besós River in Barcelona) for industrial and agricultural use, and this fact must be considered and evaluated in the preparation of clean-up schemes which affect the flow rate on certain sectors of the rivers.

In some cases, alluvial water with a high percentage of used water in the total recharge, is also utilized for supplying the population with community water, and this is a form of indirect reuse which differs little from direct use because, in some cases, the transport time between the infiltration area and the wells is only a few dozen hours. No detailed studies have been made concerning the sanitary implications of this situation, although waters are in some cases disinfected before discharging and the community water supply plants maintain a good level of chemical and bacteriologic monitoring of the water distributed.

No serious problems of subsidence caused by the extraction of groundwater have been experienced in Spain. In the delta of the Llobregat River, where confined aquifers have undergone drops of as much as 35 metres, sustained for several years, there are no clear effects of subsidence. There have been localized problems in buildings, attributable to soil drainage, as in the basilica of Teror (on the island of Gran Canaria), and depressions have been produced by subsidences in La Garrotxa (Gerona) and in the valley of the Ebro near Zaragoza, attributed to the collapse of cavities resulting from the dissolution of gypsum, although there is no clear relationship between these phenomena and present exploitation or irrigation. In fact, the Lake of Banyoles is simply but a large and deep depression of similar origin.

In some areas the flooding of basements, tunnels and underground storehouses due to rising phreatic level has occurred. That is the case in the eastern sector of the Llobregat delta and in the delta of the Besós (Barcelona); it is due to construction work done in dry areas at times when the phreatic levels were severely depressed because of extensive pumping. The cessation of such pumping for various reasons (marine intrusion, economic crisis, transformation of an industrial area to an urban area) has restored the levels to their original state, with the consequent flooding of structures for which this possibility had not been foreseen. Similar problems may occur in many other areas of Spain.

D. Administration and management

1. Research methods and teams

Studies of groundwater have made great advances in Spain during the past two decades. The objectives of these studies, and consequently their degree of detail, vary with the areas concerned but, at least at the preliminary level (scale of 1:200,000), the situation in the entire country is known to an acceptable degree.

The methodologies and the teams applying them may be classified into the following groups:

(a) Geological Service of Public Works [Servicio Geológico de Obras Públicas (SGOP)] of the General Directorate of Hydraulic Works [Dirección General de Obras Hidráulicas (DGOH)] of the Ministry of Public Works and Urban Development

In addition to a number of detailed studies aimed at solving concrete problems relating to community water supply or to irrigation, the feature which has perhaps been most characteristic of SGOP is the conduct of so-called studies of Compilation and Synthesis of Total Hydraulic Resources [Recopilación y Síntesis de Recursos Hidráulicos Totales]. These are studies aimed at the reconnaissance of relatively large areas (from 5,000 to 25,000 km^2), generally coinciding with hydrologic basins, which examine the entire hydrological cycle, as well as the modifications introduced or introduceable by man. The scale of the maps produced by these studies is usually 1:100,000 or 1:200,000. In almost every case these compilation and synthesis studies have been carried out in collaboration with basin agencies subordinate to DGOH. The outcome of these studies, initiated in 1965, may be described as positive. In more recent years SGOP has put greater emphasis on studies relating to the combined use of surface water and groundwater. The Hydrogeological Section of SGOP includes some 20 staff members with university degrees who devote their time entirely to this type of work, although recently their number has been substantially reduced because of structural problems. The number of university graduates in DGOH who are working in the various basin organizations and have had experience with groundwater is small - probably less than 15 out of the total of more than 400 university graduates. Thus, today there is a clear shortage of professionals specializing in groundwater within DGOH, still further aggravated by the transfer of many DGOH staff members to the Autonomous Communities [Comunidades Autónomas], as well as by the failure to replace staff members who change jobs or retire.

(b) Geological and Mining Institute of Spain a part of the Ministry of Industry and Energy [Ministerio de Industria y Energía (MIE)]

The Geological and Mining Institute of Spain (IGME) has a division of Groundwater and Geotechnology which has been quite active during the past two decades. In reality this continues a tradition begun in the early years of the nineteenth century, which had moments of some intensity at the beginning of the twentieth century and eventually was revived and boosted by a hydrogeologic study of the Guadalquivir basin, which was carried out during the second half of the 1960s, in cooperation with FAO. In the early 1970s IGME initiated a systematic programme of hydrogeologic reconnaissance of the entire territory of Spain, known as the Groundwater Research Plan (PIAS), which has been completed. In a sense, this involves reconnaissance studies similar to SGOP's compilation and synthesis studies, but with greater emphasis on groundwater. In the area already studied by PIAS, IGME is usually continuing a Programme of Aquifer Management and Conservation [Programa de Gestión y Conservación de Acuíferos (PGCA)], which includes various actions such as establishment of networks for observation of the levels and quality of groundwater, studies concerning protection against contamination, mathematical models and other studies. IGME is carrying out a National Plan for Community Water Supply to Urban Centres [Plan Nacional de Abastecimientos a Núcleos Urbanos (PNANU)], the purpose of which is to determine the possibility of using groundwater to provide community water to populations which are experiencing water-supply difficulties. It also publishes maps on a scale of 1:50,000 which are used for guidance in solid-waste disposal and which indicate the varying degrees of vulnerability of the aquifers. To carry out these tasks, IGME utilizes the services of 45 university graduates, more than half of whom are employed by engineering firms, collaborating on a more or less permanent basis, although the situation may change as the annual budgets vary. In fact, the operations become more difficult when the work in one area is finished because there is no adequate supply of permanently assigned staff. Many of the Government's own staff are preferentially assigned, committed to studies and tasks contracted with companies, and this is why the Government is carrying out relatively few research and development studies of its own.

(c) Agrarian Reform and Development Institute [Instituto de Reforma y Desarrollo Agrario (IRYDA)], of the Ministry of Agriculture, Fisheries and Food [Ministerio de Agricultura, Pesca y Alimentación (MAPA)]

IRYDA, or more precisely its predecessor, the National Institute of Settlement [Instituto Nacional de Colonización], introduced modern well-construction technology into Spain during the 1940s and 1950s. It has always had but few hydrogeologists, and therefore hydrogeologic studies have had to be contracted with private firms or arranged in conjunction with IGME or SGOP. The main work has been that of drilling wells and monitoring areas of special agricultural interest.

(d) Universities

The teaching of hydrogeology at universities was not initiated in practical terms until the 1970s. Research on groundwater has been conducted with some degree of dedication only since the mid-1980s, chiefly at the Complutensian University of Madrid, the Polytechnic University of Catalonia and the University of Granada.

University research has developed at a fairly rapid rate in recent years, and teams of professors dedicated preferentially to hydrogeologic research are now beginning to be trained in several universities. In addition to the three universities listed above are the Autonomous Universities of Madrid and Barcelona, the Universities of Alcalá, Zaragoza, Bilbao and San Sebastián and the Polytechnic Universities of Madrid and Valencia. There are about 20 full professors specializing on the matter. The total number of doctoral theses in hydrogeology or underground hydrology that have been updated annually in recent years is of the order of 25 to 30, and the number of master's degrees is probably two or three times as high.

Most of the master's or doctoral theses written up to the 1980s may be grouped under the heading of regional hydrogeologic studies. In recent years, although there still are perhaps a fairly large number of regional theses being written, there have been prepared - or are now being prepared - theses on a thematic basis, such as isotopic hydrogeology, mathematical models for combined use, contamination of groundwater, geostatistics applied to hydrogeology, characterization of the area of mixing between fresh and salt water in aquifers, and hydrogeochemistry.

(e) Regional and autonomous organizations

The activities of regional organizations have begun to be felt during the past 10 years. Conspicuous among them is the continued and important work being done by the Drilling Deputation of Navarre [Diputación Floral de Navarra], concentrating on springs and water resources in calcareous materials, although attention is also being given to vast alluvial areas of large development. The Generality of Catalonia maintains an agreement with the Polytechnic University of Catalonia to carry out hydrogeologic studies in areas of special interest, and its own Geological Service has initiated a number of studies. Other work has been promoted or is currently being carried on by local agencies in the Balearic and Canary Islands, in the latter case with special attention to hydrogeochemistry and isotopic hydrology.

(f) Engineering firms

Engineering firms comprise a highly diversified group. Approximately half of Spanish hydrogeologists are employed by private firms. These private firms work largely for DGOH or IGME, and consequently their studies are included among those described above.

The proportion of studies on groundwater that have been entrusted to engineering firms by autonomous regional governments or local entities has increased in recent years. It is difficult to predict how this situation will be influenced by the entry into force of the new Water Law, which declares groundwater to be public water.

Special attention should be given to the work done by the Centre for the Study, Investigation and Applications of Water [Centro de Estudios, Investigación y Aplicaciones del Agua], a foundation of the General Water Company of Barcelona [Sociedad General de Aguas de Barcelona], which between 1965 and 1975 promoted the exchange of experience and technology in the field of groundwater by holding technical meetings and publishing the Documentos de Investigación Hidrológica [Documents of Hydrological Research] and the journal Agua [Water].

(g) Other organizations

The work of the Nuclear Applications Office [Gabinete de Aplicaciones Nucleares] of the Study and Experimentation Centre [Centro de Estudios y Experimentación] of the Ministry of Public Works has been successful in improving simplified techniques for the use of artificial radioactive tracers, which are widely employed today, and promoted environmental isotope studies. The National Hydrological Plan, promoted by various official bodies under coordination of DGOH, has carried out numerous groundwater studies.

2. Inventories and data banks

Statistics on the use of water in Spain are not very accurate because, until recently, there has been no organization responsible for systematically collecting, processing and publishing statistics on the various uses of water in Spain.

In recent years a number of authors has published estimates on the uses of water, with numerical data which are more or less similar. Table 48 and figure 62 reproduce one of the latest estimates. The following facts should be noted:

In Spain irrigation accounts for 86 per cent of water use (not counting isolated industrial uses or hydroelectric use); urban water supply using groundwater accounts for only 38 per cent of the total. The use of groundwater for irrigation represents only one fourth of the total volume of water used for that purpose, although the irrigated surface area is perhaps one third of the total (2.8 million hectares), which indicates that irrigation with groundwater is more efficient than traditional irrigation with surface water. These overall data for Spain as a whole may lead to erroneous conclusions concerning the importance of groundwater use, since the situation varies considerably from one region to another; in the Mediterranean and island regions - areas with the most flourishing agriculture and tourism - groundwater is developed as much as surface water, or even more, but in the humid areas of the north-west groundwater development is minimal.

Although there are legal records on the extractions of groundwater which require a concession from DGOH or authorization from the General Directorate of Mines [Dirección General de Minas], in general these files are hard to work with or inaccessible, and the technical data they contain are usually of little value. With a view to remedying these defects, IGME began some years ago to build up a computerized national archive of water sources such as springs, wells and boreholes. At the beginning of 1985 its computerized records contained data on somewhat more than 100,000 entries, although the amount and quality of the data are generally poor. The new Water Law requires legal authorization for those water points which have a yield of more than 5,000 m^3 per year, and the number of these may be of the order of half a million.

Table 48. Spain: Use of water by source and basin

Basin	Urban use			Agricultural use			Total use		
	Surface	Ground	Total	Surface	Ground	Total	Surface	Ground	Total
				(in km^3 per year)					
Northern	240	110	350	425	153	578	665	263	928
Duero	145	35	180	2 408	465	2 873	2 553	500	3 053
Tajo	570	165	735	1 260	264	1 524	1 830	429	2 259
Guadiana	85	100	185	1 248	452	1 700	1 333	552	1 885
Guadalquivir	315	95	410	3 103	336	3 439	3 418	431	3 849
Southern	80	95	175	511	651	1 162	591	746	1 337
Segura	315	60	140	660	599	1 259	740	659	1 399
Júcar	80	240	305	1 620	810	2 430	1 685	1 050	2 735
Ebro	80	70	355	5 217	378	5 595	5 502	448	5 950
Eastern Pyrenees	65	270	645	168	264	432	543	534	1 077
Balearic Islands	285	71	79	--	158	158	8	229	237
Canary Islands	375	92	100	--	376	376	8	468	476
Total	2 216	1 403	3 659	16 620	4 906	21 526	18 876	6 309	25 185

Figure 62. Spain: Water resources utilization

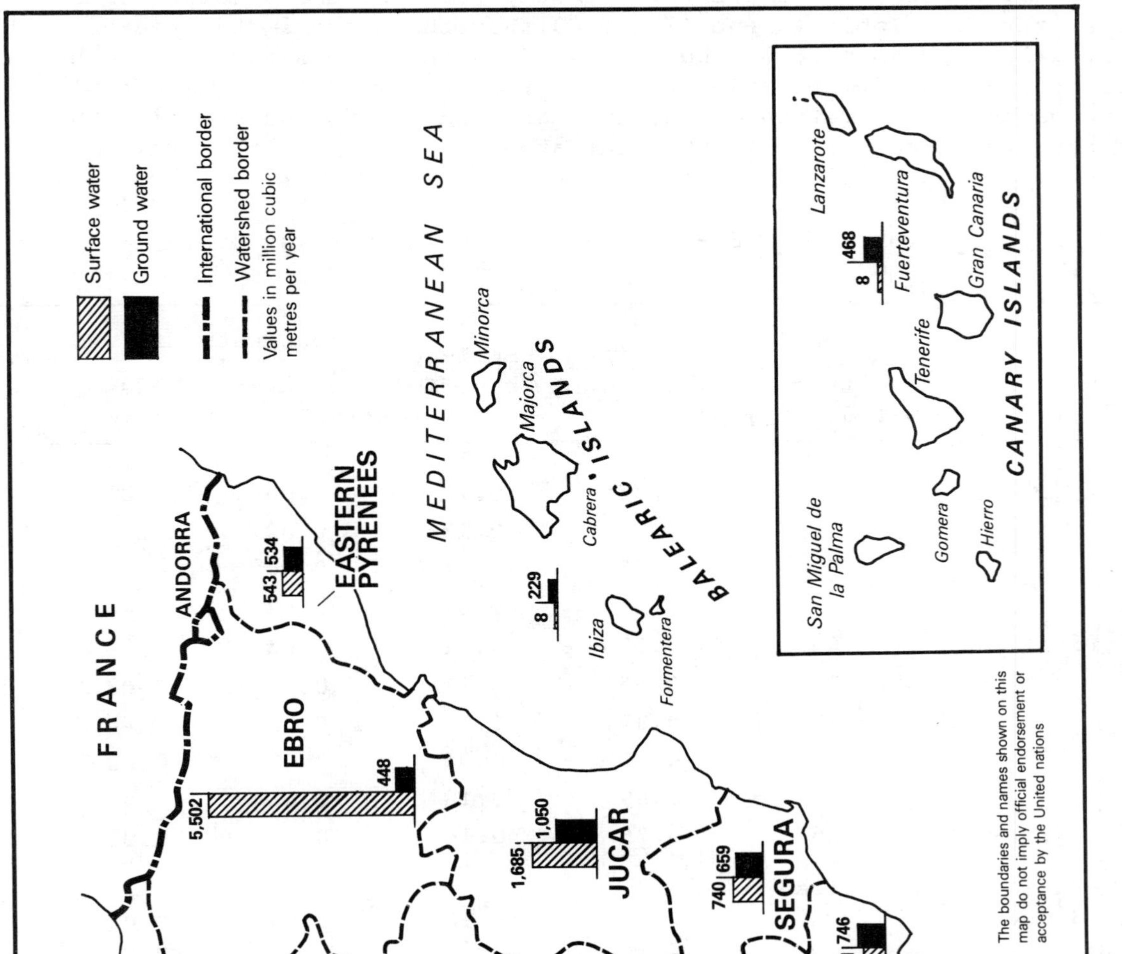

Preliminary estimates of groundwater resources, assimilating these resources to natural average recharge, have been made in each of the Spanish basins. These estimates have generally been made on the basis of conventional hydrometeorological methods or by separation of the surface component and the groundwater component in the river runoff. Table 49 and figure 63 show the approximate breakdown of the hydrological cycle in each of the basins of Spain. These data should be considered only as indicative, owing not only to the inaccuracy of the data but also to the relative value of any method based on water balances for determining the groundwater resources of any particular area.

Table 49. Spain: Water balance by river basin

Basin	Area (10^3 km^2)	Precip- itation	Surface runoff	Groundwater flow		Evapotran- spiration
				To rivers	To sea	
		(metres per year)				
Northern	54	1.35	0.72	0.08	0.03	0.52
Duero	79	0.59	0.13	0.02	..	0.44
Tajo	55	0.65	0.14	0.04	..	0.47
Guadiana	60	0.53	0.07	0.02	--	0.44
Guadalquivir	63	0.62	0.11	0.03	--	0.48
Southern	18	0.56	0.11	0.03	0.02	0.40
Segura	18	0.38	0.03	0.03	--	0.32
Júcar	43	0.52	0.04	0.03	0.03	0.42
Ebro	86	0.59	0.18	0.03	--	0.38
Eastern Pyrenees	16	0.75	0.14	0.03	0.03	0.55
Average for peninsula (493,000 km^2)		0.67	0.18	0.04	0.01	0.44

The designs of mathematical models for groundwater flow may be regarded as a technique well established in Spain since the beginning of the 1970s. The first detailed model constructed by Spanish hydrologists was the Llobregat delta model (1971).

This was followed shortly thereafter by a model of the Palma Plain (1973), and since that time many other models have been constructed, using well-established techniques, in their original form or improved, or by introducing the inverse method. Some of these, such as the model of the island of Tenerife, have presupposed an analysis of a three-dimensional flow system. Most of the models were constructed by DGOH and by IGME.

In general, all of these models have served to improve the knowledge of aquifer systems, but they have seldom proved to be permanent tools for improved exploitation of the aquifer, with the exception perhaps of the new model of the Llobregat delta, worked out by the Hydrological Plan of the Eastern Pyrenees [Plan Hidrológico del Pirineo Oriental] and subsequently equipped with a new system for processing the information.

Figure 63. Spain: Water balance for main river basins

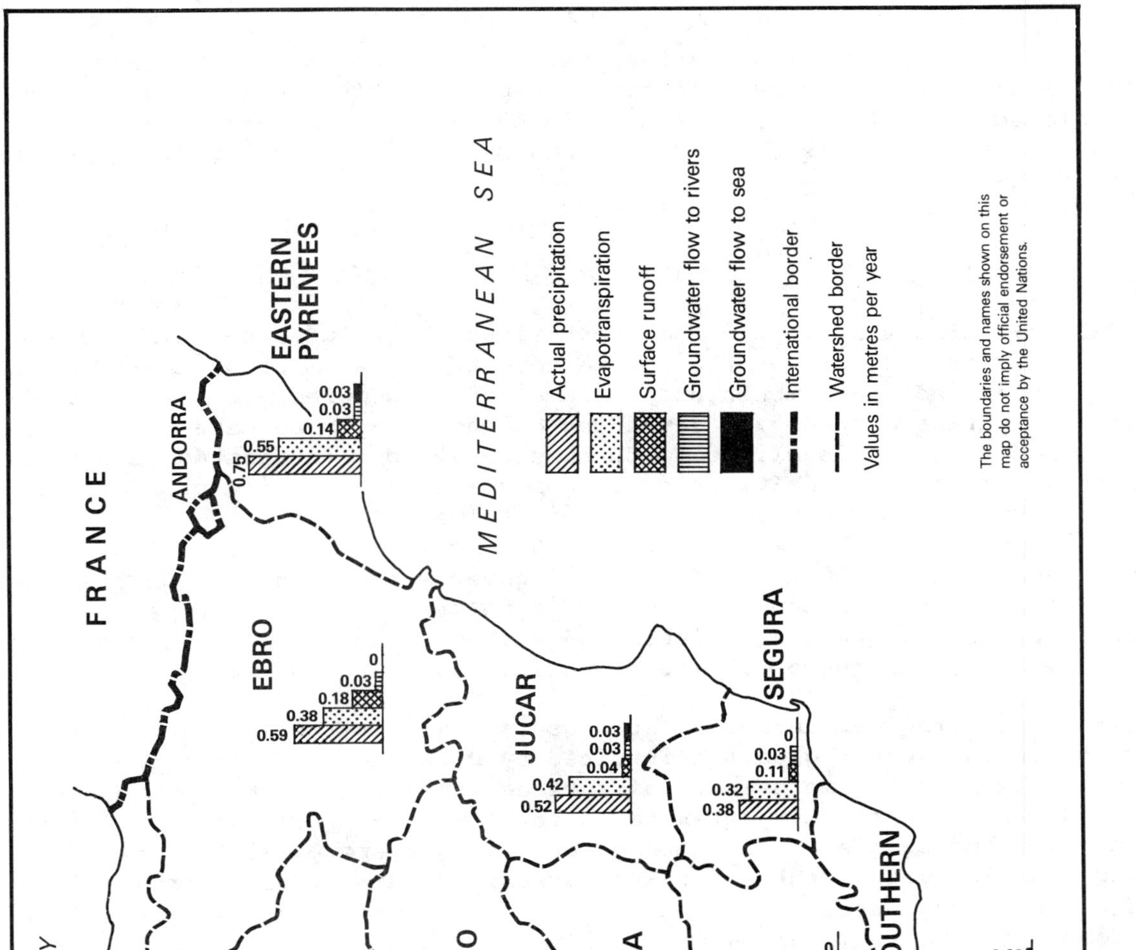

Achievements relating to the models for the transport of solutes are still in the initial stage, although some work has been done on the accumulation of nitrates (1979) and on the distribution of different types of environmental radioactive isotopes (1984).

A non-stationary model of marine intrusion with a sharp interface has been prepared (1983), and the possibilities of applying this model to real cases are now being explored, with the first attempts in the area of Valencia and Castellón. In the Canary Islands work is being done along the same lines (Ferreras, personal communication).

3. Observation networks

IGME maintains a network of piezometers which includes approximately 3,000 water points distributed over the entire territory. At most of these points, levels are measured every month and at some points samples are taken for chemical analysis, at variable intervals depending on the area. The data collected by this observation network are published with some delay, but interested persons may obtain them direct from IGME. These networks are experiencing problems of operation and aging which have not yet been solved.

Some agencies of DGOH and of regional governments also monitor piezometer networks of local or regional extent. Especially notable is the network in the Eastern Pyrenees, in operation since 1965 with more than 500 points whose level is monitored monthly, and almost 25 permanent limnigram recorders.

The hydrometric monitoring of springs is still in the incipient stage. The network of surface-water gauges maintained by DGOH, however, is designed preferentially for the improvement of surface-water regulation by means of reservoirs. Consequently it often fails to reflect adequately the relations between river and aquifer. The data of the network are published for each hydrologic basin by the Centre for Hydrographic Studies [Centro de Estudios Hidrográficos], and data not yet published may be obtained from the Offices of the Water Commissioners [Comisarías de Aguas].

4. Groundwater associations

A large proportion of Spanish hydrogeologists and groundwater specialists belong to the Hydrogeological Working Group [Grupo de Trabajo de Hidrogeología], which is part of the Association of Spanish Geologists [Asociación de Geólogos Españoles]. It holds meetings and symposia periodically and has about 150 members. The Spanish Chapter of the Spanish Association of Hydrogeologists [Asociación Española de Hidrogeólogos], with a membership of about 120, largely overlapping those of the previously mentioned groups, has been organized and also intends to hold scientific and technical meeetings. An Association of Well-Drillers [Asociación de Perforadores de Pozos] has been created.

Other associations are smaller in size, such as the Hydrogeology Section [Sección de Hidrogeología] of the University of Granada, or the Association of Former Participants in the Courses of the Hydrology Institute [Asociación de Ex-participantes en los Cursos del Instituto de Hidrologia (AICA)], and in general they conduct short courses or hold meetings. Spain has no periodical relating to groundwater, and to date it does not even have a journal of hydrology, because those which had existed in earlier years have ceased publication. A journal of a more general and popular nature, however, entitled Tecnología del Agua [Water Technology], has been created recently. The Revista de Geofísica [Review of Geophysics], which includes a section on Hydrological Sciences, is published irregularly and hard to obtain.

5. Agencies involved

(a) The new Water Law

In 1986 Spain promulgated a new Water Law which replaces the earlier Water Laws of 1879. In accordance with the previous law most groundwater was considered to be private property, provided that its extraction did not interfere with any other exploitation or with a surface watercourse or spring. In fact, groundwater had been exploited chiefly by private persons or entities or by small municipalities, and only to a slight extent by agencies of the central Government. The new Water Law declares practically all groundwater to be in the public domain. The exploitation of any new well that extracts a volume of more than 5,000 m^3 per year requires governmental authorization.

The organization of water management for all waters (both surface and groundwater) has been entrusted to a number of basin agencies known as Hydrographic Confederations [Confederaciones Hidrográficas], which are autonomous, with a juridical personality distinct from that of the State, and are the successors of the currently existing Hydrographic Confederations and Water Commissariats [Comisarías de Aguas]. These basin agencies are directly under the jurisdiction of the Ministry of Public Works, except in the case when the entire surface of the basin is within the limits of a single regional government, which occurs only in the archipelagos and in two small basins, the Eastern Pyrenees basin and the Southern basin. Participation by water users in the basin agencies is relatively low, not exceeding one third. As stated earlier, studies of the exploitation of groundwater during the past decade have been promoted chiefly by IGME and by DGOH's SGOP and, on a smaller scale, by IRYDA. With the new system of organization, it appears that the promotion of these studies may pass preferentially to MOPU's DGOH, except in the four basin agencies which are under the jurisdiction of regional governments.

The Water Law does not apply to academic or professional titles, to the training of experts in water resources or to methods for improving research on water resources.

(b) Training

The hydrogeologic training provided by university studies, not only in geology but also in civil engineering and mining engineering and other specialities, was scanty until the 1970s. In 1971 the Complutensian University of Madrid created the first official chair for the teaching of hydrogeology. At present, there are at Spanish universities approximately 20 official chairs held by experts in hydrogeology or underground hydrology. These experts carry on their teaching and research activities chiefly in Departments of Geodynamics, although some others do so in Departments of Soil Engineering, Hydraulics and Mining Geology, and occasionally in Departments of Nuclear Engineering.

However, at the level of professional training it is important to emphasize the work done by two courses that have been developed since 1977, intended for university graduates who wish to acquire training or improve their previous training in this field. The first is the International Course in Underground Hydrology [Curso Internacional de Hidrología Subterránea (CIHS)], given at the Polytechnic University of Catalonia, and the second is the Noel Llopis Hydrogeology Course at the Complutensian University of Madrid. Both of these courses are, in addition, sponsored by various State and regional entities, and even by private bodies in the case of Barcelona. The Spanish Ministry of Foreign Affairs plays an essential role in granting fellowships for attendance by foreigners, chiefly from Spanish-speaking or Portuguese-speaking countries. The courses last six months and have been conducted every year since 1977. The Barcelona course has a somewhat more technologic and quantitative focus and is longer and more intensive. The number of participants in each of these courses has been 25, maintained at an almost constant level. Approximately half of the participants have come to Spain from other countries.

In 1982 CIHS, with the collaboration of the Autonomous University of Madrid, organized an intensive course in Underground Hydrology lasting six weeks and focusing on English-speaking developing countries. This experiment has been repeated.

In addition to these, there have been various other short courses, lasting one or two weeks, on a wide variety of subjects. Prominent among these is the set of short courses organized almost every year by the School of Mining Engineers [Escuela de Ingenieros de Minas]. These courses are interrelated, and their total duration is usually three to four months.

Occasionally, special intensive courses lasting one or two weeks are organized, generally by groups working at the Complutensian and Autonomous Universities of Madrid, the Polytechnic Universities of Valencia and Catalonia, the Barcelona International Course on Underground Hydrology and the Geological Service of the Ministry of Public Works. The most common subjects are models and hydrogeochemistry.

6. International cooperation

The main lines of cooperation with developing countries - in the broad sense - are those linking Spain with Ibero-American countries. Much effort is being made to facilitate participation by graduates from those countries in hydrogeology courses in Spain, especially the courses given in Barcelona and Madrid (the International Course in Underground Hydrology and the Noel Llopis Hydrogeology Course). About 30 fellowships per year are granted in the field of groundwater, and applicants from other continents may also apply for these, with preference given to Spanish-speaking or Portuguese-speaking applicants. The total number of graduates in 19 years is of the order of 400, and less then 20 of these have been from Africa or the Middle East. The only exception is an intensive course given in English, in which 25 African and Western Asia professionals participate.

The conduct of doctoral studies in hydrogeology by foreigners has not developed substantially in Spain, partly because legislation is not sufficiently encouraging, partly because of Spain's own deficiencies in making this type of instruction official in the country, as mentioned before, and partly because there are few financial endowments for scholarships.

The exchange and attendance of professors and experts is organized largely at the personal level, since there are not sufficient practical arrangements at the official level. The advantages of linguistic and idiosyncratic ties, and in many cases ties based on a similarity of problems, technology and circumstances, have not been fully utilized, and to some extent the work has been obstructed by lack of specialized Spanish university personnel available for large-scale exchanges.

At the entrepreneurial level, relations have developed on an occasional basis, although work is being done to establish more permanent relationships. In the field of drilling equipment, a firm has already concluded export agreements and production agreements with several Ibero-American countries, and successful collaboration with Africa has been initiated. Activity in the field of groundwater has been less than in the field of surface water, although the trend is towards a reversal of the relative importance of these two.

Relations with European countries involve various facets and characteristics. In general, most of these relations are maintained with neighbouring countries, and France occupies a prominent place, helped by its particularly favourable policy of welcoming Spanish students and professionals for additional training and by the visits of French specialists to Spain. This latter aspect has been a constant in connection with the two principal underground-hydrology courses in Spain.

There are agreements for the exchange of professors with Portugal, France and the United Kingdom, although in the field of hydrogeology there have been few agreements.

A number of students from European countries have participated in underground-hydrology courses in Spain, including students from Romania, Italy, Switzerland and Finland. In addition, a number of European students, chiefly French and German, have done or are doing doctoral research in Spain, with the support of Spanish organizations and institutions and scholarships from their respective countries. Their freedom of action and the facilitation of their work have traditionally been considerable.

From the entrepreneurial point of view, a number of studies have been conducted by consortiums of Spanish firms with other European firms, both in Spain and abroad, and it is not unusual to find various European firms competing with Spanish firms in public bidding on studies relating to groundwater.

Relations involving cooperation and exchange with the United States of America are fruitful and frequent, many of them being channelled through the Hispano-American Joint Committee for Scientific and Technological Cooperation [Comité Conjunto Hispanoamericano para la Cooperación Científica y Tecnológica]. Agreements have been concluded and are continuing to be concluded with various public agencies and various universities, and a good many Spanish activities have been initiated through the channels opened by such cooperation.

In recent years there has been good collaboration with the International Association of Hydrogeologists, with the International Association of Water Resources, with the European Water Institute, and through them with UNESCO, the United Nations and the Council of Europe.

References

Aragonés, J. M., L. López and M. R. Llamas. Informe para la realización de un modelo digital simplificado para el estudio de la explotación de los acui'feros del Llano de Palma de Mallorca (Report on construction of a simplified digital model for study of exploitation of aquifers of the Plain of Palma on Majorca), Bol. Servicio Geológico de Obras Públicas, No. 38, 75 p., Madrid, 1973.

Ardemans, T. Fluencias de la tierra y curso subterráneo de las aguas. Imprenta de Francisco del Hierro, Madrid, 1972.

Batista, E., A. Bayó, E. Custodio and J. Doménech. Some aspects of sea water intrusion in Catalonia (Spain). Geologia Applicata e Idrogeologia, vol. 18, part 2, Bari, p. 299-314, 1983.

Castiella, J., J. Solé', S. Niñerola and A. Otamendi. Las aguas subterráneas en Navarra (Groundwater in Navarre), Servicio Geológico de la Diputación Foral de Navarra, Pamplona, 229 p., 1982.

CIHS. Jornadas sobre análisis y evolución de la contaminación de las aguas subterráneas en España (Discussions on the analysis and evolution of groundwater pollution in Spain), Curso Internacional de Hidrología Subterránea (CIHS), Barcelona, 2 vols., 1,350 p., 1981.

Coma, J. E. Las aguas subterráneas en el Plan Nacional de la Minería (Groundwater in the National Mining Plan). Hidrología, No. 4, p. 61-75, Madrid, 1971.

Custodio, E. Las aguas subterráneas en España (Groundwater in Spain). Revista de Obras Públicas, April 1977, Madrid, p. 321-334, April 1977.

___________. Geohidrología de terrenos e islas volcánicas (Geohydrology of volcanic terranes and islands). Centro de Estudios Hidrográficos, Pub. 128, Madrid, 303 p., 1978.

___________. Utilización del agua subterránea en España: reconocimiento y aprovechamiento (Utilization of groundwater in Spain: reconnaissance and exploitation). Quaderns d'Enginyeria, No. 3 (1981) 2, p. 370, and No. 5 (1984) 1, p. 145-172, Barcelona.

___________. Utilisation des réserves d'eau souterraine en Catalogne et aux îles Canaries (Utilization of groundwater reserves in Catalonia and the Canary Islands). 4ème Conférence Internationale sur la Planification et la Gestion des Eaux, Marseille, vol. 1, Comm. Europ. Méditerranéenne de Planification des Eaux, p. 693-700, 1982(a).

___________. Nitrate build-up in Catalonia coastal aquifers. In Impact of Agricultural Activities in Groundwater, Internat. Assoc. Hydrogeologists, p. 171-181, 1982(b).

___________. Model of the aquifers in the Llobregat delta (Catalonia, Spain). In Groundwater Models - Studies and Reports in Hydrology, No. 34, UNESCO, Paris, p. 39-51, 1982(c).

Custodio, E. Los estudios hidrológicos del Pirineo Oriental (Hydrologic studies in the East Pyrenees). Seminario del Agua en Aragón (Water Seminar in Aragon), Department of Geomorphology, Hydrogeological Section, University of Zaragoza, vol. 8, p. 828-851, 1984.

___________. Low permeability volcanics in the Canary Islands (Spain). In Hydrogeology of Rocks of Low Permeability, published by Internat. Assoc. Hydrogeologists, p. 533-544, 1985.

___________. Recarga artificial de acuíferos: avances y realizaciones (Artificial recharge of aquifers: advances and achievements), Boletín del Servicio Geológico de Obras Públicas, No. 48, approx. 350 p. (in press), 1986.

Custodio, E. and M. R. Llamas. Hidrología Subterránea, Ediciones Omega, Barcelona, 2 vols., 2,350 p., 1976, 1983.

Custodio, E., A. Bayó and E. Batista. Sea water encroachment in Catalonian coastal aquifers, Mem. Internat. Assoc. Hydrogeologists, vol. 13, part 1, p. F.01-F.22, 1977.

Custodio E., F. Cacho, M. Suárez and others. Combined use of surface and groundwater in the Barcelona Metropolitan Area (Spain). Mem. Internat. Assoc. Hydrogeologists, vol. 13, part 1, p. C.14-C.32, 1977.

Custodio, E., J. Isamat and J. Miralles. Twenty-five years of groundwater recharge in Barcelona (Spain). In Artificial Groundwater Recharge, vol. 1, DVWK Bull. 11, Verlag Paul Parey, Hamburg-Berlin, p. 171-192, 1982.

Custodio, E., J. A. Guerra, J. Jiménez and others. Effects of agriculture on volcanic aquifers of the Canary Islands. Environmental Geology, vol. 5, No. 4, p. 225-231, 1984.

Davis, S. N. Teodoro Ardemans, pioneer water supply engineer of Spain. Water Resources Bull., vol. 9, p. 1,028-1,034, 1973.

Fernández-Rubio, R. Objetivos y actividades del Grupo de Trabajo de Hidrogeología de la Universidad de Granada (Objectives and activities of the Hydrogeology Working Group of the University of Granada). Hidrogeología y Recursos Hidráulicos, vol. 2, p. 903-911, Madrid.

Fernandopullé, D., R. Heras, A. Sahuquillo and E. Custodio. Estudio de los recursos de agua de las Islas Canarias, SPA-15, Spanish Government and UNESCO, 4 vols., Centro de Estudios Hidrográficos, Madrid, 1972.

Gutiérrez, D. Sobreexplotación de acuíferos, cuenca del Segura (Overexploitation of aquifers: the Segura basin). Tecnología del Agua, Year 5, No. 21, p. 59-71, Barcelona, 1985.

Iglesias, A. Usos del Agua en España. Report published by Instituto Geológico y Minero de España, 87 p., Madrid, 1984.

Llamas, M. R. Influence of failure of groundwater supply to Madrid on the National Water Policy of Spain. International Symposium on Groundwater in Water Resources Planning, Koblenz, UNESCO and International Association of Hydrogeologists, p. 421-427, 1983.

Llamas, M. R. La hidrogeología del terciario de Madrid: hechos generalmente admitidos, incertidumbres, posibles líneas de acción (Hydrogeology of the Tertiary in the Madrid area: generally agreed facts, uncertainties and possible lines of action). Jornadas sobre la Explotación de las Aguas Subterráneas de Madrid, Canal de Isabel II, Madrid, preprint, 20 p., 1985.

Llamas, M. R. and I. Coleto. Nuevas consideraciones sobre la estadística española de usos de agua (New considerations on Spanish statistics concerning water use). I Congreso Español de Geología, vol. 4, p. 215-224, 1984.

Llamas, M. R., E. Custodio and G. Ariño. El Proyecto de Ley de Aguas, Informe, Instituto de Estudios Económicos, 166 p., Madrid, 1985.

Martínez-Gil, J. Notas sobre la historia de la Hidrogeología española (Notes on history of Spanish hydrogeology), Agua, No. 67, p. 35-43, and No. 68, p. 30-48, Centro de Estudios, Investigación y Aplicaciones del Agua, Barcelona, 1971.

MOP. Estudio de los recursos hidráulicos totales del Besós y del Bajo Llobregat, published by Comisaría de Aguas del Pirineo Oriental y Servicio Geológico de Obras Públicas (MOP), 4 vols., Barcelona, 1966.

Moragas, G. Corrientes subálveas: estudio general sobre el régimen de las aguas contenidas en terrenos permeables e influencia que ejercen los alumbramientos por galerías o pozos y especial del régimen o corriente subterránea del delta acuífero del Besós. In Anales de la Revista de Obras Públicas, Madrid, 133 p., 1896.

REPO. Estudio de los recursos hidráulicos totales del Pirineo Oriental, published by Comisaría de Aguas del Pirineo Oriental (REPO) y Servicio Geológico de Obras Públicas, Barcelona, 22 vols., plan, general summary, 1971.

Sahuquillo, A. La utilización conjunta de aguas superficiales y subterráneas en España (Joint use of surface and groundwater in Spain), Revista de Obras Hidráulicas, Madrid, March 1985.

SGOP-UPV. Utilización conjunta de aguas superficiales y subterráneas, Geological Service of Public Works (SGOP) and Polytechnic University of Valencia (UPV), Madrid-Valencia, 420 p., 1983.

XIV. SWITZERLAND

Area: 41,293 km^2 Population: 6.6 million

A. General geography

1. Physiography and geology

The Swiss Confederation for most part of its territory is a mountainous country. The highest point is Pointe Dufour on Mount Rosa (4,634 metres) and the lowest point is the bank of Lake Maggiore (193 metres). From the plain of the Po in the south, the land surface rises almost without interruption to the summit of the Alps (figure 64). The northern side of this massif gradually slopes down to the Swiss plateau, a gently sloping basin, drained and occupied by several great lakes, among them Geneva and Constance. The Jura then develops its undulating terrain towards France. In the north, the foothills of the Black Forest rise on the right bank of the Rhine. Switzerland includes parts of the basins of the main rivers of central Europe (table 50).

Like the relief, the vegetation of Switzerland is highly varied. It ranges from species of the Arctic and high mountains to species of the subtropical zone. The forests of conifers, beech, oak, ash and alder cover roughly more than a quarter of the total area of Switzerland (table 51).

Switzerland has a complex geologic structure (figure 65). To the north, the Hercynian crystalline massif of the Black Forest plunges under the tabular Jura of Mesozoic age. The Jura mountains are folded over the basement; they form a continuous chain from the region of Baden near Gex to the west of Geneva and even beyond, as far as Chambéry, France.

Between the Jura and the Alps, the Swiss plateau consists of a molasse basin containing detrital sediments stripped from the outlying chain during the uplifting which started during the Oligocene period. The molasse of the plateau is slightly deformed while the broken segments (scales) of the sub-Alpine molasse has piled up in front of the Alps over a thickness that may attain 5,000 to 6,000 metres.

The Alpine structure is formed of autochthonous and parautochthonous crystalline cores (Aiguilles Rouges, Arpille, Mont Blanc, Aar) and their folded sedimentary cover, which is partly pushed towards the north-east. These formations are overlain by Helvetic and ultra-Helvetic layers as well as by external Pennine layers.

To the south there are outcrops of internal Pennine layers. These were imbricated and carried towards the north-west along with parts of the crystalline substratum. Strongly marked by regional metamorphism, they are themselves covered by Austro-Alpine layers and separated from the southern Alps by the Insubrian line or the Tonale event which represents a considerable vertical throwback since its northern lip was raised from 12 to 15 km.

Several glaciations shaped the Swiss landscape during the Quaternary. The last glaciation covered the greatest area about 15,000 years ago. Many moraines were deposited during the withdrawals and the glacial meltwater deposited gravel in the open plains. This alluvial buildup has continued until the present in the Alpine valleys.

Figure 64. Switzerland: Hypsometry

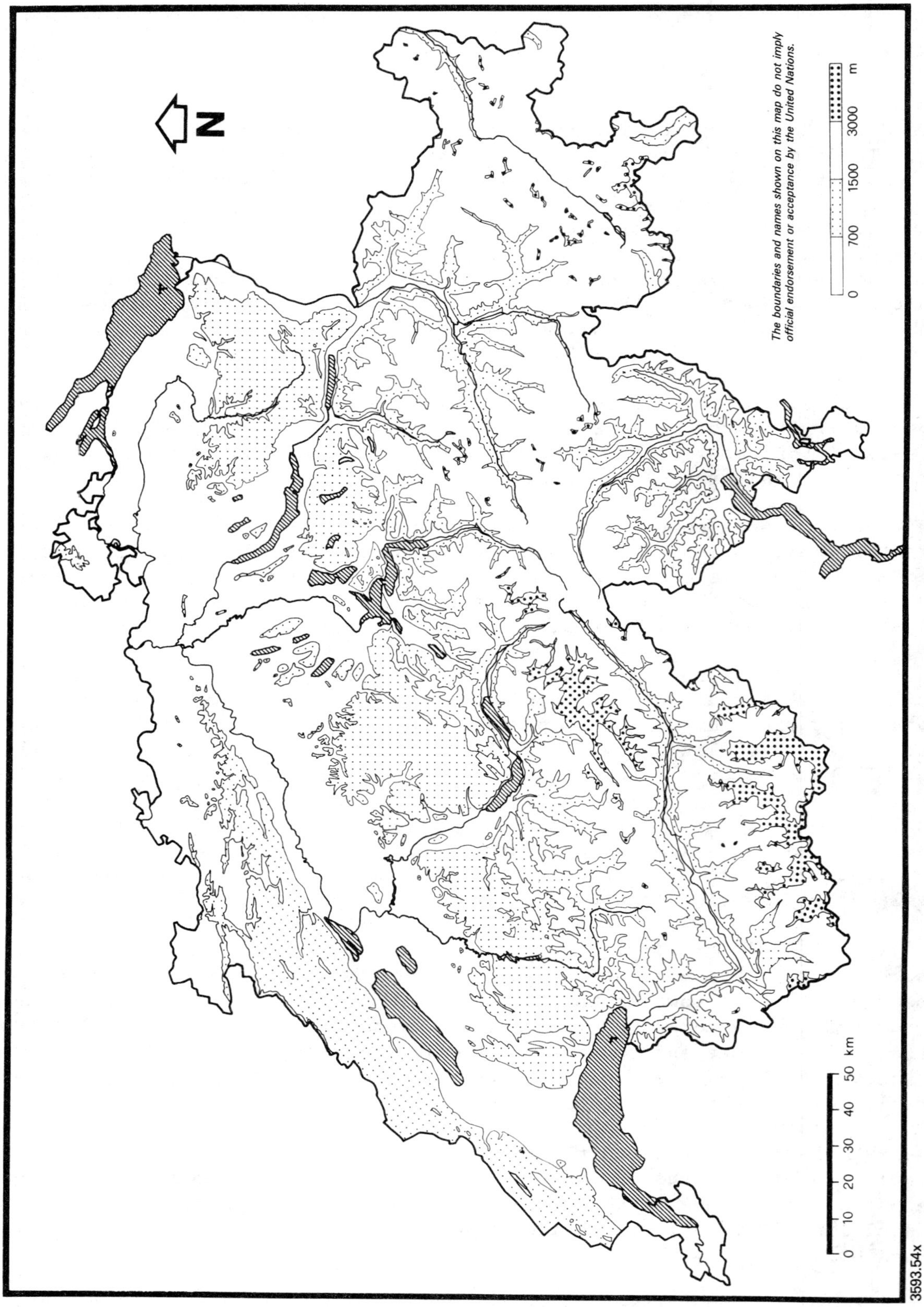

Figure 65. Switzerland: Tectonics

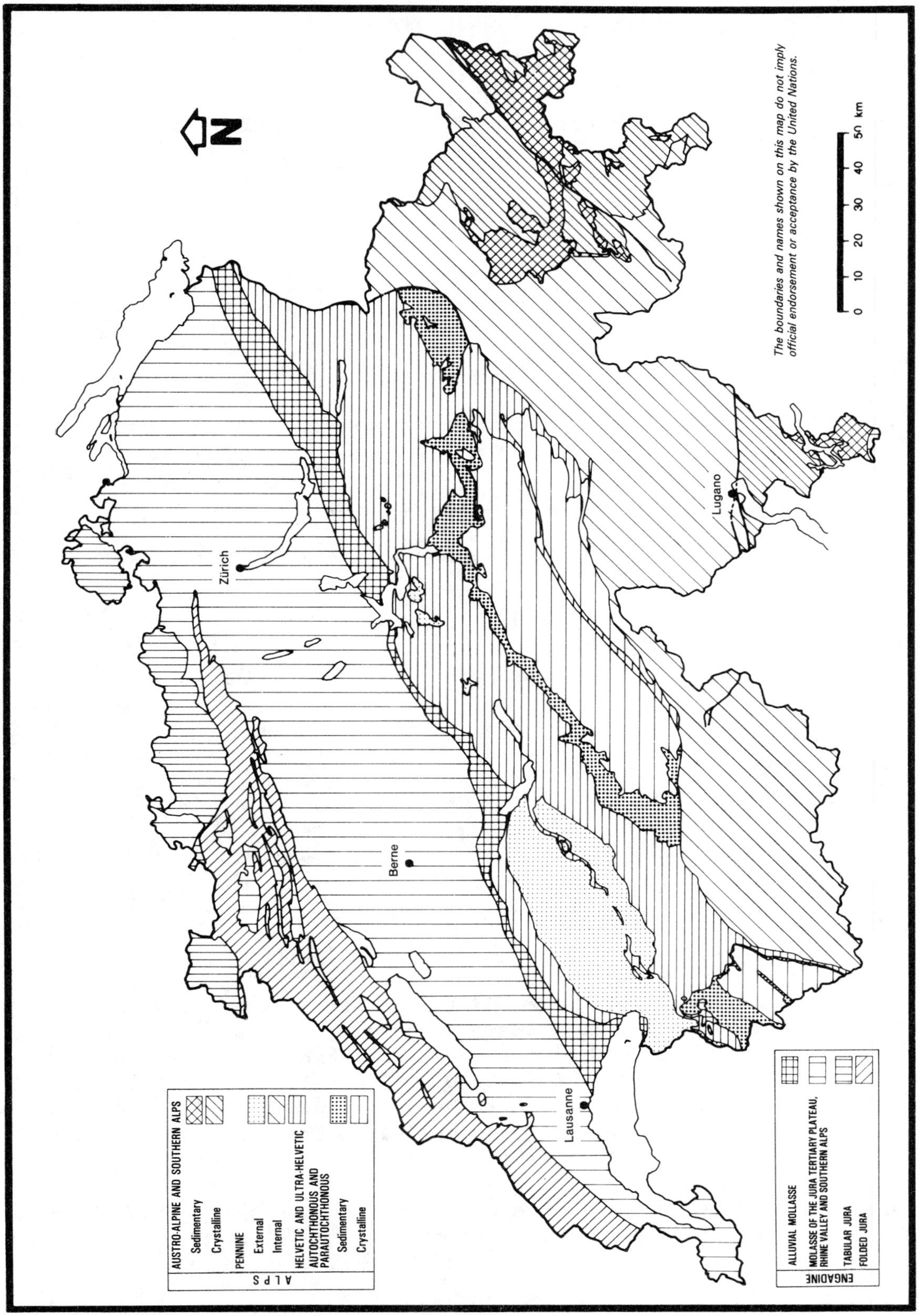

Table 50. Switzerland: Principal river basins

River	Outlet	Area in Switzerland (km^2)	(%)
Rhine	North Sea	27 973	67.8
Rhône	Mediterranean	7 522	18.2
Po	Adriatic	3 852	9.3
Adige	Adriatic	132	0.3
Danube	Black Sea	1 814	4.4
Total		41 293	100.0

Table 51. Switzerland: Land use, 1972

Use	Area (km^2)	(%)
Cultivated farmland	11 685	28.3
Pasture	8 510	20.6
Forest	10 520	25.5
Uncultivated (rocks, glaciers)	7 273	17.6
Lakes	1 416	3.4
Rivers	111	0.3
Built-up areas	1 687	4.1
Industrial complex	57	0.1
Communication installations	34	0.1
Total	41 293	100.0

2. Climate and water balance

Being of a continental type, the climate of Switzerland is strongly influenced by the barrier of the Alps and to a lesser degree by the Jura. During the summer, the temperature decreases from 0.5° to 0.6° C for each 100-metre rise in altitude; in winter this difference is only 0.4° to 0.5° C per 100 metres. The annual average temperature varies from -8° C at the Jungfraujoch (3,579 metres above sealevel), to 10.3° C in the Montreux region (408 metres) or 11.7° C in the region of Lugano (276 metres) south of the Alps.

The geographic distribution of rainfall is also irregular since it depends on the relief and prevailing winds. The influence of the Alps is felt over a great distance right to the foothill region. Annual precipitation may exceed 3 metres in the high mountains; it is close to 1 metre on the Swiss plateau but decreases sharply in the transversal valleys of the Alps which are more protected from orographic effects. For example, precipitation is 0.6 metre along the plains of the Rhône in the Sion region.

Evapotranspiration remains the least known factor of the water balance. Estimates are of annual averages below 400 mm in the Jura and 200 mm at the summit of the Alps. Table 52 shows the relative importance of the various elements of the water cycle in Switzerland.

Table 52. Switzerland: Annual water cycle

Direction	Form	Amount (mm)	
Gain	Precipitation		1 460
Loss	Evapotranspiration	480	
	Runoff, both surface and underground	980	
	Total	1 460	

B. Hydrogeology

1. Background

Switzerland is rich in groundwater. It has a large number of aquifers in loose soil and in fractured rocks (figure 66). The main areas with permeable unconsolidated water bearing formation include:

a. Glaciofluvial and fluviatile coarse sediments deposited in large alluvial plains

b. Glaciofluvial gravel of terraces and trenches on the plateau beyond the large valleys

c. Cones of dejection low in clay

d. Landslides and talus

e. Moraines which form aquifers of local interest

Among the fractured rocks, limestone and dolomite constitute the most important reservoirs because of their large areal extent, their degree of fracturing and their karstification potential.

In the Jura, the main carbonate aquifers are as follows:

a. Cretaceous

i. Upper Hauterivian (yellow stone of Neuchatel)

ii. Lower Valanginian (Berriasian)

Figure 66. Switzerland: Principal aquifers

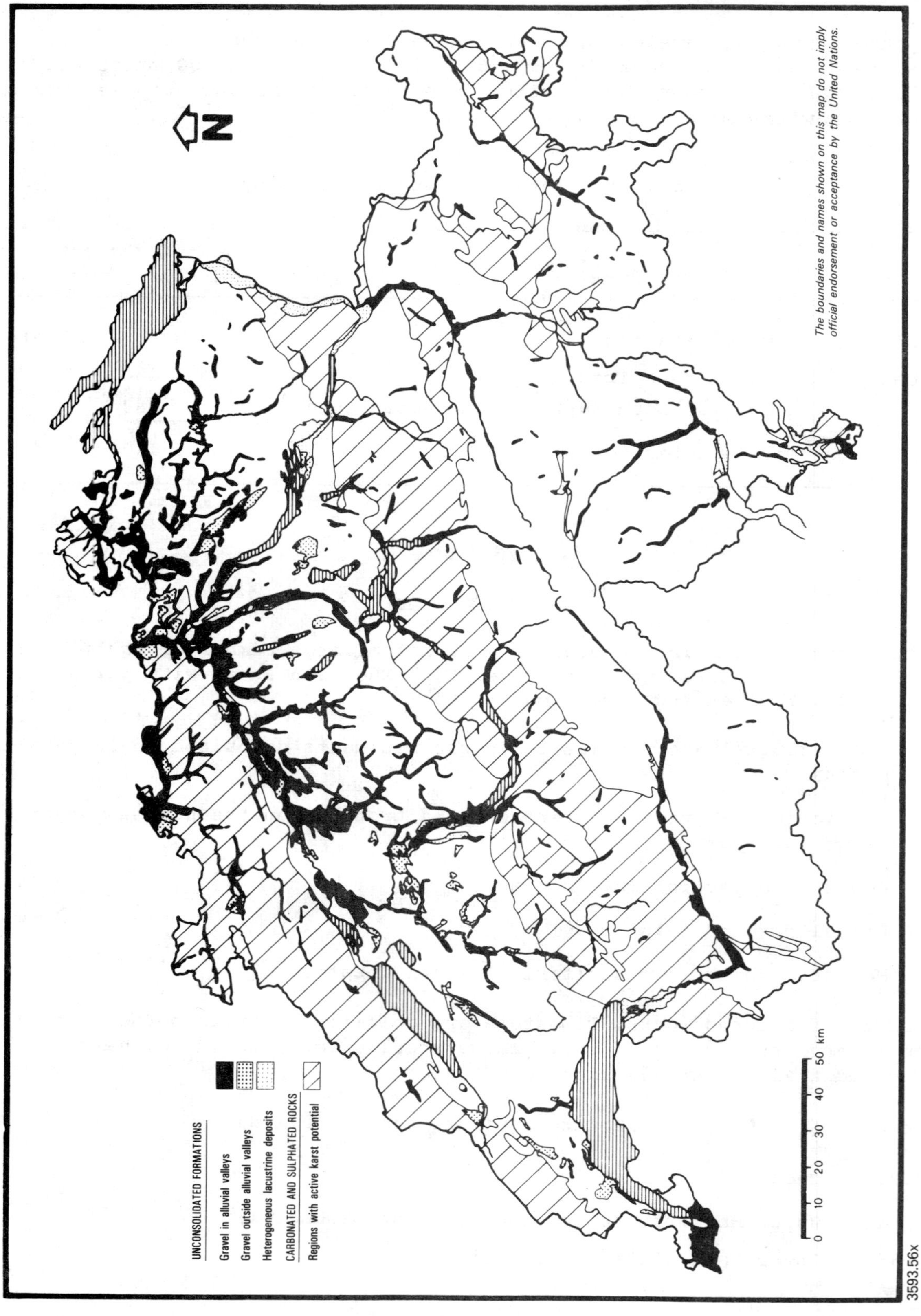

3593.56x

b. Jurassic

i. Malm (middle and upper Oxfordian, Kimmeridgian and Portlandian)

ii. Dogger (Bajocian-Bathonian: Hauptrogenstein)

c. Triassic (upper Muschelkalk, dolomite)

In the Alps, active karsts develop in different calcareous or dolomitic series (table 53). Both in the Jura and in the Alps, Karst phenomena affect the thick Triassic beds of gypsum which may also conceal important reserves of sulphated water.

Other fissured rocks may form aquifers. This happens especially in the sandstone and conglomerate of the plateau molasse and of the Alpine flysch which constitute reservoirs and discharge in a multitude of small springs. There also may be water circulating in the crystalline rocks, whether endogenous or fissured metamorphic, but its importance is limited.

Finally, Switzerland has many thermo-mineral sources, some used since the time of ancient Rome. They come mainly from deep aquifers which are not well known. Today, the water of some 10 of these natural springs is used for curative purposes or as bottled mineral water, and about a dozen thermal zones are tapped by boreholes.

2. Principal unconsolidated aquifers

The depressions carved by glaciers into the Alps and through the Swiss plateau have been partially filled by thick series of gravel and sand (figure 66) which hold groundwater.

These aquifers have been exploited for more than a century with ordinary shallow wells. Their systematic prospection began around the years 1960-1970 because of a growing demand for good-quality water and because of the increasing risks of overexploitation or pollution. The following are some characteristic examples of such aquifers.

(a) Alpine valley of the Rhine

From Disentis on the upper Rhine at Lake Constance and Thusis on the lower Rhine, the valley of the Alpine Rhine is filled with glaciofluvial and fluviatile sediments, some lacustrine or morainal, interrupted locally by the dejection cones of tributaries and by large masses of talus (Flims, Kunkels, Sennwald). The granulometry and permeability of these loose sediments decrease with altitude and the fine lacustrine deposits become preponderant in depth from Bad Ragaz onwards. From here to Lake Constance the main aquifer is represented by Recent alluvium of the Rhine the thickness of which can attain several tens of metres near the river and its former meanders.

The depth to bedrock is still not clearly known (table 54). It is 120-140 m or more north of Thusis, is 170 m at Coire, and 250 m at Salez. It is shallower towards the north as indicated by the inselbergs of the Sennwald region and the Ruthi borehole which reached Cretaceous limestone (Schrattenkalk) at a depth of 125 m.

Table 53. Switzerland: Principal aquifers of the Swiss Alps

Tectonic unit	Period	Epoch	Stage	Lithology
Autochthonous and parautochthonous (including Morcles aquifer)	Cretaceous	Urgonian	Lower Aptian Upper Barremian	Massive limestone (80-100 m)
	Jurassic	Upper Malm	Sequanian Portlandian	Compact limestone (200-300 m)
	Triassic		..	Cargneule
Helvetic and ultra-Helvetic	Cretaceous	Turonian		Limestone (0-10 m)
		Lower Urgonian	Barremian	Reef limestone (Schrattenkalk) (100-200 m)
	Jurassic	Upper Malm	Sequanian Portlandian	Compact limestone (150-200 m)
	Triassic		..	Cargneule and dolomitic limestone
Pre-Alps	Jurassic	Upper Malm	Sequanian-Portlandian	Calcareous massifs (100-300 m)
		Lias	Toarcian	Calcareous spathic (70-100 m)
	Triassic		Norian	Cargneule and white dolomite
Pennine	Jurassic	Upper Malm	..	Marbles (Barrhorn series)
		Lias	..	Limestone
	Triassic		..	Limestone, sand-stone, dolomite, marble (500 m)
Austro-Alpine	Jurassic	Lias	..	Limestone
	Triassic		Anisian Norian	Dolomite
Southern Alps	Jurassic	Lias	Carixian Hettangian	Kieselkalk
	Triassic		Anisian Norian	Limestone, dolomite (500-600 m)

Table 54. Switzerland: Characteristics of principal Quaternary aquifers, Alpine Rhine

	Bonaduz	Felsberg	Rossboden Chur	Salufer-feld Chur	Viertel-loeser, Zizers	Malans	Maienfeld	Flasch	Vitters	Heuwiesen	Ruthi
Maximum depth (m)	172	120	168	120	170	226	141	100	110	72	130
Depth of substratum (m)	..	..	..	..	..	..	..	..	125	..	125
Aquifer position 1 (m)	0-57	1-16	4-153	27-120	2-123	1-80	0-115	3-74	1-91	72	2-21
Piezometric level (m)	..	..	5-6	24-27	3-4	25	3	5-6	3	4	..
Permeability (X 10^{-4} m/s)	<0.5	..	20-90	2-30	4-100	0.5-2	50	>5	>0-6	20	..
Aquifer position 2	77-166	41-90	-	-	-	142-189	-	-	-	-	-
Piezometric level (m)	42-43	9	-	-	-	24	-	-	-	-	-
Permeability	1	>6	-	-	-	2	-	-	-	-	-

The Alpine valley of the Rhine has been the subject of systematic hydrogeologic study for about a dozen years especially as regards the geometry of the aquifer formations, the groundwater flow and its relation to the channelled and embanked Rhine, with its tributaries and lateral canals; today these studies make it possible to simulate the behaviour of the unconfined aquifers known above Bad Ragaz (table 55).

Table 55. Switzerland: Flow of unconfined aquifers, Alpine Rhine and of the river

Profile	Maximum flow of aquifers (m^3/s)	Average flow of the Rhine (years measured) (m^3/s)
Reichenau	0.28	-
Domat-Ems	0.90	-
Felsberg	0.40	116 (1962-1981)
Chur	0.45	-
Haldenstein	1.20	-
Untervaz	0.80	-
Zizers-Igis	1.10	-
Bad Ragaz	1.25	173 (1978-1981)

According to the developed mathematical model, these aquifers are fed essentially by groundwater from the slopes (65 per cent), by infiltration of surface water (30 per cent) and by precipitation. From Reichenau to Bad Ragaz the natural resources, exploitable over the long term with no diminution of the reserve, are calculated on the average at 4,000 l/s. Total withdrawals are now less than 2,500 l/s over this part of the Rhine valley.

From Bad Ragaz to Lake Constance, groundwater reserves are also large because of an abundant inflow from mountain slopes and from the Rhine and its tributaries. Between Oberriet and St. Margrethen, a stretch of less than 15 km, the Swiss communes and industries withdraw almost 900 l/s. The drilling of new wells is possible in this region near the Rhine, where the infiltration represents an available potential of 3,000 l/s in winter to 4,500 l/s in summer.

The results of chemical and bacteriological analyses show that the groundwater used in the upper valley of the Alpine Rhine are naturally of good to very good quality, except for certain confined deep aquifers which are poor in oxygen. Conversely reduction potential is in many cases significant from Bad Ragaz onwards in the parts of the aquifer which are surrounded by peat or which are not fed by surface infiltration. Such adverse conditions usually do not occur near the Rhine.

(b) Valley of the Alpine Rhône

For nearly 115 km, from Brigue to Lake Geneva, the valley of the Rhône River is filled with fluviatile and glaciofluviatile deposits, stratified or deposited in lenticular groups which may include morainal and glacio-lacustrine sediments. These become abundant and even predominant below Martigny. Many dejection cones and landslides (Sierre, Chessel-Noville) complicate the geohydrology of the permeable sand and gravel aquifers.

The position of the bedrock is not known precisely (table 56). However, with the help of geoelectric, gravimetric and seismic studies, we can estimate the depth at more than 200 m above Martigny and above 500 m between St. Maurice and Lake Geneva. With an average thickness of the permeable earth equal to 100 m, the total storage capacity of the aquifer is calculated to be more than 4 km^3.

The aquifers of the Rhône plain are fed by the river and its tributaries, precipation and runoff. The runoff is probably large but like other elements of the hydrologic balance it cannot be quantified in the present state of knowledge.

The connections existing between the embanked Rhône, the canals and groundwater will soon be better known. The surface aquifers of the alluvial plain are now the subject of systematic studies as part of a project for the utilization of the Rhône for the production of power.

Unfortunately, the chemical and bacteriological quality of the groundwater makes much of it unsuitable for consumption. For example, a heavy concentration of sulphates is observed where the Quaternary aquifers are in hydraulic continuity with evaporite beds of Triassic age. There is also a lack of oxygen and a marked reducing power where the loose sediments contain peat (table 56). The situation generally improves near the Rhône and in the ancient gravel beds which extend over the plain and which the river recharges through infiltrations.

(c) Valley of the Emme

The Emme rises in the massif of Hohgant south-east of Berne and north of Interlaken. It crosses the whole of the Swiss plateau and joins the Aar near Soleure.

For more than 35 km the lower valley of the Emme contains glaciofluvial gravel of the Wurm stage of the Pleistocene, resting on molasse sandstone and marl or on the deep limonitic-argillaceous moraine.

These permeable series are characteristic of most of the aquifers in the unconsolidated sediments of the plateau (figure 66). The unconfined aquifers that they contain are some of the best studied in Switzerland (table 57). A precise hydraulic balance has been established, the relations between the rivers and the aquifers are well known and management of the aquifer can be based on a developed model simulating steady or transient flow conditions.

Table 56. Switzerland: Characteristics of principal Quaternary aquifers, Alpine Rhône

	Vouvry	Massongex	Martigny	Fully	Saxon	Bramois	Sierre-Chippis	Loeche	Brigue-Clis
Maximum depth (m)	31	64	60	31	60	61	60	60	59
Aquifer position 1 (m)	5-9	5-38	6-20	6-17	3-16	4-30	10-32	19-24	4-29
Piezometric level (m)	0.4-1.0	2.4-4.5	2.1-3.2	0.7-1.6	1.8-2.3	1.4-3.5	2.0-7.6	0.5-1.0	0.0-1.0
Permeability (X 10^{-4} m/s)	1.4	1.8	3-4	0.3-0.7	1	7-10	1-8	0.7	8-11
Oxygen (mg/l)	0.3	2.6-4.8	2.6-7.7	0.6	2.8-4.5	2.9-4.3	6.2	2.2	4.9-9.0
Sulphates (mg/l)	153	56-95	26-59	49	72-82	108-110	95	7.0	115-12
Nitrates (mg/l)	--	2.0-2.9	0.0-0.3	0.2	0.0-0.7	3.5-5.9	22	--	6.9-7.0
Aquifer position 2 (m)	11-21	42-57	29-59	25-31	20-23	44-57	38-53	36-40	41-59
Piezometric level	0.3-1.0	2.4-4.4	2.0-3.4	0.0-1.3	..	1.5-3.5	2.0-7.6	Artesian	0.0-1.0
Permeability	2-4	2.9	1-10	0.6	..	3.8	0.9	1.4	0.5-2.0
Oxygen	0.2	0.7-6.2	2.0-6.3	0.2	..	0.0-3.4	1.9-6.4	5.1	3.7-9.0
Sulphates	207	143-197	40-52	43	..	205-221	224-395	--	230-33
Nitrates	--	0.0-0.3	--	0.3	..	5.7-5.8	4.3-7.9	--	6.7-8.0
Aquifer position 3 (m)	22-31	-	-	-	37-60	-	-	46-60	-
Piezometric level	0.2-1.0	-	-	-	1.7-2.4	-	-	Artesian	-
Permeability	0.9	-	-	-	0.4	-	-	..	-
Oxygen	0.2	-	-	-	2.9-7.8	-	-	7.5	-
Sulphates	99	-	-	-	64-83	-	-	8.0	-
Nitrates	--	-	-	-	0.0-0.5	-	-	8.5	-

Table 57. Switzerland: Characteristics of Quaternary aquifers, Emme Valley

Aspect	Emme Valley		
	Upper (Schupbach-Ramsel)	Middle (Ramsel-Burgdorf)	Lower (Burgdorf-Gerlafingen)
Length (km)	8	7	20
Aquifer thickness (m)	29-54	45	50
Annual rainfall (mm)	1,000	970	1,070
Permeability (10^{-3} m/s)	5.8-11	3-9.5	2-13
Effective porosity (%)	14-18	3-21	8-23
Average gradient (o/oo)	7	5.1-8.5	2.6-8.9
Velocity maximum (m/d)	260-500	..	200
Velocity speed average (m/d)	24-55	..	30-40
Reserves (10^6 m^3)	29	29	200
Maximum flow (l/s)	2,550	1,100	3,600
Contribution by rain (l/s)	..	115	740
Withdrawals (l/s)	200	23	630

3. Karstic aquifers

Figure 66 shows the extent of karstic aquifers in the Jura in the north-west and north of Switzerland. Infiltration amounts to 70 per cent of annual precipitation in the limestone regions of the Jura above 1,000 metres and 60 per cent in the lower basins. This water supply is often reinforced by surface runoff. The limestone formations of the Cretaceous, the Malm and Dogger (Jurassic) and the limestone and dolomite formations of the Muschelkalk (Triassic) form large karstic reservoirs.

The karstic areas are anisotropic and heterogeneous. The storage function is helped by porosity and microfissuration while the cavities and open communicating conduits ensure good transmissivity. In a massif as a whole, permeability may range from 10^{-6} to 10^{-7} m/s and the permeability of the unfilled fractured zones may exceed 10^{-2} m/s (table 58). In the Jura as a whole the effective porosity of the main karstic reservoirs varies from 1 to 20 per mille with higher values, of the order 0.5 to 1 per cent in the deeply fractured dolomitic limestone of the Muschelkalk.

Many karstic springs occur in the folded Jura mountains. Some give rise to rivers (Venoge, Nozon, Orbe, Areuse, Birse) whereas others have long been used for community water supplies and for industrial use.

Table 58. Switzerland: Characteristics of karstic aquifers, Swiss Jura

Aspect		La Brevine	Les Rosiers-Malleray	Le Puet-Moutier	Wahlen	Grellingen	Hurstet-Oberhof	Densburen-ASP
Maximum drilling depth (m)		450	486	402	210	213	62	237
Reservoirs	Age	Malm	Malm	Malm	Malm	Middle Dogger Hauptrogenstein	Upper Muschelkalk	Upper Muschelkalk
	Lithology	Limestone	Limestone	Limestone	Limestone	Oolitic limestone	Limestone and dolomite	Limestone and dolomite
	Aquifers	Unconfined	Confined flowing borehole	Confined	Confined	Confined	Confined	Confined
Thickness (m)	Total	500	350-400	350-400	200	140	70	70
	Apparent, exploited	390	325	330	131	104	37	86.5
Permeability (10^{-7} m/s)	By pumping	2.1	..	0.9-1.6	0.05-0.18	4-6	170	19
	By injection	15.1	..	..	..	..	..	..
Coefficient of storage (o/oo)		2-4	0.1-1.0	0.5	4.5	3	-	-
Maximum yield (l/min)		40	1,200	75	25	180	180	420

Exploration of the karstic aquifers is well advanced in the central and northern part of the Jura where several deep boreholes have been drilled during the last 20 years (table 58). The Muschelkalk of the tabular Jura, for example, is already being exploited on a large scale on the Rhine east of Bâle. Systematic prospecting, however, is only just beginning in Ajoie and in the south-west part of the Jurassic arc.

4. Deep aquifers under Swiss plateau

Four principal aquifers underlie the Swiss plateau. In declining order of stratigraphic position they are:

a. Upper marine molasse (Burdigalian-Helvetian)

b. Karstified Mesozoic limestone (Cretaceous and Jurassic)

c. Upper dolomitic Muschelkalk (Triassic)

d. Lower Buntsandstein (Triassic) and altered surface of the crystalline basement

The first of these aquifers has been utilized in recent times for the production of mineral and thermal water at Zurich and on the Swiss bank of Lake Constance. Oil exploration drillings carried out on the Swiss plateau have revealed active karstic circulation in the limestone series of the Cretaceous and Malm (Jurassic) which continue the permeable series of the Jura under the molasse. This confined aquifer is utilized at the thermal spa of Yverdon-les-Bains.

The thermo-mineral springs of Baden, Schinznach and Lostorf are of the Muschelkalk epoch whereas the waters of Buntsandstein and the basement rock are extracted at Zurzach. The development of the geothermic energy of these reservoirs is being studied.

These different deep aquifers are still poorly known. In the north-west of Switzerland they are being systematically explored as part of a project to dispose of radioactive wastes in the crystalline basement.

C. Groundwater development

1. Exploration techniques

Hydrogeologic prospecting has greatly benefited from recent progress in the fields of geophysics, tracers and the drilling technology. This development is clearly revealed by studies carried out from 1976 to 1983 under the sponsorship of the National Swiss Fund for Scientific Research on the subject "Basic problems of the water cycle in Switzerland".

Random and empiric at the beginning, the studies have become regional and sytematic in character. The use of computers makes it possible to establish water balances and forecasts by various simulations with the help, for example, of the method of finite elements in two or three dimensions. Nevertheless, knowledge of the geology of the aquifer system, of its structure, and of its relations with surface water is still of primary importance.

The exploration phase is reaching its end in the main alluvial valleys (1986). There is here a substantial demand for groundwater for domestic use. Various observation networks have been developed in the cantons. At the federal level, the National Hydrological Service maintains some 30 piezometers in Switzerland as a whole; its network should finally include some 100 stations serving as a reference point for the dense systems established by the cantons.

2. Current status

For a long time, spring water met the drinking water needs of the Swiss population without it being necessary to use pumps to extract groundwater. This situation has changed with higher living standards and industrial development. Thus, water consumption has more than quintupled since the beginning of the century. It is now stabilized at between 450 and 500 litres per inhabitant per day on the average, including part of the needs of artisan workers and industry. However, it may double during prolonged drought, as in the summer of 1976, particularly following exceptional demand for agricultural and domestic use.

In 1980, the communal services used more than 1.1 km^3 (1,100 million m^3) of water for distribution to households, to artisan workers and to industry (table 59).

Table 59. Switzerland: Water consumed, all sources, 1980

Source	Quantities distributed (10^6 m^3)		(%)	
Lake water		181		16
Groundwater				
Springs	473		43	
Wells	449		41	
Subtotal		922		84
Total		1 103		100

Withdrawals from aquifers are, in fact, much greater; industry and agriculture exploit directly certain aquifers such as those of the Rhône valley and the Emme valley. Also in recent years there has been considerable development of boreholes feeding heat pumps. The total of these withdrawals probably represents more than a cubic kilometre of groundwater per year.

3. Quality and quantity control

With the Federal Law of October 1971 on the protection of water against pollution, Switzerland has a powerful and effective instrument. The measures taken during recent years have made it possible to improve substantially the quality of water in rivers which feed the aquifers to limit the runoff and infiltration of used water into the subsoil, to collect solid waste and to eliminate it by incineration or by deposit in controlled discharge, in a way which protects the environment. Thus the danger of accidents during the course of transport or storage of polluting liquids, such as petroleum products, can be avoided.

Entered into force on 1 January 1985, the Federal Law of 7 October 1983 on the protection of the environment has reinforced control over dangerous substances and waste; it also makes provision for better protection of the soil. This law should also have a positive effect on the maintenance of the quality of groundwater in Switzerland.

Many problems remain to be solved. They include the risk of excessive cooling of aquifers used for heat withdrawals, the cleansing of aquifers contaminated by liquid petroleum products and by former uncontrolled discharges and pollution by such chlorine compounds as trichloroethylene or tetrachlorethylene (perchlorethylene) used in the chemical cleaning industry and, for example, in workshops to clean metal surfaces.

On the Swiss plateau, the most serious type of pollution is nitrates. Heavy and widespread use of chemical fertilizer and the maintenance of fallow land during winter may be cited as the main factors in this general contamination. Locally, other factors may play a role, such as the mineralization of freshly drained soils or infiltration of surface water high in nitrate.

A revision of federal legislation on water protection is under way. It aims at ensuring the conservation of groundwater and preserving its source of supply, indeed of improving it. In past years, the water services and industries have been confronted with problems of overexploitation of groundwater. Several of them have already built installations for the artificial recharge of aquifers (table 60).

Table 60. Switzerland: Artificial recharge of groundwater, 1978

User	Location	Year of first service	Consumption of crude water (m^3/s)	Nature of installation
GWA Lange Erlen	Bâle/Richen BS	1920	1.5	Submerged wooded areas, canals
Hardwasser AC	Pratteen BS	1956	2.0	Canals, ponds
Metallworks AG	Dornach BL	1957	..	Basins
Service des eaux La Chaux-de-Fonds	Gorges de l'Areuse NE	1958	..	Sprinkling
Verzasca	Tenero TI	1965	..	Basins in the river bed
WV Oberwinterthur	Winterthur ZH	1965, 1971	0.033	Basins
GWA Seeland	Aarberg/Gimmiz BE	1971*	0.2	Basins
GWA Aesch	Aesch BL	1976	0.5	Basins and injection wells
GWA Breitfeld	St-Gall SG	1976	0.2	Basins
SI Geneve	Vesey/Geneve GE	1979	0.6	Infiltration pipes
GWW Hardhof	Zurich ZH	1980	0.926	Injection wells

* Closed from 1979.

4. Agencies involved

Because of the federal political structure of Switzerland, hydrogeologic exploration, and the management and protection of aquifers are primarily the responsibility of the cantons. Usually, the Confederation merely monitors the application of federal legislation on water protection and coordinates the activities of the 26 cantonal establishments concerned. These establishments may also delegate an important part of their responsibility to the communes and to the major water services. Furthermore, private organizations play an important role in the preparation and publication of geologic and hydrogeologic maps or, for example, in the updating of water supply statistics.

The federal polytechnic schools of Lausanne and Zurich share the teaching of geology in Switzerland with the universities of Bâle, Berne, Fribourg, Geneva Lausanne, Neuchatel and Zurich. The Federal Polytechnic School of Lausanne also provides postgraduate training in operational and applied hydrology, while the Centre for Hydrology of the University of Neuchatel organizes each year a third-cycle course in hydrogeology open to holders of a geology diploma from a Swiss school or an equivalent foreign diploma.

Switzerland participates in several groundwater research programmes, mainly in the Sahelian zone of the Niger and Mali. This cooperation may be direct, with the assignment of experts in the field and teachers from universities; or it may be in the form of financial participation in projects supported by Governments or humanitarian organizations: in 1984 only about 50 million Swiss francs have been allocated by the Federal Department for foreign affairs as a contribution to solving water problems in west Africa.

References

Blau, R. V. Grundlagen fûr Schutz und Bewirtschaftung der Grundwasser des Kantons Berne. Hydrogeologie Emmental, volume 4, Modellstudie zur Bestimmung des Grundwasserdargebotes im Testgebiet Emmental. Wasser- und Energiewirtschaftsamt des Kantons, Berne, 1983.

__________. Quantitative Erkundung von Lockergesteinsgrundwasserleitern am Beispiel Emmental. Manual, Gaz-Eaux-Eaux usées, No. 5, 1984.

Emmeneger, Ch. Le reseau federal de stations d'observation des debits et d'autres paramètres physiques et chimiques des Cours d'eau. Gaz-Eaux-Eaux usées, No. 9, 1980.

Fonds national suisse de la recherche scientifique. Programme national de recherche No. 2 "Problèmes fondamentaux du cycle de l'eau en Suisse". Programme partiel 2 "Les ressources en eaux souterraines; aspects quantitatifs et qualitatifs". Final report, Gaz-Eaux-Eaux usées, No. 9, 1983.

Hantke, R. Eiszeitalter. Die jüngste Erdgeschichte der Schweiz und ihrer Nachbachgebiete. Bände 1-3. Ott Verlag, Thun 1978, 1980, 1983.

Högl, O. Die Mineral- und Heilquellen der Schweiz. Haupt, Bern und Stuttgart. 1980.

Hug, J. Die Grundwasservorkomnisse der Schweiz. Annalen der Schweizer, Landeshydrographie, 1918.

Imhof, Ed. Atlas de la Suisse. Edition du Service topographique fédéral. Wabern-Berne, 1965-1984.

Jackli, H. Die hydrogeologische Karte der Schweiz 1:500,000. Eclogae Geologicae Helvetiae, vol. 60, No. 2, 1967.

Kindle, Th. Die Grundwasserverhältnisse im Liechtensteiner Rheintal. Sonderdruck aus der Jahreschrift "Bergheimat" 1984 des LAV.

Lang, H. and others. Die Verdunstung in der Schweiz. Beiträge zur Geologie der Schweiz - Hydrologie. No. 25. Geographischer Verlag Kummerly and Frey, Berne, 1978.

Mornod, L. Recherches hydrogéologiques dans la plaine du Rhône entre Brigue et le Léman (résumé). Rapport ASPEE, No. 124/2, Gaz-Eaux-Eaux usées, No. 11, 1975.

Nänny, P. Die Bedeutung der Grundwasservorkommen für die Versorgung unseres Landes mit Trink- und Brauchwasser. Wasser-Boden-Luft, 1966.

Office fédéral de la statistique. Annuaire statistique de la Suisse 1984. Editions Birkhauser, Bâle, 1984.

__________. Das Grundwasser in der Schweiz; seine Bedeutung für die Wasserversorgung; seine Erforschung und Nutzung. Siedlungswasherwirtschaft in der Schweiz. Eine Standortbestimmung der achtziger Jahre. Société suisse de l'industrie du gaz et des eaux. 7 November 1984.

Société suisse de l'industrie du gaz et des eaux (SSIGE). Résultats statistiques des services des eaux en Suisse 1983. SSIGE, Zurich, September 1984.

Trüeb, E. Grundwasser, Eckpfeiler der Trinkwasserversorgung unseres Landes. Schweizer Maschinemarkt, Nr. 22, 1971.

__________. Struktur und Besonderheiten der Schweizer Wasserwerke. gwf-wasser/abwasser, 123, H. 8, 1982.

Trümpy, R. <u>and others</u>. <u>Geology of Switzerland: A guide-book</u>. Part A: Outline of the geology of Switzerland, Part B: Geological excursions. Commission géologique suisse. Wepf and Co. Publishers, Basel and New York, 1980.

Weber, E. <u>and others</u>. Der Grundwasserstrom des Alpenrheins. Wasser-Energie-Luft, H. 5, 1978.

Zahner, Ph. La protection des eaux souterraines en Suisse: Bases légales et experiences (Groundwater protection. Strategies and practices). Seminar of the United Nations and Economic Commission for Europe, Athens, October 1983.

XV. UNITED KINGDOM OF GREAT BRITAIN AND NORTHERN IRELAND

Area: 244,100 km^2 Population: 56.4 million

A. General geography

1. Physiography and climate

The United Kingdom of Great Britain and Northern Ireland in the British Isles includes the island of Great Britain (England, Scotland and Wales), Northern Ireland (Western County) and several smaller islands.

The highest relief in the United Kingdom is in the north and west in the mountains of the North West Highlands, the Grampians, and the Southern Uplands of Scotland and in the Cumbrian Mountains of the Lake District massif, the Cambrian Mountains of Wales and the mountains of Northern Ireland (figure 67). Much of England, except for the Lake District, the Pennines and the south-west peninsula is relatively low-lying, undulating country characterized in the east by alternating scarp and vale topography. Considered as a whole the United Kingdom is topographically subdued, with a maximum elevation of only 1,343 m above sealevel on Ben Nevis in Scotland.

The main watershed extends from north to south; it lies nearer the west coast and consequently rivers draining west generally have smaller catchments, although exceptions are the Clyde in Scotland and the Severn and Wye in England and Wales (figure 67). The largest river basin is the Severn with an area of more than 11,000 km^2. The Tay is the largest basin in Scotland and the Bann in Northern Ireland, both more than 5,000 km^2. The Bann expands in mid-course into Lough Neagh, the largest natural lake in the United Kingdom although only 397 km^2 in area. The overriding feature of the country as a whole is that the drainage basins are small, reflecting the short distances between source and mouth of the rivers; no point in the country is farther from the coast than 130 km. The upper reaches of many rivers in upland regions have been impounded for water supply.

Britain has a mild temperate climate, not subject to extremes, because of its position on the north-west margin of Europe in the extra-tropical westerly air stream. This mild oceanic airstream can be interrupted by weather patterns originating over the European continent, and the two conflicting patterns give rise to a typically variable climate. The mean monthly temperature in the coldest month (January or February) is about 4.5° C while that for July and August ranges from 13° C in the north to 17° C in the south. Significant variations in individual years are not uncommon.

The average annual rainfall is about 1,100 mm with some 870 mm over England, 1,300 mm over Wales, 1,400 mm over Scotland and 1,100 mm over Northern Ireland, the figures reflecting the higher incidence over upland regions in the north and west. Annual amounts vary from less than 500 mm on the lowlands of eastern England to more than 2,500 mm in highland Britain. Values vary from year to year and the coefficient of variation ranges from 8 to 18 per cent. In general terms, the rainfall is evenly distributed throughout the year although the winter tends to be wetter.

Figure 67. United Kingdom: Physical features

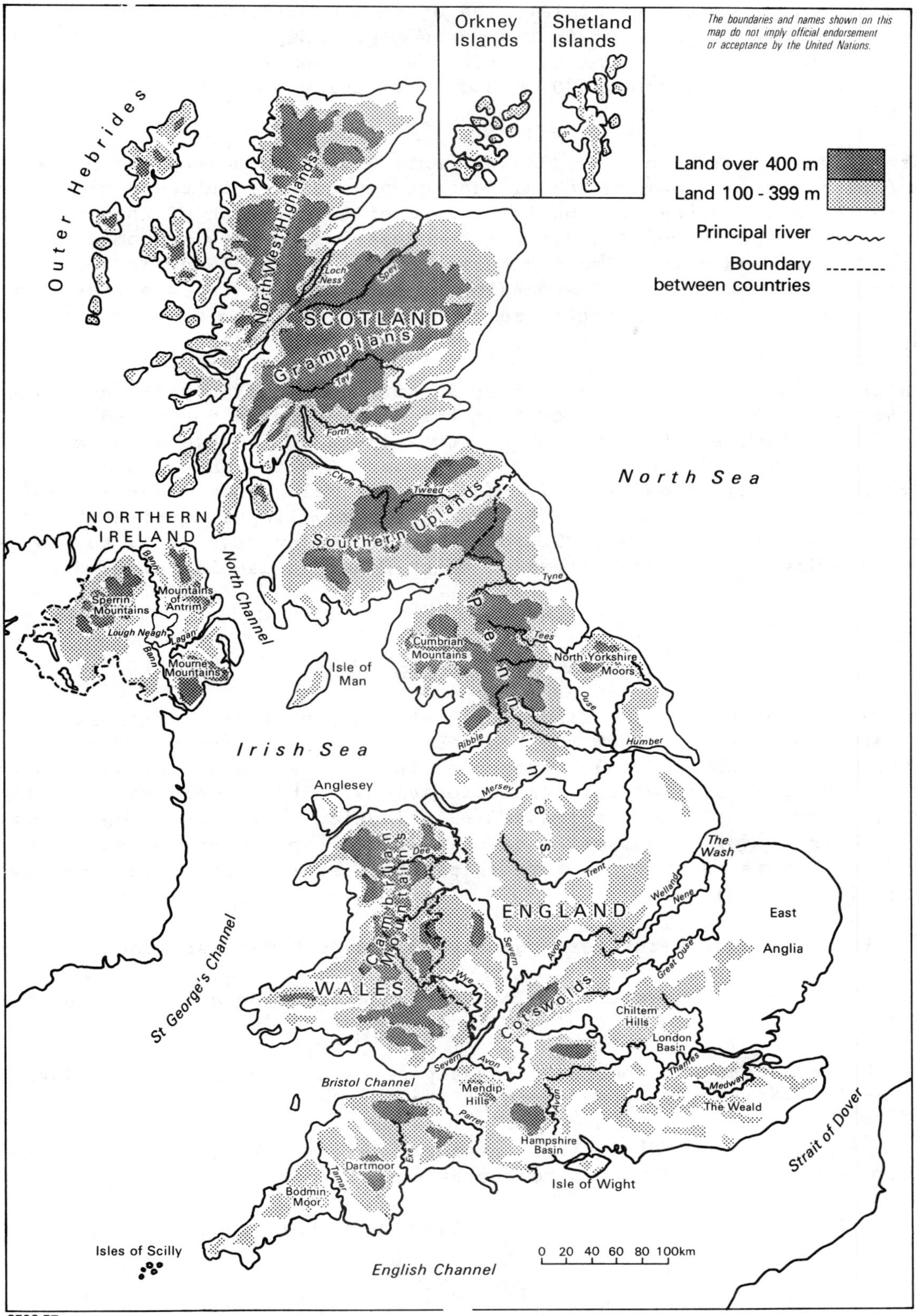

3593.57x

During the six warmer months, evaporation commonly exceeds rainfall in areas where the total rainfall is less than 700 mm per year. In eastern and southern England, soil moisture deficits often exceed 100 mm in the summer and there is need for irrigation in 5 years out of 10. The average annual evaporation over Britain is 470 mm with some 75 per cent in the period April-September, consequently runoff is seasonal. Minimum flow-rates occur in the late summer or early autumn. The average annual runoff is about 630 mm but it is twice as high in Scotland as it is in England and Wales.

Most of the population (some 83 per cent) lives in England where the average density (361 per km^2) is among the highest of any country in the world. This is related to the high degree of urbanization; over 80 per cent of the population live in urban areas. The United Kingdom is an industrial country economically dependent on the production and sale of manufactured goods. The tendency is for diversification of products and expansion of service industries as the traditional heavy industries decline. About three quarters of the land area is used for agriculture, the rest is mountainous, forest-covered or urban.

A high proportion of the population is served by publicly financed water supply and sewer systems; 99 per cent of the population is connected to public water supply systems and 95 per cent to sewer systems. Public water supplies provide some 7.1 km^3 (7,100 million m^3) per year of which about 25 per cent goes to industry and commerce (Agnew, 1987). England and Wales uses about 6.1 km^3 per year (85 per cent) of the total, Scotland 800 million m^3 (11 per cent) and Northern Ireland 250 million m^3 per year (3.5 per cent). The average daily supply per head of the population is some 350 litres of which domestic use is believed to amount to 130 litres.

2. Regional geology

The geology of the United Kingdom is extremely varied; surface outcrops include formations ranging from the Precambrian to the Holocene. The geologic structure is the outcome of major orogenies in the Precambrian as well as the Caledonian, Variscan and Tertiary tectonic events. Structural and geomorphologic patterns conform closely to the age and distribution of the rocks which may be divided into the major groups of Precambrian, lower and upper Paleozoic, Mesozoic and Cenozoic. In terms that are more relevant to hydrogeology, they may be classified (figure 68) as:

a. Younger cover sequences: gently folded post-Carboniferous sediments;

b. Older cover sequences: relatively undeformed Devonian and Carboniferous rocks;

c. Variscan orogenic belt: intensely deformed strata with east-west fold axes of late Paleozoic age lying south of the Variscan Front;

d. Caledonian orogenic belt: intensely deformed lower Paleozoic and Precambrian rocks with north-east to south- west fold axes;

e. Precambrian basement: crystalline metamorphic rocks.

Figure 68. United Kingdom: Basic geologic structure

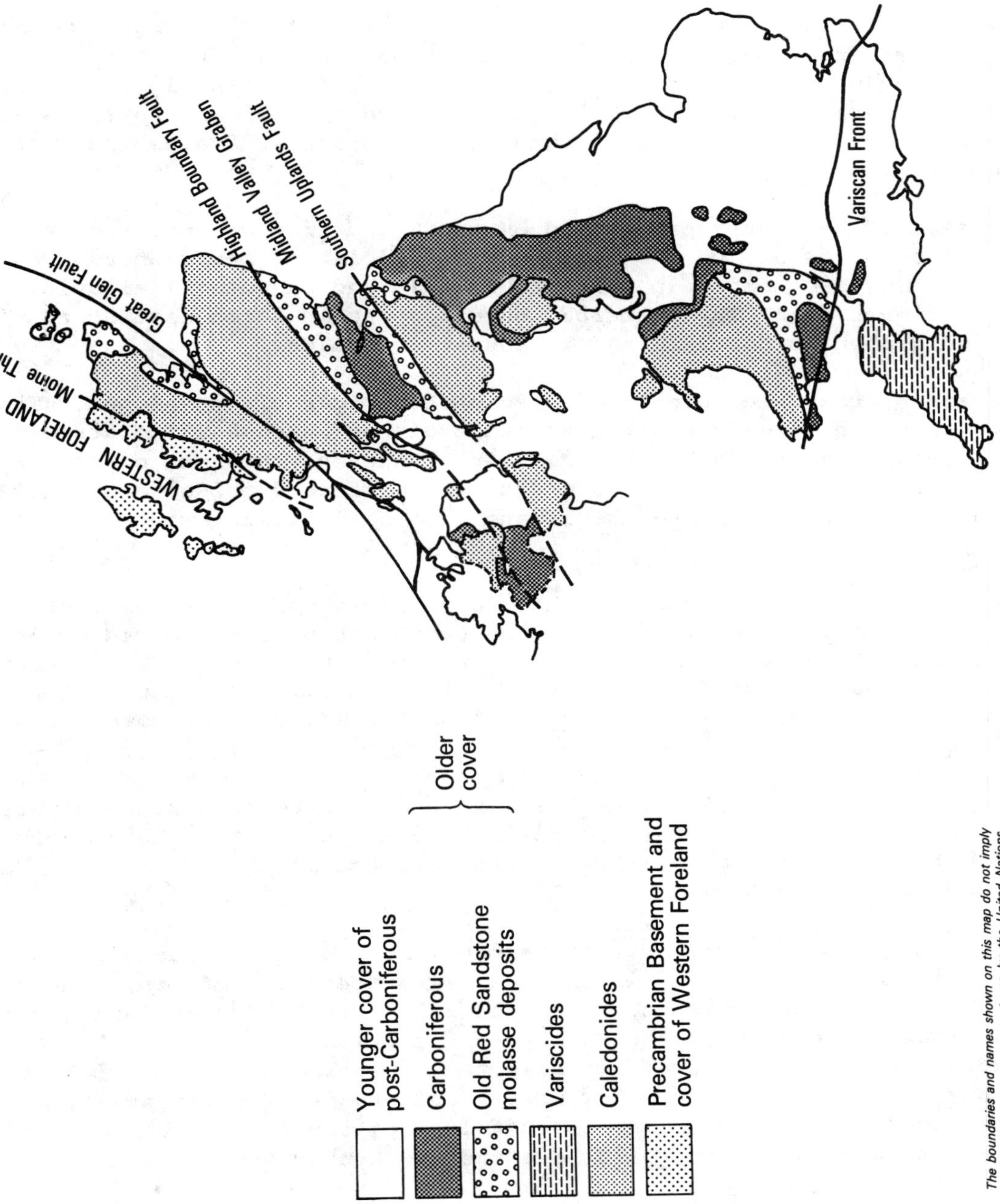

The boundaries and names shown on this map do not imply official endorsement or acceptance by the United Nations.

3593.58x

The oldest rocks in Britain are of Precambrian age occurring at the surface over much of the Scottish Highlands. They are high-grade metamorphic rocks that have been repeatedly deformed to give a complex structure. Precambrian rocks probably also form the deep basement in southern Britain but the few isolated outcrops indicate they are different in character from the rocks of northern Scotland.

The oldest known rocks beneath a large part of Britain form part of the complex Caledonian orogenic belt which includes a wide variety of sediments and volcanic rocks ranging in age from Precambrian to Silurian. They were folded at the end of the early Paleozoic and now outcrop in the Scottish Highlands, Northern Ireland, the Southern Uplands, the Lake District and central and north Wales. In central England the lower Paleozoic accumulated on a stable Precambrian basement and, consequently, is only gently folded.

The erosion of the continent formed by the Caledonian orogeny produced the thick detrital deposits of the Old Red Sandstone facies of Devonian age which formed internal molasse basins in northern Britain and on coastal alluvial plains in the south (figure 68). The principal outcrops are in north-east Scotland, the Midland Valley graben in Scotland and in south-east Wales.

At the end of the Paleozoic the Variscan orogenic cycle developed. The Variscan Front, which marks the northern limit of the intense structural deformation, extends east-south-east from South Wales to the Kent coast. South of the Front, Devonian and Carboniferous rocks were intensely folded and faulted. They outcrop in the south-west peninsula and have been encountered at depth in boreholes across southern England.

On the foreland north of the deformed zone, Lower Carboniferous rocks were deposited in fault-controlled basins. These sediments were succeeded by a widespread blanket of fluvio-deltaic Silesian (stage) sediments. The Carboniferous sequences contain many beds of sandstone and limestone, which are potential aquifers on the margins of the structural basins formed by earth movements during the Late Carboniferous.

Granite intruded the crystalline basement and fold-belt rocks in several phases, the most notable being the Caledonian granite bodies of the eastern Highlands of Scotland, the Southern Uplands and northern England, and the Variscan granite of south-west England.

The post-Variscan younger cover of Permian, Mesozoic and Cenozoic continental and marine sediments occur mainly in the low-lying areas of eastern, southern and western England. They form sequences several kilometres thick in fault-controlled basins that began to develop in the Permian and continued throughout the Mesozoic and Cenozoic. These basins are essentially the on-shore extensions of major off-shore basins such as the North Sea, Irish Sea and Anglo-Paris basins. In southern England, the younger cover rocks have been folded by the Alpine compression to give the principal structures of the Weald Dome and the London and Hampshire basins.

Finally, Pleistocene glaciation influenced the geomorphology of much of Britain except in the extreme south. The complex sequence of glacial sand and boulder clay have had considerable bearing on the geomorphology and form of the drainage pattern and on the rate of recharge to underlying bedrock aquifers. At least three major and several minor ice advances extended across lowland Britain reaching, at their maximum, as far south as the Thames Valley.

B. Hydrogeology

1. Background

Groundwater development in the modern sense can be related to the advent of the Industrial Revolution in the seventeenth and eighteenth centuries. With the development of industry and the associated trend towards urbanization, the increasing need for a good quality water supply, at a site where it was to be used, focused attention on groundwater. Where major aquifers occurred and the demand for water existed, groundwater resources were developed at an expanding rate. In London the first deep wells were drilled through the Tertiary deposits to the Chalk in 1823 and by 1900 abstraction from the aquifer in and around London was more than 60 million m^3 per year. Similar developments took place in other towns and cities, including Birmingham, Nottingham and Liverpool. Coastal towns on the Chalk benefited by their ability to economically develop good quality groundwater.

In time, undesirable consequences became apparent in the form of significant falls of water level and saline intrusion. By 1900, groundwater levels had fallen more than 35 m in central London and similar declines or reduction in quality were recorded in other areas of high abstraction. These trends eventually led in 1945 to legislation and the control of groundwater abstraction by a licensing system, thus providing for the first time the essential basis for the sound management of aquifers as practised today.

2. Regional flow systems

The form of groundwater flow systems is closely related to topography, geology and depth of the comparatively impermeable basement that defines the base of regional flow systems. In the United Kingdom the geologic horizon of the basement differs in different tectonic provinces. South of the Variscan Front, the base of the Mesozoic is the limit of significant groundwater flow but north of the Front it is the base of the Carboniferous, except in Scotland where the base of the Devonian is more appropriate.

By recognizing an impermeable basement below which flow is negligible, the combined upper Paleozoic, Mesozoic and Cenozoic cover sequences can be viewed as Groundwater Provinces, or Basins (figure 69) on the basis that groundwater in these provinces is incorporated in major regional flow systems flowing to a common outlet (Downing and others, 1987).

The principal aquifers occur in the post-Carboniferous Younger Cover sequences of England and Wales and include the Cretaceous Chalk, Middle Jurassic limestone and Permian-Triassic sandstone (figure 70). Secondary aquifers are also found in these sequences as well as in the Older Cover sequences of the Carboniferous and Devonian (table 61 and figure 68).

The main features of the aquifers are discussed below in the context of the Younger and Older Cover sequences after an initial comment on the water-bearing character of the Superficial Deposits as represented by alluvial deposits and glacial sand and gravel.

Figure 69. United Kingdom: Groundwater Provinces

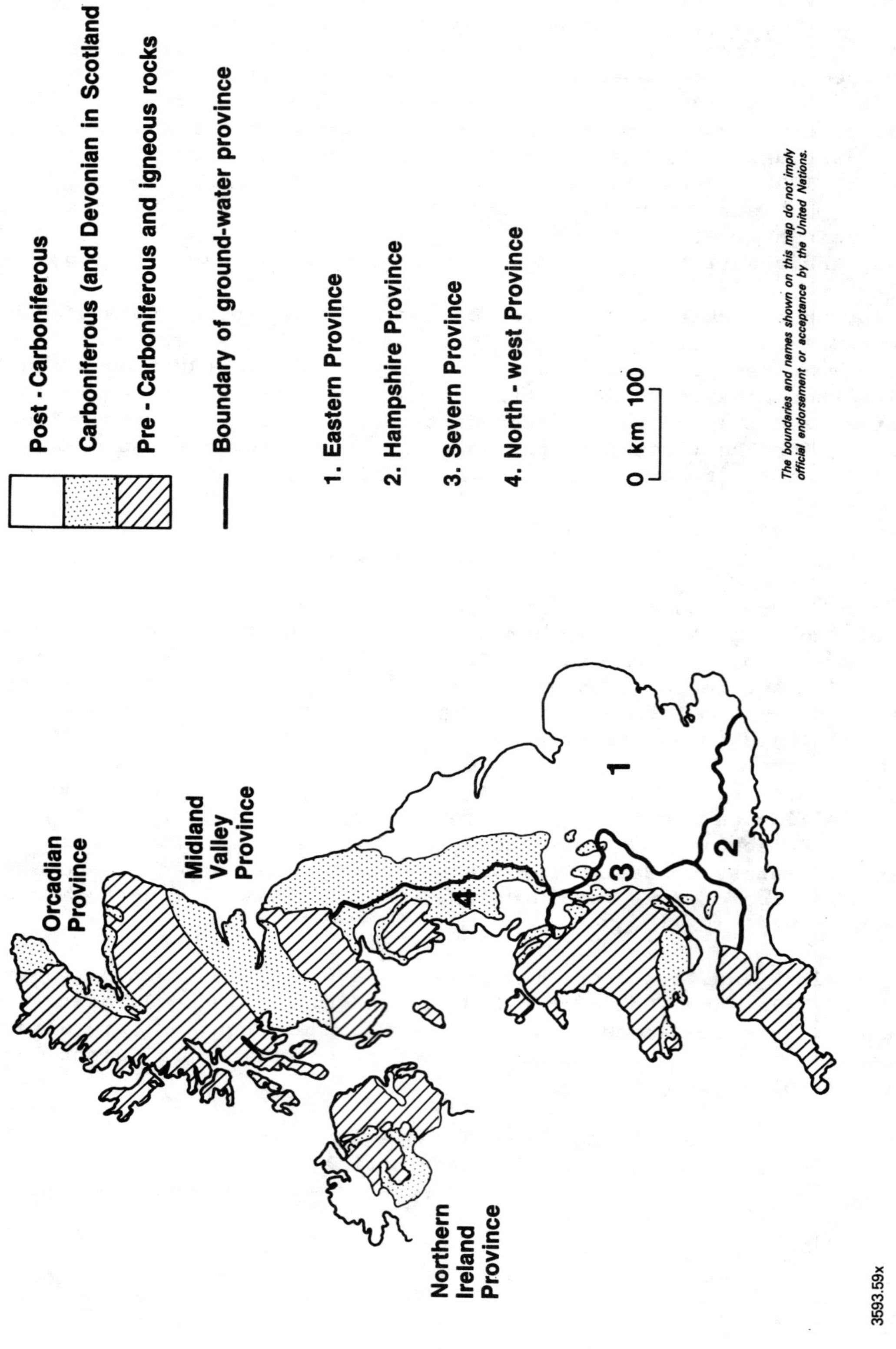

Figure 70. United Kingdom: Principal aquifers and groundwater use in England and Wales

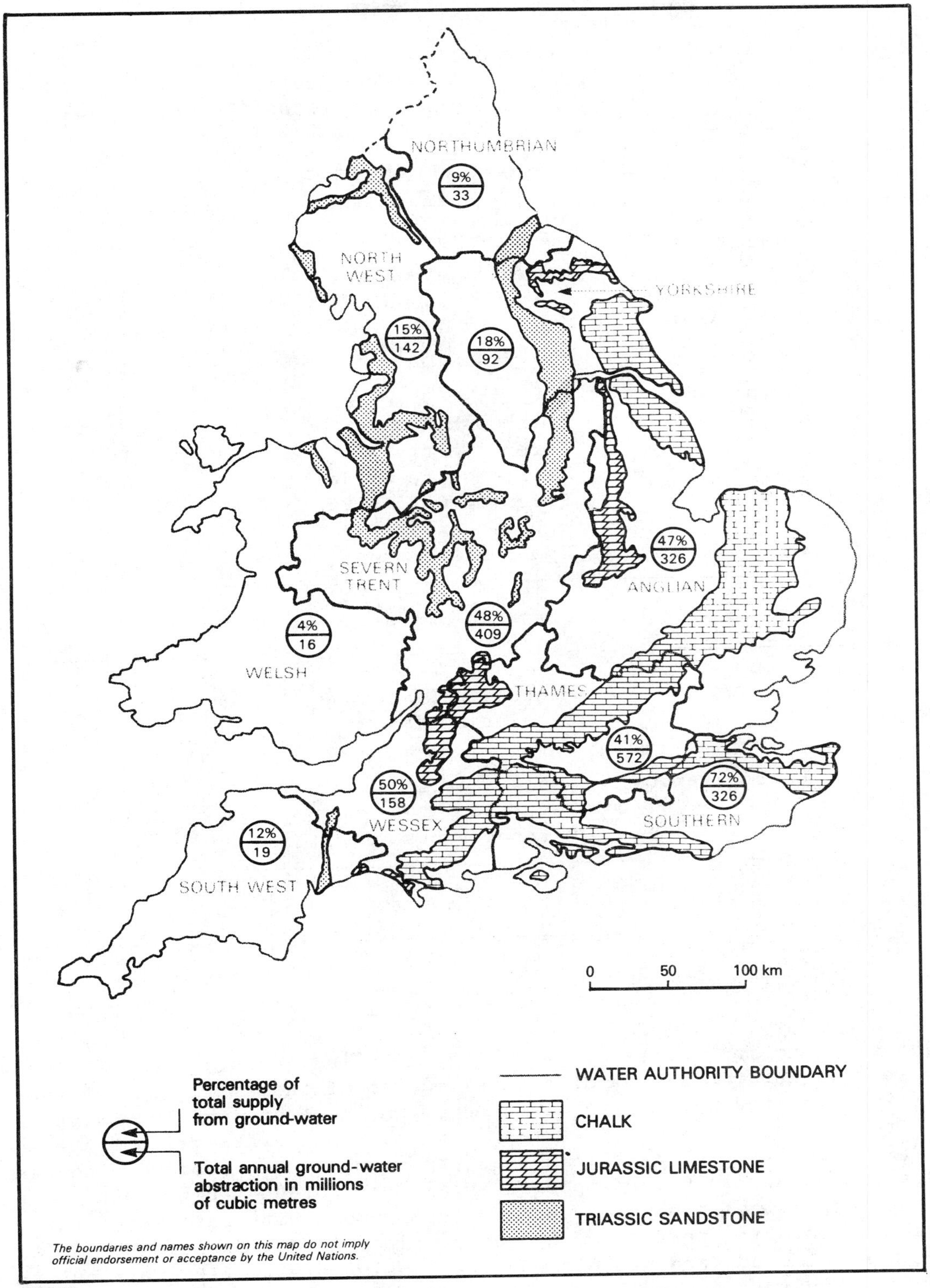

3593.60x

Table 61. United Kingdom: Principal aquifers

Era	Period	Formation
Cenozoic	Quaternary	Alluvial gravel Glacial sand and gravel Lower Pleistocene sand (Crags)
	Tertiary	Lower Tertiary sand
Mesozoic	Cretaceous	Chalk Upper Greensand Lower Greensand Hastings Beds (Tunbridge Wells Sand, Ashdown Beds) Spilsby Sandstone
	Jurassic	Corallian Cornbrash Forest Marble Great Oolite Series Inferior Oolite Series (including Lincolnshire Limestone) Upper Lias Sands
	Triassic	Sherwood Sandstone Group
Upper Paleozoic	Permian	Magnesian Limestone Permian sandstone
	Carboniferous	Coal Measures Namurian Carboniferous limestone
	Devonian	Old Red Sandstone facies

3. Alluvial aquifers

In contrast to many parts of the world, alluvial deposits are not important aquifers in Britain. This is a direct reflection of the small size of the river systems. Spreads of alluvial gravel do occur in the Trent and Thames valleys, in Northern Ireland and, to a lesser extent, in other main valleys, but the average thickness is only 5 m. Where thicknesses are greater than average, the generally high hydraulic conductivity and specific yield (50 to 400 m per day and 15 to 25 per cent respectively) have encouraged development with large-diameter wells, sometimes incorporating perforated collector pipes, to give yields of 50 to 200 l/s. Glacial sands and gravels cover extensive areas of Britain but they are not important aquifers and provide only small local supplies.

4. Post-Carboniferous aquifers

The youngest cover rocks are the lower Pleistocene sands (referred to as Crags) which occur in the eastern part of East Anglia attaining maximum thicknesses of 30 to 60 m. They tend to be unconsolidated, ferruginous sands that provide up to 25 l/s in favourable situations.

The Tertiary sequences are confined to the London and Hampshire basins. They contain a number of relatively thin sandstone formations that yield 5 to 10 l/s and locally up to 20 l/s. Of these the Bagshot Beds of the Hampshire Basin and the sands of the Lower Eocene of the London Basin are the most important.

The Cretaceous Chalk is the principal aquifer in the United Kingdom. It is a soft, white, microporous rock comprising two fine powders. The coarser of the two consists of shell debris and foraminifera, and the finer is composed of minute calcareous shells (coccoliths) that are the skeletons of plankton. The Chalk underlies much of eastern and southern England and also occurs in Northern Ireland although there it is harder and less porous and permeable because of the metamorphic effect of an extensive cover of Tertiary basalt lavas.

The aquifer has a high porosity of 30 to 40 per cent although generally less than 20 per cent for the harder bands of Chalk. The majority of pore diameters are less than a micron (Price and others, 1976). Consequently, the bulk of the water in the matrix is immobile, being held in the rock by capillary forces, resulting in a specific yield of only 1 per cent. The Chalk owes its position as an aquifer to an extensive system of secondary fissures which give typical permeabilities of 1 to 100 metres per day; the permeability of the matrix is no more than 10^{-4} to 10^{-3} m per day. Although the Chalk attains maximum thicknesses of the order of 400 m, only the top 50 to 60 m are usually fissured. The more permeable zones in the aquifer are closely related to the alignment of the major drainage pattern with typical transmissivities along valleys in the range 1,500 to 3,000 m^2 per day decreasing to no more than 10 to 20 m^2 per day below higher ground.

The Chalk is a classic example of a dual porosity system, with one porosity represented by the matrix and the other by the fissures. Water flows rapidly through the main fissure system and more slowly through the fine micro-fissure network and larger intergranular pore spaces of the matrix. The fissures act as a water distribution system gaining water from the matrix.

Groundwater is developed by wells up to 1 metre in diameter and up to 150 m deep. Such wells can yield more than 150 l/s while 50 l/s are quite common from sites in valleys. In the past, yields from large-diameter wells were supplemented by driving adits (or tunnels) over 1 metre wide and nearly 2 metres high, often for considerable distances in different directions from the well. In part of east London, the Thames Water Authority has a total of 18 km of adits connected to 48 pumping stations and similar tunnels are found in other parts of eastern England.

In addition to the Chalk, the Cretaceous contains three other aquifers: the upper Greensand, the lower Greensand and sandstone of the Hastings Beds.

The upper Greensand underlies the Chalk and may supplement the yields of wells that are primarily in Chalk. However, where it crops out in western England it is a poor aquifer.

The lower Greensand is important in south-east England, comprising beds of both sandstone and limestone. Typical values for the aquifer properties of the sand facies are 10 to 20 per cent for specific yield and 5 to 20 m per day for hydraulic conductivity. Yields can be of the order of 10 to 25 l/s and in favourable situations 35 l/s. The calcareous facies tends to have a higher permeability because of a more extensive fissure system and higher yields, of the order of 35 l/s and exceptionally 100 l/s, can be anticipated.

The Hastings Beds, found in the Weald of south-east England, include two sandstone aquifers - the Tunbridge Wells Sand and the Ashdown Beds. Of the two, the Ashdown Beds is the better aquifer giving typical yields of 10 to 25 l/s and sometimes 50 l/s in comparison to 5 to 10 l/s from the Tunbridge Wells Sand.

In eastern England, the Spilsby Sandstone, which straddles the Lower Cretaceous and Upper Jurassic boundary, is an aquifer of local significance, providing typical yields of 10 to 25 l/s and in some circumstances as much as 40 to 50 l/s.

The Jurassic is represented by a wide range of predominantly marine sediments which include several aquifers of which the Inferior Oolite and Great Oolite series of the Middle Jurassic and the Corallian of the Upper Jurassic are the most important (figure 70).

The Inferior Oolite is well developed in the Cotswold Hills of central England where it attains a maximum thickness of 100 m. An aquifer of equivalent age in eastern England, south of the Humber, is known as the Lincolnshire Limestone. Upper Lias Sands are commonly in hydraulic continuity with the Inferior Oolite in the west of England.

The Great Oolite is also important in the Cotswold Hills. In places it is distinct from the underlying Inferior Oolite but elsewhere they form a single aquifer. In the same area, limestones of the overlying Forest Marble and Cornbrash, although not important aquifers in their own right, supplement yields of wells drilled to the Great and Inferior Oolites and the entire sequence may be regarded as a single aquifer system.

The Corallian is of some significance in the northern part of eastern England, where it is more than 100 m thick. Other Jurassic limestones and sandstones are also important locally, for example the Marlstone, in Northamptonshire, which is a ferruginous sandstone.

The transmissivity of the Jurassic limestones is dependent upon the extent of the fissure systems. In many places values exceed 1,500 m^2 per day and may be as high as 4,000 to 4,500 m^2 per day. Where extensive fissure systems exist yields can be prolific. Artesian flows of 200 l/s have been recorded from the Lincolnshire Limestone and individual wells in this aquifer provide some of the largest yields in Britain. Large capacity wells in the Inferior and Great Oolites produce 50 to 100 l/s and in some cases as much as 200 l/s.

The Permian and Triassic rocks contain a complex sequence of continental sandstone beds that pass laterally into limestone and evaporites in eastern and northern England and in Northern Ireland. The sandstone, referred to as the Permo-Triassic sandstones (and which incorporate the Sherwood Sandstone Group of the Triassic), are the second most important group of water-bearing formations in the United Kingdom. They are predominantly fine to medium-grained sandstone but include pebble beds, coarse sandstone, siltstone and mudstone. Maximum thickness exceeds 1,000 m. They occur in the Midlands of England, extending into northern England, along the east and west flanks of the Pennines, as well as into south-west England. They are also important aquifers in Northern Ireland and parts of Scotland.

The sandstones have typical porosities of about 30 per cent and specific yields of betwen 20 and 30 per cent. Their permeability can be 1 to 2 m per day and, although this gives a significant matrix transmissivity, the total transmissivity of up to 1,500 m^2 per day is related to the presence of fractures. These fractures are commonly sub-horizontal and related to lithologic discontinuities. The extent of fracturing can be accentuated by mining in the underlying Coal Measures. The ratio of the intergranular to fracture transmissivity varies considerably depending upon the extent to which the rock is cemented, which is a major factor controlling yield from wells. In the soft, uncemented sandstones of eastern and central England yields are generally 50 to 100 l/s. The aquifer is commonly developed by pumping stations comprising several wells providing a total of 150 or more litres per second. Where the sandstone is more cemented 30 to 50 l/s are more usual.

East of the Pennines, there is an important sequence of Permian dolomitic limestones, referred to as the Magnesian Limestone, with a series of sands at their base known in northern England as the Yellow Sands and in Yorkshire and Nottinghamshire as the Basal Sands and Breccia. These sands, of variable thickness but ranging up to more than 50 m, tend to be unconsolidated, giving problems with well construction; they yield at least 5 to 10 l/s. The Magnesian Limestone is most important in north-east England where the thickness is as much as 200 m and yields from the fissured limestone are of the order of 50 l/s. Farther south, yields are commonly less than 10 l/s.

5. Upper Paleozoic aquifers

The Older Cover sequences may be considered in terms of the Upper and Lower Carboniferous and the Devonian (figure 68 and table 61). As mentioned earlier, the Devonian sandstones tend to have low permeabilities in England and only in Scotland can they be regarded as significant aquifers.

The Coal Measures comprise a thick sequence of mudstone, sandstone and coal beds. The sandstone tends to be well-cemented and fracture permeability is the rule. Yields from individual sandstone beds are usually no more than a few litres per second although shafts in the more arenaceous Upper Coal Measures of the Midlands of England do yield 20 to 50 l/s. Considerable volumes of water are pumped from the Coal Measures to drain the mines.

Although the lithology of the Namurian is similar to the Coal Measures, individual sandstone beds are thicker and more extensive and where the Millstone Grit facies occurs in northern England, yields are commonly 15 to 20 l/s and exceptionally 50 l/s. In Scotland, sandstone beds of the same age are also important aquifers locally.

The dominant aquifer of the Lower Carboniferous is the Carboniferous Limestone. It is a massive limestone, well-developed in several areas of England including Derbyshire, the Mendip Hills and north and south Wales; thicknesses are up to 1,000 m. It is a compact rock with a low porosity and a negligible specific yield. It is, however, extensively karstified and groundwater flow is entirely along joints and fissures that have been greatly enlarged by solution. Drilling for water is speculative because of the negligible specific yield of the rock and the fact that water flows in discrete, widely spaced channels that are not well interconnected. Large flows of the order of 1,000 l/s have been obtained when fissures have been intersected by tunnelling carried out for engineering purposes. Because of the high risk that wells will fail to give a supply, the groundwater from the limestone is commonly used at spring outlets. The extreme range in the flow from springs (for example 90 to 4,550 l/s at Cheddar in the Mendip Hills) means that development of springs for water supply must be in conjunction with reservoirs.

The Devonian Old Red Sandstone facies of England and Wales typically comprises hard, compact rocks giving only small yields. Farther north in Scotland, however, individual formations are more permeable, in particular the Knox Pulpit Formation which, together with overlying and underlying sandstone beds, constitutes an aquifer group 300 m thick. It has a porosity of more than 20 per cent but fissure flow is the main contributor to transmissivity. Yields of 40 l/s are common with specific capacities of about 1.5 l/s per metre of drawdown.

6. Basement rocks

The basement complex comprises, primarily, the Precambrian crystalline metamorphic rocks and the Caledonian and Variscan fold-belts and also the Devonian, except in Scotland. These rocks have low porosity and permeability and water flow through them is limited. Where flow does occur it tends to be restricted to preferred routes controlled by the alignment of major fractures and faults. At outcrop, fractures formed by weathering are more common in the top 100 m and small amounts of water can be obtained from shallow wells but generally yields are no more than a litre per second. Small springs are important sources of water supply in the upland areas formed by these rocks.

7. Water quality

Aquifers at outcrop in the United Kingdom generally contain a calcium bicarbonate groundwater as a consequence of acidic soil water reacting with carbonates in limestone or in the cement in sandstone. Such waters have a high carbonate hardness. Examples are common in the Chalk, Jurassic limestones and cemented Triassic sandstones (table 62). If a sandy soil with a low organic content, and hence low partial pressure of carbon dioxide in the soil air, overlies a carbonate aquifer, the carbonate hardness will be relatively low, as in the Chalk of west Norfolk and the Triassic sandstones of Nottinghamshire. In such situations the calcium content may be so low that the water is a sodium bicarbonate type.

Unconsolidated sand overlain by an organic-rich soil will also contain water with a low hardness, in this case because of the absence of carbonates, but it will be corrosive as a result of a relatively high free carbon dioxide content. Examples are found in glacial sand, the Upper Greensand and the Lower Greensand.

Table 62. United Kingdom: Hydrochemistry of the principal aquifers

(in mg/l)

	Chalk a/	Chalk b/	Chalk c/	Lower Greensand d/	Lower Greensand e/	Jurassic limestone a/	Triassic sandstone e/
Type	Ca-HCO_3	Ca-HCO_3	Na-HCO_3	Ca-HCO_3	Ca-HCO_3	Ca-HCO_3	Ca-Mg-HCO_3
Calcium (Ca)	100	188	6	21	85	109	37
Magnesium (Mg)	4	14	3	4	2	8	16
Sodium (Na) and Potassium (K)	2	40	247	11	11	52	23
Bicarbonate (HCO_3)	276	398	352	35	209	311	157
Sulphate (SO_4)	21	200	92	17	35	93	20
Chloride (Cl)	14	36	123	20	19	36	21
Nitrate (NO_3)	7	7	2	12	13	--	15
Total dissolved solids	320	710	655	150	260	465	225
Carbonate hardness, as $CaCO_3$	230	325	27	28	171	255	128
Non-carbonate hardness	37	205	--	40	49	51	43

a/ Typical groundwater from limestone at outcrop.

b/ Chalk overlain by boulder clay, composition influenced by high carbon dioxide and sulphate contents of the boulder clay.

c/ Chalk overlain by Tertiary, composition influenced by natural ion exchange.

d/ Sandstone at outcrop containing only limited carbonate cement.

e/ Sandstone at outcrop containing carbonate cement.

The composition of groundwater is modified, as it flows laterally through aquifers below confining beds, from a calcium bicarbonate to a sodium bicarbonate type as a result of ion exchange (table 62). With increasing distance from the outcrop the water becomes increasingly saline and eventually non-potable. This sequence can be recognized in many aquifers in Britain, including the Chalk, Jurassic limestones, Lower Greensand and sandstone of the Coal Measures.

Glacial deposits overlying aquifers can exert a considerable influence on the quality of their water. For example, boulder clay cover tends to produce a hard water with a high sulphate content in underlying aquifers.

Saline intrusion has occurred in a number of areas, particularly where the Chalk and Permian-Triassic sandstones have been extensively developed near coasts or estuaries. Many of the wells causing the problem are close to the coast and the amount of water pumped is limited by the chloride content. If the rate of increase in the salinity cannot be controlled, the well is eventually abandoned. To this extent the problem is partly self-correcting.

The principal areas where saline intrusion has occurred are the estuaries of the Mersey, Humber and Thames and along the south coast of England. The latter example is near Brighton where the Chalk is intensively developed for public water supply. The aquifer is carefully managed to limit the extent of the intrusion by restricting abstraction from coastal wells in the summer while increasing abstraction from wells more remote from the coast. The policy is reversed in winter (Headworth and Fox, 1986).

Considerable concern has been expressed in recent years about the steady pollution of groundwater from agricultural fertilizer and pesticides, landfill, seepage of industrial effluent or accidental spillage, and leaking sewers. The problem of rising concentrations of nitrate in groundwater derived from fertilizer has been recognized since the early 1970s. By 1984, 10 per cent of the licensed yield of groundwater in England and Wales exceeded the guideline nitrate value of 50 mg/l. The most seriously affected areas are the Chalk and Permian-Triassic sandstones in central and eastern England where effective rainfall (i.e., rainfall less evapotranspiration) is less than 200 mm per year (Skinner, 1987). There is some evidence that low concentrations of pesticide residue exist in groundwater associated with high nitrate content in eastern England (Aldrick and others, 1986; Foster and others, 1986).

The principal point sources of pollution are landfills. Although much of the waste disposed of in this manner is inert, about 20 per cent is municipal solid waste containing soluble and degradable substances that produce a highly polluting leachate (Aldrick and others, 1986).

The steady decline in groundwater quality as a consequence of environmental pollution is not in question. This is particularly disturbing in view of the difficulty, if not impossibility, of rehabilitating an aquifer to its original pure state, once it has been polluted.

C. Groundwater development

1. Agencies involved

Groundwater planning and development is in the hands of the water industry, and in the four entities that constitute the United Kingdom, the industry operates under three different administrative systems. In England and Wales it is controlled by 10 independent water authorities, in Scotland responsibility rests with local government in the form of regional (and island) councils, while in Northern Ireland, which is virtually one water authority, control is exercised through the Department of the Environment of Northern Ireland (figure 71). Despite these differing administrative structures, the overall objective and the underlying basic theme are the same in each entity, namely river basin management - the comprehensive management of all aspects of the water cycle in major river basins. The boundaries of the authorities are governed by hydrologic features, they are watersheds of natural river basins and not political boundaries.

The water authorities are responsible for:

a. Water resource planning

b. Control of abstraction through licensing

c. Pollution monitoring and control (including licensing industrial discharge)

d. Prevention of environmental pollution of rivers, estuaries, tidal waters and groundwater

e. Land drainage

f. Flood prevention

g. Certain aspects of water recreation and navigation

h. Provision of water supplies, water treatment and distribution

i. Provision of sewer systems; sewage and effluent treatment and disposal

The Government has changed the organization of the water industry in England and Wales. It has created a National Rivers Authority that assumes responsibility for the first seven functions listed above, that is planning, river-basin management and regulatory functions. The water authorities will become private companies responsible for providing water supplies, sewer systems and sewage treatment and disposal.

In Scotland, prevention of pollution of surface and groundwater is a function of seven river purification boards which have similar boundaries to and work closely with the regional councils.

Groundwater management readily fits into a system involving the comprehensive management of river basins. Groundwater policy is concerned with efficient management of the resource, bearing in mind the relationship with other facets of water management. It is concerned not only with providing a water supply and maintaining water quality but also with the consequences of aquifer development on the environment. Social and environmental factors as well as economic studies are thus necessarily included in the consideration of development proposals.

Figure 71. United Kingdom: Water authorities

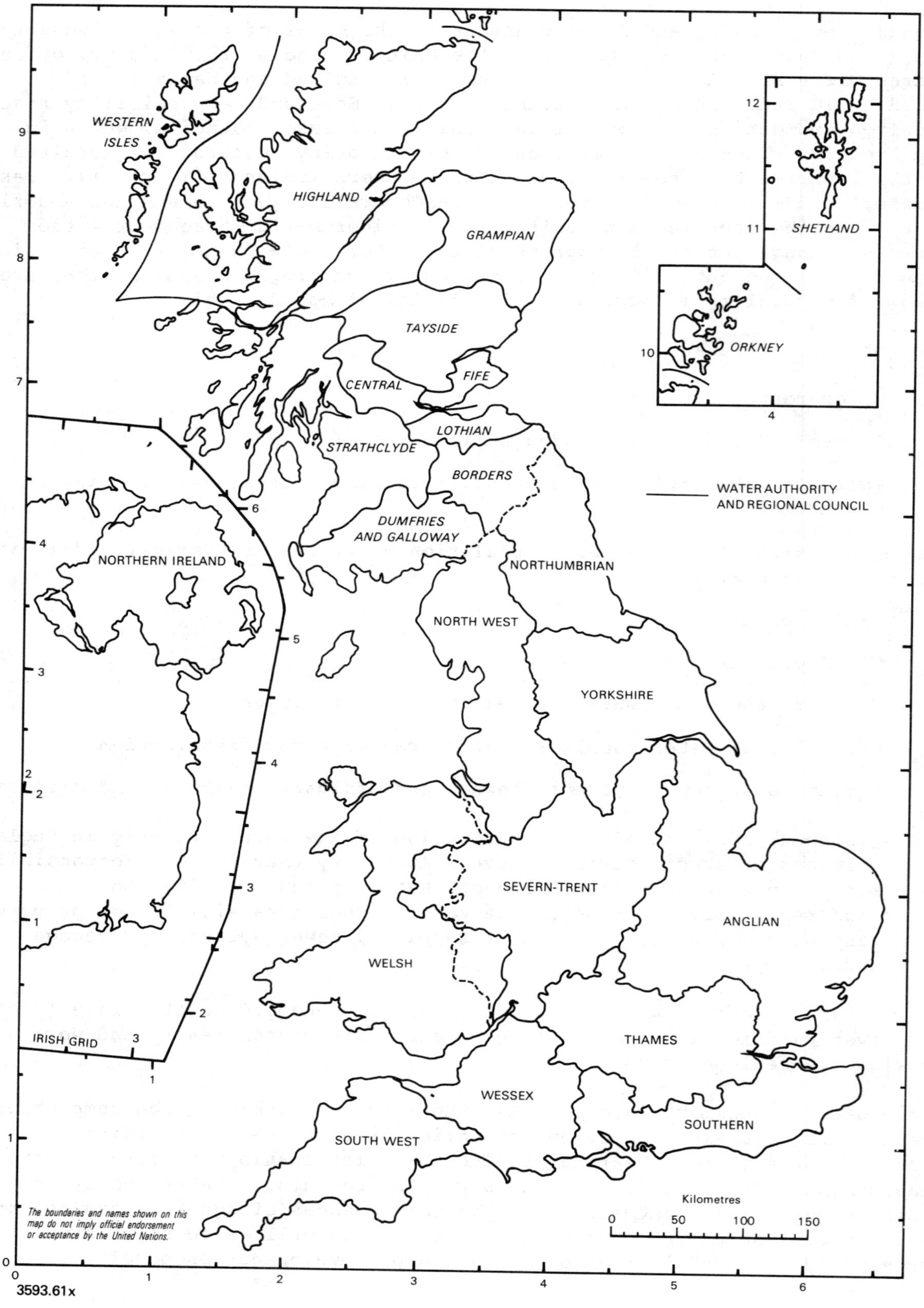

2. Current status

(a) General

In 1986, the abstraction of groundwater in England and Wales was 2.5 km^3, approximately 80 per cent of which was for public water supply, 15 per cent (excluding water used for cooling purposes) was used by industry and 3 per cent for agriculture, including spray irrigation. Groundwater provides one third of the public water supply but there is wide variation from region to region with many areas, especially in the south and east, entirely dependent on groundwater. Figure 70 shows the 10 water authority catchments in England and Wales and the amount of groundwater abstracted within each for public supply, both by volume and as a proportion of the total public supply.

The major aquifers in England and Wales are the Chalk, the Permian-Triassic sandstones and the Jurassic limestones which together provide more than 90 per cent of the abstracted groundwater. Apart from public supply, groundwater contributes significantly to industrial and agricultural water supplies in many areas. The volumes abstracted by industry have not increased in recent years and in some places there has been a significant decline caused partly by contraction in the traditionally high water-consuming industries and also because of changes to industrial processes. There has been some increase in agricultural abstraction caused by greater irrigation from groundwater in the Midlands and east of the country. In much of the United Kingdom, irrigation is a marginal activity and thus actual abstraction varies significantly from year to year and was low over the period 1985-1987.

As a consequence of the nature of the geology, with a predominance of older, hard, compact rocks, Scotland and Northern Ireland are not so dependent on groundwater for supply. In Scotland only about 3 per cent of the public water supply is derived from groundwater sources; in Northern Ireland the figure is 12 per cent.

Most groundwater used for public supply is supplied directly and continuously, that is, one or more boreholes provide a continuous supply to a centre of demand. This is the traditional form of groundwater use and it provides an economic source, reliable in both quality and quantity. The only form of treatment which is normally provided is disinfection, usually by chlorine. Some sandstone aquifers have naturally higher iron and manganese concentrations and these are also removed by treatment, usually pressure filtration.

The organization of the water industry in England and Wales into multi-functional catchment-based authorities has encouraged the planning and operation of water resources on a catchment-wide basis in an integrated manner. This has influenced the way in which groundwater is used and led to groundwater developments other than for continuous direct supply. Three different modes of use of groundwater as part of a river basin resource system have been commissioned in various parts of England and Wales (Skinner, 1983):

i. Groundwater used in conjunction with surface sources,

ii. Groundwater used to augment river flow, and

iii. Artificial recharge of groundwater.

All these modes of operation take advantage of the fact that the overall yield for water supply can be enhanced if the storage characteristics of aquifers are more fully exploited than is possible if abstraction is at a fixed and constant rate. Intermittent high-rate abstraction from groundwater, possibly further augmented by artificial replenishment, can provide a valuable component of water supply at times when surface-water systems are under pressure. The high storage capacity of aquifers allows abstraction to be balanced with recharge over a long period. In the United Kingdom, authorization to abstract groundwater as part of a combined surface- and groundwater system will typically require a balance over a five-year rolling period and acceptance of short-term depletion of storage in the interim.

(b) Groundwater used in conjunction with surface sources

In a conjunctive use system, ground- and surface-water abstractions supplying a common demand point can be balanced so that while the full demand is always met, the groundwater output is reduced at times of readily available surface water and then "overdrawn" for short periods to compensate for deficiencies in surface supplies during periods of dry weather. A higher total yield is obtained than by using the two sources individually and in the long term the abstraction from groundwater is kept in balance with replenishment. The most substantial scheme of this type in the United Kingdom is the Lancashire Conjunctive Use Scheme operated by the North West Water Authority.

(c) Groundwater used to augment river flow

The same principles can be applied by using groundwater to directly augment river flows from which water is subsequently abstracted for public supply. This type of groundwater development has been commonly developed in the United Kingdom and there are four major schemes (Thames, Anglian, Southern and Severn-Trent Water Authority areas) and a number of smaller ones.

(d) Artificial recharge of groundwater

The integrated use of surface- and groundwater resources can be further extended by using surface water available at times of surplus to augment groundwater storage; a number of schemes exist which operate on this principle recharging both the Chalk and sandstone aquifers. The largest scheme involves recharging the Chalk and lower Tertiary sands in the London Basin with water from the Thames and Lee rivers. The water is first treated to potable standards and recharge takes place when the water treatment plants are not used to their full capacity. The scheme should ultimately yield 90,000 m^3 per day.

3. Projected needs

In many parts of the United Kingdom existing groundwater resources are fully exploited, and in some cases overexploited. The policy in these areas is to prevent further development through application of an abstraction licensing system and, where possible, to reduce abstraction. The scope for further development of groundwater to meet projected demands in urban centres is small and it is expected that most new development will take the form of components of regional resource schemes. There is also increasing interest in using minor aquifers for rural public supplies in areas that have hitherto traditionally relied on surface water, particularly in Scotland and Wales. There may be a reduction in groundwater use in some areas if such supplies are taken out of use in favour of other sources because of contamination by nitrate or other diffuse pollutants.

4. Conflicting interests

Groundwater as part of the hydrologic cycle is capable of both affecting and being affected by "users" of other components of the cycle. There are many such issues but by scale of impact the greatest of these is the effect of groundwater abstraction on surface water, including wetlands, and the effects on groundwater of pollution, particularly from agricultural practice.

In many parts of central, eastern and southern England, where the main aquifers crop out, natural groundwater discharge to the river systems amounts to over 75 per cent of the total flow. Even where less permeable rocks occur in other parts of the United Kingdom it can amount to 30 or 40 per cent. Many rivers that include an appreciable groundwater component in their flow are also committed to water supply schemes that involve direct abstraction of river water. Consequently, all groundwater abstraction proposals must take into account potential effect upon surface water, both to ensure that surface abstraction rights are not prejudiced and also that environmental criteria in terms of preservation of flow for fisheries or maintenance of wetlands are not unacceptably prejudiced. There is an important link with water quality in that maintenance of stream flows might also be important to provide adequate dilution for surface-water discharges.

The effect of nitrate pollution from agriculture is a major concern and the maintenance of groundwater as an important source of supply in many areas depends upon effective control of land-use being implemented around critical boreholes. Strategic water resources are being placed permanently at risk by the continued leaching of nitrate from agricultural land in borehole catchment areas.

A further feature is the declining use of groundwater in urban areas because of changing industrial need. This is causing problems of groundwater "flooding" in some cities as the water tables rise. The groundwater in many of these areas has suffered from urban pollution and so the matter cannot be resolved by increasing abstraction for public supply.

D. Training and international cooperation

Hydrogeologic research is carried out by the British Geological Survey, the Water Research Centre and many universities, but in particular University College of the University of London, and the universities of Birmingham, Bath and Newcastle-upon-Tyne. These organizations undertake research for government, the water industry, international organizations and the private sector as well as initiating relevant projects themselves. Considerable emphasis is currently placed on quality problems. The water authorities carry out or commission research and investigations concerned with the assessment and development of groundwater resources and the protection of its quality.

Postgraduate education in hydrogeology is mainly provided by University College, London and the University of Birmingham. The universities of Birmingham, Durham and Newcastle-upon-Tyne and Imperial College, London, provide postgraduate courses in engineering hydrology, and Birmingham and Newcastle courses in water resources engineering.

Technical cooperation projects in developing countries, funded by the Overseas Development Administration (ODA) of the British Government through the British Geological Survey, commonly involve on-the-job training and the training of staff at Survey laboratories. The British Council, through grants and scholarships, supports the training of individuals from developing countries at universities in the United Kingdom.

The water industry and research institutes in the United Kingdom are involved in cooperative groundwater studies with many international agencies and developing countries. These studies are concerned with resource investigations or development as well as research into the principles of the subject. The British Geological Survey, as advisers to ODA, is active on a world-wide basis. Current studies, funded by ODA, include groundwater resources investigations in Botswana, the South Pacific, Kenya, Sri Lanka and Costa Rica. The problem of using thin fresh-water lenses above saline groundwater (the Ghyben-Herzberg principle) is being investigated in Pakistan in association with consulting engineers. The optimum design of wells in alluvial aquifers is being evaluated in Bangladesh, both in the field and the laboratory.

Geological Survey staff have been seconded to government hydrogeology units in Fiji, Vanuatu, Honduras, Zimbabwe and Botswana. Workshops have been organized in Sri Lanka, Botswana, Mexico, Peru and Zimbabwe to deal with a wide range of groundwater problems. The development of groundwater in basement aquifers in Africa is a major research programme of the Survey, and a workshop in Zimbabwe, in 1987, was an international meeting that addressed the problems of using groundwater in basement rocks.

<u>References</u>

Agnew, S. C. Infrastructure planning, water overview. <u>Proceedings</u>, Institution Civil Engineers, Part 1, 82, p. 381-396, 1987.

Aldrick, R. J., K. J. Edworthy <u>and</u> C. P. Young. Groundwater pollution and protection, p. 543-590, in <u>Groundwater occurrence, development and protection</u> (T. W. Brandon, ed.). Institution Water Engineers and Scientists, London, 1986.

Anonymous. <u>Survey of British hydrogeology 1980</u>. The Royal Society, London, 1981.

Brandon, T. W. (editor). <u>Groundwater occurrence, development and protection</u>. Institution Water Engineers and Scientists, London, 1986.

Downing, R. A., W. M. Edmunds <u>and</u> I. N. Gale. Regional groundwater flow in sedimentary basins in the UK, p. 105-125, in <u>Fluid flow in sedimentary basins and aquifers</u> (J. C. Goff and B. P. J. Williams, eds.). Geological Society of London, Special Publication, No. 34, 1987.

Foster, S. S. D., L. R. Bridge, A. K. Geake <u>and others</u>. The groundwater nitrate problem. <u>Hydrogeological Report, No. 86/2, British Geological Survey</u>, 1986.

Headworth, H. G. <u>and</u> G. B. Fox. The South Downs aquifer: its development and management. <u>Journal Institution Water Engineers and Scientists</u>, 40, p. 345-361, 1986.

Price, M., M. J. Bird <u>and</u> S. S. D. Foster. Chalk pore-size measurements and their significance. <u>Water Services</u>, 80, p. 596-600, 1976.

Robins, N. S. <u>and</u> I. B. Harrison. Private groundwater use and consumption in Scotland. <u>Journal Institution Water Engineers and Scientists</u>, 40, p. 362-368, 1986.

Rodda, J. C., R. A. Downing <u>and</u> F. M. Law. <u>Systematic hydrology</u>. Newnes-Butterworth, London, 1976.

Skinner, A. C. Groundwater development as an integral part of river basin resource systems, p. 641-650, in <u>International Symposium on Groundwater in water resource planning</u>, Koblenz, pub. by UNESCO, Paris, 1983.

__________. Groundwater protection and agriculture in the United Kingdom, p. 317-335, in <u>Proceedings of Conference on Impact of Agriculture on Water Resources</u>, Deutscher Verein des Gas und Wasserfaches, Frankfurt, 1987.

Toth, J. Theoretical analysis of groundwater flow in small drainage basins. <u>Journal Geophysical Research</u>, 68, p. 4,795-4,812, 1963.

Annex I

GROUNDWATER PROJECTS IN WESTERN AND CENTRAL EUROPE SPONSORED BY THE UNITED NATIONS DEVELOPMENT PROGRAMME

Country and project	Project symbol	Executing agency	Duration	Groundwater component
		1. Malta		
Waste disposal and water supply	MAT-66-505	WHO	1966-1975	Substantial
Development and conservation of water resources	MAT-78-006	DTCD	1978-1981	Substantial
		2. Portugal		
Etude des ressources en eau de la Péninsule de Setubal	POR-77-015	UNESCO	1978-1981	Substantial
Evaluation des ressources en eau des systèmes aquifères de l'Algarve	POR-79-001	UNESCO	1980-1982	Exclusive
Plan National de l'Eau	POR-82-004	UNESCO	1982-1987	Substantial
		3. Spain		
Hydrogeological investigation in the Guadalquivir River basin	SPA-64-009	FAO	1965-1969	Exclusive
Pilot project of groundwater utilization for agricultural development in the Guadalquivir River basin	SPA-69-016	FAO	1969-1972	Exclusive
Scientific study of water resources in the Canary Islands	SPA-69-515	UNESCO	1969-1976	Exclusive
Groundwater utilization of the Guadalquivir River basin	SPA-71-003	FAO	1971-1978	Exclusive
National Programme for underground water study	SPA-73-002	FAO	1974-1976	Exclusive

Annex II

UNITED NATIONS PUBLICATIONS ON WATER RESOURCES

(Listed in chronologic order)

Report of United Nations Interregional Seminar on Flood Damage Prevention Measures and Management, Tbilisi, USSR, 1969, 117 p. English only. (ST/TAO/SER.C/144)

Integrated River Basin Development. Report of panel of experts. Revised edition, 80 p. English (French and Spanish out of print). Sales No. E.70.II.A.4.

Solar Distillation as Means of Meeting Small-scale Water Demands, 86 p. English (French out of print). Sales No. E.70.II.B.1.

Triennial Report on Water Resources Development, 1968-1970, 202 p. English. Sales No. E.71.II.A.15.

Abstraction and Use of Water: A Comparison of Legal Regimes. 254 p. French and Spanish (English out of print). Sales No. 72.II.A.10.

Proceedings of Interregional Seminar on Water Resources Administration, New Delhi, 1973. 204 p. English. (DP/UN/INT-70-371)

National Systems of Water Administration. 183 p. English. Sales No. E.74.II.A.10.

Groundwater Storage and Artificial Recharge. Natural Resources/Water Series No. 2, 270 p. English, French, Spanish. Sales No. 74.II.A.11.

Management of International Water Resources: Institutional and Legal Aspects. Report of panel of experts. Natural Resources/Water Series No. 1. 271 p. English, French and Spanish. Sales No. 75.II.A.2.

Demand for Water: Procedures and Methodologies for Projecting Water Demand in the Context of Regional and National Planning. Natural Resources/Water Series No. 3. 240 p. English and Spanish (French out of print). Sales No. 76.II.A.1.

Groundwater in the Western Hemisphere. Natural Resources/Water Series No. 4. 337 p. English, French and Spanish. Sales No. 76.II.A.5.

Guidelines for Flood Loss Prevention and Management in Developing Countries. Natural Resources/Water Series No. 5. 183 p. English. Sales No. E.76.II.A.7.

Report of United Nations Water Conference, Mar del Plata, 14-25 March 1977. 181 p. English and French (Spanish out of print). Sales No. 77.II.A.12.

Review of the United Nations Groundwater Exploration and Development Programme in the Developing Countries, 1962-1977. Natural Resources/Water Series No. 7. 84 p. English, French and Spanish. Sales No. 79.II.A.4.

Efficiency and Distributional Equity in the Use and Treatment of Water: Guidelines for Pricing and Regulations. Natural Resources/Water Series No. 8. 175 p. English (French and Spanish out of print). Sales No. E.80.II.A.11.

Water Resources Planning: Experiences in a National and Regional Context. Report of United Nations Workshop Convened in Cooperation with the Government of Italy. 293 p. English. (TCD/SEM.80/1)

Rural Water Supply. Report of United Nations Interregional Seminar Convened in Cooperation with the Government of Sweden. Uppsala, Sweden, 6-17 October 1980. 44 p. English. (TCD/SEM.81/1)

Groundwater in the Eastern Mediterranean and Western Asia. Natural Resources/Water Series No. 9. 230 p. Arabic, English and French. Sales No. 82.II.A.8.

Experiences in the Development and Management of International River and Lake Basins. Natural Resources/Water Series No. 10. Proceedings of United Nations Interregional Meeting of International River Organizations, Dakar, Senegal, 5-14 May 1981. 424 p. English, French and Spanish. Sales No. 82.II.A.17.

Flood Damage Prevention and Control in China. Report of Study and Workshop in the People's Republic of China, 16-31 October 1980. Natural Resources/Water Series No. 11. 121 p. English. Sales No. 82.II.A.13.

Groundwater in the Pacific Region. Natural Resources/Water Series No. 12. 289 p. English. Sales No. E.83.II.A.12.

Technical Cooperation Among Developing Countries in Groundwater Resources Development: Report of United Nations International Colloquium Convened in Cooperation with the Government of Yugoslavia in Zagreb, 23-28 May 1983. 123 p. English. (TCD/SEM.83/1)

Treaties Concerning the Utilization of International Watercourses for Other Purposes than Navigation (Africa). Natural Resources/Water Series No. 13. 98 p. English/French. Sales No. E.F.84.II.A.7.

Use of Non-Conventional Water Resources in Developing Countries. Natural Resources/Water Series No. 14. 278 p. English. Sales No. E.84.II.A.14.

Groundwater in Continental Asia. Natural Resources/Water Series No. 15. 391 p. English. Sales No. E.86.II.A.2.

Water Resources Legislation and Administration in Selected Caribbean Countries. Natural Resources/Water Series No. 16. 171 p. English, French and Spanish. Sales No. 86.II.H.2.

Follow-Up to the Mar del Plata Action Plan: Report of Meeting. Interregional Symposium on Improved Efficiency in the Management of Water Resources, 5-9 January 1987. 47 p. English, French and Spanish. (TCD/SEM.87/1)

Institutional Issues in the Management of International River Basins: Financial and Contractual Considerations. Natural Resources/Water Series No. 17. 111 p. English, French and Spanish. Sales No. 87.II.A.16.

Application of Computer Technology for Water Resources Development and Management in Developing Countries, 1987. 95 p. English. (TCD/WATER/1)

Groundwater in North and West Africa. Natural Resources/Water Series No. 18. 415 p. English and French. Sales No. 87.II.A.8.

Groundwater in East, Central and Southern Africa. Natural Resources/Water Series No. 19. 341 p. English and French. Sales No. 88.II.A.5.

Water Resources Planning to Meet Long-Term Demand: Guidelines for Developing Countries. Natural Resources/Water Series No. 21. 117 p. English, French and Spanish. Sales No. 88.II.A.17.

Non-Conventional Water Resources Use in Developing Countries: Proceedings of Interregional Seminar, Willemstad, Curacao, Netherlands Antilles, 22-28 April 1985. Natural Resources/Water Series No. 22. 515 p. English. Sales No. E.87.II.A.20.

Groundwater Economics. Report of United Nations International Symposium and Workshop Convened in Cooperation with the Government of Spain. Barcelona, Spain, 19-23 October 1987. 65 p. English. (TCD/SEM.88/2)

Assessment of Multiple Objective Water Resources Projects: Approaches for Developing Countries. 125 p. English. United Nations/United Nations Environment Programme. (TCD/SEM.88/3)

Legal and Institutional Factors Affecting the Implementation of the International Drinking Water Supply and Sanitation Decade. Natural Resources/Water Series No. 23. 121 p. English. Sales No. E.88.II.A.21.

كيفية الحصول على منشورات الأمم المتحدة

يمكن الحصول على منشورات الأمم المتحدة من المكتبات ودور التوزيع في جميع أنحاء العالم . استعلم عنها من المكتبة التي تتعامل معها أو اكتب إلى : الأمم المتحدة ، قسم البيع في نيويورك أو في جنيف .

如何购取联合国出版物

联合国出版物在全世界各地的书店和经售处均有发售。请向书店询问或写信到纽约或日内瓦的联合国销售组。

HOW TO OBTAIN UNITED NATIONS PUBLICATIONS

United Nations publications may be obtained from bookstores and distributors throughout the world. Consult your bookstore or write to: United Nations, Sales Section, New York or Geneva.

COMMENT SE PROCURER LES PUBLICATIONS DES NATIONS UNIES

Les publications des Nations Unies sont en vente dans les librairies et les agences dépositaires du monde entier. Informez-vous auprès de votre libraire ou adressez-vous à : Nations Unies, Section des ventes, New York ou Genève.

КАК ПОЛУЧИТЬ ИЗДАНИЯ ОРГАНИЗАЦИИ ОБЪЕДИНЕННЫХ НАЦИЙ

Издания Организации Объединенных Наций можно купить в книжных магазинах и агентствах во всех районах мира. Наводите справки об изданиях в вашем книжном магазине или пишите по адресу: Организация Объединенных Наций, Секция по продаже изданий, Нью-Йорк или Женева.

COMO CONSEGUIR PUBLICACIONES DE LAS NACIONES UNIDAS

Las publicaciones de las Naciones Unidas están en venta en librerías y casas distribuidoras en todas partes del mundo. Consulte a su librero o diríjase a: Naciones Unidas, Sección de Ventas, Nueva York o Ginebra.

Litho in United Nations, New York
16582-May 1991-2,580
ISBN 92-1-104355-7

United Nations publication
Sales No. E.90.II.A.25
ST/TCD/12